Denise

AN INTRODUCTION TO THE THEORY OF MECHANICS

AN INTRODUCTION TO THE
THEORY OF MECHANICS

by

K. E. BULLEN

M.A., Sc.D., F.R.S.

*Professor of Applied Mathematics
in the University of Sydney*

EIGHTH EDITION

CAMBRIDGE
AT THE UNIVERSITY PRESS
1971

By the same author

AN INTRODUCTION TO THE THEORY OF SEISMOLOGY

Cambridge University Press, Third Edition, 1965

SEISMOLOGY

Methuen, 1954

© SCIENCE PRESS 1971

ISBN 0 521 082919

PRINTED IN AUSTRALIA

*Printed in Australia by Halstead Press, Sydney
for the publishers, Science Press, Marrickville, N.S.W.
Registered in Australia for transmission by post as a book.*

PREFACE TO THE EIGHTH EDITION

A number of corrections and improvements have been made, but the general structure of the seventh edition has been preserved fairly intact. The m.k.s. units in the book conform with the SI (Système International) units which are now being widely used.

I am indebted to past and present colleagues for various forms of help in preparing the book. These include Professors E. S. Barnes, M. Brearley, T. M. Cherry, F. Chong, H. David and K. C. Westfold; Mr J. M. Allen, Mr J. Guest, Mr L. Karlov, Mr R. P. Loh, Mr J. B. Miller, Mr A. E. Minty, Mr P. R. Pettit, Mr M. Ritter, Mr W. B. Smith-White, Mr R. J. Storer and Mr M. L. Urquhart.

K. E. BULLEN

SYDNEY
January, 1971

PREFACE TO THE SEVENTH EDITION

This edition is the first in which a substantial revision of the original book has been undertaken. The immediate reason for the revision has been the demand by many teachers of Mechanics for a change-over from c.g.s. to m.k.s. units. The demand has been heeded to the extent that m.k.s. units are now given pride of place throughout the book, except in the chapters on Statics and Hydrostatics. At the same time, full provision has been made for readers who wish to become competent in the use of c.g.s. units. This was felt to be desirable because c.g.s. units continue to be used, not only in many school and university courses, but also in important branches of research. F.p.s. units are included with the same degree of emphasis as in earlier editions.

In the worked and set exercises, pains have been taken to ensure that the reader who wishes may gain a mastery of all three unit systems, or, if he prefers, may concentrate on one or other of the m.k.s. and c.g.s. systems. This has required a substantial re-wording of many of the numerical exercises, with occasional minor numerical changes. With rare exceptions, the revised exercises correspond closely to those in earlier editions.

The changes in units have made it necessary to re-set the entire book, and advantage has been taken of the occasion to improve the exposition in many places. In order to reduce inconvenience to readers, pains have been taken, however, to preserve intact the original structure of the first sixteen chapters; in particular, the original section headings and numbers remain practically unaltered.

Chapter XVII has been appreciably enlarged, principally to include an introduction to some of the more immediate problems of extra-terrestrial mechanics—orbits of space vehicles, weightlessness, etc. It is hoped that the account will be found somewhat more reliable than accounts which have lately appeared in many elementary physics texts on these topics. Because of the increased size of Chapter XVII, it was decided to drop the Chapter XVIII of earlier editions.

PREFACE

The following remarks are adapted from prefaces to earlier editions:

The book assumes no prior knowledge of Mechanics or Calculus, but assumes that the reader will at least be starting to learn the calculus as he reads the book. Most of the earlier exercises can be solved without the calculus, but simple calculus is intermittently required in the theory and in later exercises.

Vector methods up to scalar products and related topics are introduced in Chapter VI and applied in the following chapters on Dynamics. Vector products are introduced in Chapter XVII. The book contains all the vector theory needed for the topics it treats.

The main emphasis is on Dynamics, which occupies most of Chapters III to XIII, but the book is designed so that readers who wish may start with Statics and Hydrostatics. These readers may proceed to Chapters XV and XVI after reading only a small number of the sections in earlier chapters; directions are given in §§ 10·4, 15·1A.

Special efforts have been made to present the subject of Dynamics as Applied Mathematics—i.e. as a branch of Science in which mathematics is auxiliary to the central purpose of describing of a body of observational evidence about the natural world. Thus, so far as seems reasonable in an elementary book, attention is given to questions of inductive inference as well as logical soundness. The former questions are disregarded by many mathematical writers on Mechanics, and the latter, often blatantly, by physicist and other writers. I am of course aware that very much more can be written on the inductive side than appears in this book, and in particular on Newton's laws of motion, but there are limits to what can be reasonably attempted in a first course.

I should like to record my appreciation of the work of Mr A. Boden, Mr A. D. Castleman, and all associated with the Science Press in enabling the book to be so well produced.

K.E.B.

TOKYO
January, 1965

"Reason is but an item in the mystery; and behind the proudest consciousness that ever reigned, reason and wonder blushed face to face."

CONTENTS

PREFACE *page* v

CHAPTER I SOME REMARKS ABOUT APPLIED MATHEMATICS
1·1 Science 1
1·2 Logic ; deduction and induction 1
1·3 Experiment 2
1·4 Mathematics, pure and applied 2
1·5 The scope of Applied Mathematics 4
1·6 Remarks 5
Examples I 7

CHAPTER II MOTION OF A POINT IN A STRAIGHT LINE
2·1 Units and measures 9
2·2 Position coordinate 10
2·3 Velocity 11
2·4 Acceleration 11
2·5 Motion in a straight line ; uniform acceleration 12
2·6 Vertical motion under gravity 17
2·7 Motion in a straight line ; variable acceleration 19
2·8 Change of reference origin 20
2·9 Velocity-time graph 22
Examples II 24

CHAPTER III DYNAMICS OF A PARTICLE MOVING IN A STRAIGHT LINE
3·1 The concepts of force and mass 28
    Particle 29
    Momentum 29
    Fundamental equation of particle dynamics 30
    Addition of forces 30
    Equilibrium 30
3·2 Dimensions and change of units 33
3·3 Weight 34
3·4 Worked examples 36
3·5 Principle of momentum 39
    Impulse ; instantaneous impulse 39, 40
3·6 Work 41
3·7 Power 43

CONTENTS

3·8	Energy	page 44
	Kinetic energy	44
	Principle of energy	45
	Conservative forces and potential energy	46
	Potential energy associated with weight	47
	Alternative form of principle of energy	47
	Conservation of energy	47
3·9	Hooke's law	51
Examples III		54

CHAPTER IV DYNAMICS OF A SYSTEM OF PARTICLES MOVING IN A STRAIGHT LINE

4·1	System of particles	58
	Centre of mass	58
	Momentum	59
4·2	Action and reaction forces	59
4·3	Principle of momentum	60
	Internal and external forces	60
	Motion of centre of mass	61
	Conservation of momentum	62
4·4	Worked examples	62
4·5	Energy	65
	Kinetic energy	65
	Potential energy	65
	Principle of energy	65
4·6	Tension and thrust	68
4·7	Problems involving pulleys	70
4·8	Elastic impacts	75
4·9	Problems where the mass changes	79
Examples IV		80

CHAPTER V INTERLUDE

5·1–5·5	Recapitulation and foreword	85
Examples V		88

CHAPTER VI PARTICLE DYNAMICS IN MORE THAN ONE DIMENSION

6·1	Introduction to the use of vectors	93
	Free and localised vectors	94
	Addition and subtraction	94
	Components of a vector	95

CONTENTS

	Product of scalar and vector	page 95
	Scalar product of two vectors	96
	Differentiation and integration of vectors	97
	Line-integral	99
	Vector product of two vectors	99
6·2	Motion of a point	101
	Position vector, velocity and acceleration	101
6·3	Fundamental equation of particle dynamics	102
	Addition of forces	103
	Principle of momentum for a particle	103
6·4	The method of resolved parts	103
6·5	Applications	104
	Particle on smooth fixed inclined plane	104
6·6	Friction	106
6·7	Work and power	109
6·8	Energy	112
	Kinetic energy	112
	Principle of energy	112
	Conservative forces and potential energy	112
	Alternative form of principle of energy	113
	Conservation of energy	113
	Particle sliding on smooth surface under gravity	113
	Energy in problems involving strings	114
6·9	Change of reference origin	116
	Relative velocity	116
	Relative acceleration	119
Examples VI		120

CHAPTER VII DYNAMICS OF A PARTICLE MOVING IN A CIRCLE

7·1	Angular velocity and acceleration of a point	125
7·2	Circular motion : preliminary relations	126
7·3	Circular motion : acceleration components	127
7·4	Uniform circular motion	128
7·5	Circular motion : dynamical equations	128
7·6	Worked examples	128
Examples VII		133

CHAPTER VIII SIMPLE HARMONIC MOTION

8·1	Definitions ; amplitude, phase, epoch	137
8·2	Velocity and acceleration at any instant	138
8·3	Period and frequency	138
8·4	Speed and acceleration in any position	139
8·5	Alternative investigation of simple harmonic motion	139

8·6	Simple harmonic motion as projection of uniform circular motion	page 139
8·7	Worked examples	140
8·8	Oscillations of particle attached to light perfectly elastic string	141
8·9	Simple pendulum	144
Examples VIII		145

CHAPTER IX FREE MOTION OF A PARTICLE UNDER CONSTANT GRAVITY

9·1	Preliminary remarks	148
9·2	Path of projectile	148
	Equations of motion	148
	Velocity and position at time t	149
	Equation and properties of path	149, 150
9·3	Estimation of range	150
	Range on inclined plane	150
	Maximum range	150
	Range on horizontal plane	151
9·4	The enveloping parabola	151
9·5	Worked examples	152
Examples IX		155

CHAPTER X INTERLUDE

10·1–10·5	Recapitulation and foreword	158
Examples X		160

CHAPTER XI CENTRES OF MASS AND MOMENTS OF INERTIA

11·1	Centre of mass	165
	Position, velocity and acceleration	165
	Grouping theorem	166
	Cartesian formulae	167
11·2	Continuously distributed mass-systems	167
	Centroid	168
11·3	Centres of mass in particular cases	169
	Uniform symmetrical bodies	169
	Uniform triangular lamina	169
	Uniform pyramid and cone	170
	Uniform wire in shape of circular arc	171
	Uniform lamina in shape of sector of circle	171
	Uniform semicircular lamina	172
	Uniform distribution over surface of sphere	172
	Uniform hemispherical shell	173
	Uniform solid hemisphere	173

CONTENTS   xiii

11·4	Theorems of Pappus	page 174
11·5	Worked Examples	175
11·6	Moment of inertia and radius of gyration	178
11·7	Parallel axes theorem	178
	Perpendicular axes theorem for lamina	179
11·8	Moments of inertia in particular cases	180
	Uniform rod	180
	Uniform rectangular lamina and block	180
	Uniform thin ring	180
	Uniform disc	181
	Uniform spherical shell	181
	Uniform solid sphere	181
	Uniform solid and hollow cylinder	182
11·9	Routh's rule	183
11·(10)	Worked examples	183
11·(11)	Equimomental systems	186
Examples XI		187

CHAPTER XII   TWO-DIMENSIONAL DYNAMICS OF A SYSTEM OF PARTICLES

12·1	Localised vectors	190
	Moment of localised vector about a point	190
	Cartesian formula for moment	191
	Addition of localised vectors	192
	Parallel localised vectors	193
	Vector couple	193
	Resultant of a set of localised vectors	194, 5
	Theorem of moments	194
	'Equivalence' of sets of localised vectors	195, 6
12·2	Systems of particles in more than one dimension	196
	Configuration and momentum	196
	Forces on systems of particles	196
	Weight ; centre of gravity	197
12·3	Principle of (linear) momentum	198
	Internal and external forces	198
	Motion of centre of mass	198
	Conservation of momentum	199
12·4	Principle of angular momentum	199
	First form of principle	200
	Conservation of angular momentum	200
	Extended forms of principle	200, 1

12·5	Energy	page 202
	Kinetic energy	202
	Translational and internal kinetic energy	202
	Potential energy	203
	Potential energy associated with weight	203
	Principle of energy	203
12·6	Applications and worked examples	203
	Elastic impacts of spheres	203
Examples XII		210

CHAPTER XIII TWO-DIMENSIONAL DYNAMICS OF A RIGID BODY

13·1	Rigid body	214
13·2	Kinematical considerations	214
	Degrees of freedom	214
	Angular velocity and acceleration of rigid body	215, 6
	Velocity of a point of a rigid body	216
13·3	Pin-joints	218
13·4	Kinetic energy of a rigid body	218
	Case of fixed axis of rotation	218
	General two-dimensional case	219
13·5	Potential energy of a rigid body	219
13·6	Use of principle of energy	219
	Work of internal forces in a rigid body	220
	Form of energy principle for rigid body	221
	Case of conservative forces	221
	Work done by forces at contacts	221
13·7	Use of (linear) momentum principle	224
13·8	Use of angular momentum principle; case of fixed axis	225
	Basic formulae	225, 6
	Work done during a rotation	226
	Work of friction couple	226
13·9	The rigid-body pendulum	231
13·(10)	Motion of rigid body when there is no fixed axis	236
13·(11)	Work done by external forces during displacement of rigid body	240
13·(12)	Effect of impulsive forces on a rigid body	241
Examples XIII		245

CONTENTS

CHAPTER XIV INTERLUDE
14·1–14·3 Recapitulation and foreword .... page 250
Examples XIV 252

CHAPTER XV TWO-DIMENSIONAL STATICS
15·1 Basic theory of statics of a single rigid body 258
15·2 Statics of a particle 260
15·3 Problems involving a single rigid body 263
      Theorem of three forces 263
15·4 Problems involving friction 266
15·5 Problems involving several rigid bodies 270
15·6 Problems on frameworks 271
15·7 Graphical statics 278
15·8 Stability of equilibrium 279
15·9 Machines 282
15·(10) The method of virtual work 287
      The method of stationary potential energy 290
      Testing stability of equilibrium 291
15·(11) Miscellaneous examples 292
Examples XV 296

CHAPTER XVI HYDROSTATICS
16·1 Stress and pressure 305
      Solids and fluids 306
      Stress and pressure in fluid at rest 306, 7
      Compressibility ; liquids and gases 307
16·2 Variation of pressure in fluid under gravity 308
      Atmospheric pressure 309
16·3 Thrusts on immersed surfaces 312
      Magnitude and line of action of thrust on plane surface .. 313
      Centres of pressure in particular cases 315
      The principle of Archimedes 318
      Thrust on curved surface 321
16·4 Work and energy considerations 323
      Work of boundary forces 323
      Use of principle of energy 324
16·5 Equilibrium of floating solids 325
      Centre and force of buoyancy 325
      Stability for vertical displacements 325
      Stability for angular displacements ; metacentre 326

16·6	Special consideration of gases	page 329
	Temperature	329
	Ideal gas relations	330
	Mixtures of gases	331
	Isothermal and adiabatic conditions	331
	Use of barometer to determine altitude	332
16·7	Applications	333
Examples XVI		337

CHAPTER XVII APPENDIX

17·1	Vector product	343
17·2	Angular velocity of a rigid body	344
17·3	Rotating frames and time-differentiation	346
17·4	Centrifugal force, Coriolis force, etc.	347
17·5	Newtonian principle of relativity	350
17·6	Newton's law of gravitation	351
17·7	Motion under inverse-square law	352
	Kepler's laws	354
17·8	Motion of artificial satellites	354
17·9	Weightlessness	355

INDEX 359

CHAPTER I

SOME REMARKS ABOUT APPLIED MATHEMATICS

"What's in a name?"

Applied Mathematics is variously defined in various places. The viewpoint taken in this book is that Applied Mathematics is primarily concerned with applying mathematical methods to the investigation of Nature. On this broad interpretation, Applied Mathematics is intimately connected with many parts of Science and plays a key role in the theory of scientific method.

1·1. Science. Science in the narrow sense consists of studies such as Physics, Chemistry, Geology, Biology, Anthropology, etc. In a wider sense, *Science* is knowledge of any form which is acquired by exercise of what is called the scientific method. The common feature of all sciences is this scientific method. (Studies such as 'Political Science' may qualify as sciences if the method is followed sufficiently closely.)

Scientific method involves *observation*, *experiment* and *reasoning*, all of which contribute to experience and knowledge.

1·2. Logic. The reasoning aspect of scientific method is the concern of Logic. One does not read far into a book on logic without discerning that logical questions run deep. Here, brief superficial comment will be made on the two important matters of logical *deduction* and *induction*.

1·21. Deduction. Examples of deductive arguments are:
(i) All readers of this book are intelligent persons.
I am a reader of this book.
Therefore, I am an intelligent person.
(ii) $x^3 + y^3$ contains the factor $x + y$.
$729343 = 90^3 + 7^3$.
∴ 729343 contains the factor 97.

In such arguments, there is ordinarily some interest in the 'truth' of the *conclusion*. This involves two distinct questions: (*a*) the correctness or otherwise of the process of deduction; (*b*) the correctness of the data (*premises*) taken (e.g., the first premise of (i) could conceivably be doubted).

In regard to (*a*), there may be (1) a breach of the accepted rules of deduction; or (2) the rules themselves may be open to question. An argument is said to be *valid* if there is no breach of the form (1); otherwise, invalid. The arguments (i) and (ii) are both valid. The case (2) raises deeper questions which are among those studied in Mathematical Logic.

In regard to (*b*), the premises are frequently conclusions derived in earlier deductive arguments. But evidently there must in any chain of deductive arguments be some initial set of premises not derived in a purely deductive way.

Thus deduction, while it is a very significant part of scientific method, does not establish the 'truth' of any conclusion.

1·22. (Logical) induction.* The inductive side of logic is concerned with setting up generalisations from past experience. Any inductive statement has an element of uncertainty in it, and is granted recognition only as being about the most probable among competing statements, in the light of the available experience. The associated probability is sometimes relatively high, sometimes not so high, and does not reach certainty; the probability is liable to change as experience grows.

Inductive statements are commonly taken as initial premises in chains of deductive arguments. The conclusions in such arguments are evidently also subject to uncertainty.

1·3. Experiment. Experiments provide a good part of the background of experience from which inductive inferences are drawn. An experimenter aims to control so far as he can some of the circumstances attending an observation. He may seek to test, directly or indirectly, some earlier premise, or to set up a new premise.

1·4. Mathematics. A branch of *Mathematics* always contains a deductive structure developed from some specific set of initial premises. The premises selected are as simple, as few in number, as little redundant and as self-consistent as it is possible to make them in the attending circumstances. There is the aesthetic aim of making the structure as little cumbersome as may be, keeping strictly all the while to the allowed rules of deduction.

Statements (*propositions*) validly deduced from the initial premises are commonly said to be 'proved'. But it is well to realise (see §§ 1·21,

---

* Not to be confused with *mathematical induction*.

1·22) that the term *proof* means only that deductive rules have been applied without error.

In most branches of mathematics considerable use is made of *symbols*. These are in large part labour-saving devices; they also help in purging the structure of ambiguities and irrelevancies.

Also in the interests of labour-saving, *definitions* are introduced, whereby brief new names or symbols are assigned to larger groups of words or symbols which would otherwise occur frequently in the structure. Of related character are *conventions*, an example of which is denoting $(a + b) + c$ as $a + b + c$.

The more important deductions in a mathematical structure are called *theorems*. A *corollary* is a fairly immediate deduction from a *theorem*. A *lemma* is an auxiliary statement which is deduced as a preliminary to proving one or a number of theorems.

1·41. Pure Mathematics. In *Pure Mathematics* there is special interest in the deductive structure. In the oft-quoted epigram of Bertrand Russell, pure mathematics 'is the subject in which we never know what we are talking about, nor whether what we are saying is true'. By this, it is meant in part that a pure mathematician aims to select his initial premises independently of happenings in the external world around him. (It is not certain how far he succeeds in this.)

1·42. Applied Mathematics. An applied mathematician is consciously interested in setting up a mathematical structure that relates to his experience of the external world. The initial premises, or *postulates*, are selected with this in view.

It may be remarked that, from a strict point of view, ordinary elementary geometry is a branch of applied mathematics in so far as the postulates are selected with a view to fitting properties of space as observed in the external world.

It needs to be stressed that the postulates in a branch of applied mathematics are of the nature of inductive statements, which are based on experiment and observation. The theoretical structure erected on the premises is a *mathematical model* whose purpose is to represent some field of observations.

In practice, particularly at the elementary stages of study, the postulates are artificially simplified *working hypotheses*, and do not accord (see e.g. § 2·63) to the fullest degree with the available experimental evidence.

This simplification is made for psychological reasons. The postulates of the Relativity Theory embody a richer set of experience than those constituting Newton's laws of motion; but it would be a bold lecturer who would try to impart a knowledge of Mechanics to a normal first year University class through the Relativity postulates.

Later, as the student's experience (including mathematical experience) develops, some of the working hypotheses may need to be amended or replaced. (In this book, that will be done only to a very limited extent.) Moreover, any mathematical model, even if accepted as the best on present evidence, is likely to be improved in the future.

There is thus an accumulation of reasons why the 'answer' to a problem in applied mathematics can never be 'right' in the sense of fitting precisely some feature of the external world. (We have glibly spoken of 'the external world'; we do not here pursue the question as to how far there exists a world outside an individual's imagination.)

On the other hand, we can produce answers in applied mathematics which are 'valid' within the deductive framework constructed on the basis of the adopted postulates, and which we judge to be of great value in helping us to understand the workings of the world into which, somehow or other, we have stumbled.

1·5. The scope of Applied Mathematics. In theory, there is no limit to the observational fields in which the procedures of Applied Mathematics can serve. Applied Mathematics becomes relevant immediately measurements are taken and thought about. Mathematical models abound in the fields of Astronomy, Physics, Engineering and Chemistry, and are being set up increasingly in Geology and Biology. Mathematical methods are starting to be used (with varying degrees of elegance) in studies such as Economics, Psychology, Education, etc.

In practice, however, it has been traditional for those who are called applied mathematicians to confine themselves largely to applications in the physical sciences—Astronomy, Physics, Geophysics, Engineering and Chemistry. At the same time, some applied mathematicians are nowadays venturing further afield.

1·51. Classical Mechanics. This book is confined to one small part of Applied Mathematics, namely the introductory theory of *Classical Mechanics*, i.e. theory developed principally from postulates on the behaviour of matter that had been made up to the time of Newton. The term 'classical' distinguishes this theory from theory based on the postulates of Relativity and Quantum Mechanics.

For several reasons, classical mechanics is normally the first branch of Applied Mathematics to be studied as such. First, it introduces ideas on *force*, *momentum*, *work*, *energy*, etc., which are an important part of general cultural knowledge in this modern world. Secondly, when suitably taught, it exhibits the spirit of applied mathematics very clearly. Thirdly, a sound understanding of the mathematical structure of classical mechanics is a pre-requisite to the understanding of many other aspects of Nature. Fourthly, the subject is of immediate utilitarian value in engineering and other contexts.

In spite of the philosophic emphasis in the present chapter, attention will be paid in the following chapters to the utilitarian as well as the scientific aspects of the subject.

1·6. Further remarks.

1·61. Applied mathematics uses mathematics. Care in argument is required no less than in pure mathematics.

1·62. Applied mathematics involves much more than pure mathematics. It is not merely formal mathematics in which the symbols are largely devoid of meaning in the sense of § 1·41. The solution to a problem generally involves discussion in the English (or other verbal) language, in addition to the use of mathematical language. This discussion is an essential part of the applied mathematical processes, and the part which is commonly worst done.

Often the process of getting a solution will consist of: (i) abstracting from an English statement in a given problem to obtain a set of algebraic or differential equations; (ii) solving these equations; (iii) interpreting the solution found in (ii). The student is warned against ignoring the step (iii).

In addition, applied mathematics is a subject where verbal precision is demanded. For example, the presence or absence of words such as 'constant', 'uniform', 'light', 'about a fixed point' can be highly significant.

1·63. It has already been emphasised (§ 1·42) that answers to problems in Applied Mathematics do not fit *precisely* any feature of the external world. The more modest aim is to set up a mathematical model which has some *correspondence* with a field of observations; the aim is to make the correspondence the *best* possible. There is no pretence of being 'right' in any absolute sense.

1·64. There will be a small number of occasions in this book where the average student's experience in pure mathematics will be insufficient for him to follow a valid deduction of a needed theorem; (this should not apply with most of the set problems). Where this happens, the reader is asked to accept our assertion that the result can be validly deduced, and is advised to substantiate this when his mathematical knowledge becomes adequate. This procedure is thought preferable to the habit (rife in accounts of mechanics in many physics text-books) of deluding the reader with an invalid proof or one insufficiently general for the purpose in hand.

1·65. Some statements and formulae are widely applicable, while others are limited to special contexts. When a formula is applied beyond the context for which it has been proved or stated to apply, the result cannot be held to be validly derived.

1·66. As already pointed out, however, Applied Mathematics involves much more than valid deductive reasoning. A simple early example is the application of the mathematical model theory for a 'particle' (see § 3·13) to physical objects such as trains and cars. Why can theory for a single particle be reasonably applied to a train sometimes, but not always? This is where many students find difficulty in solving problems at the early stages, since judgments resting on practical experience have to be made.

It is not practicable to reduce all the procedures to a short list of formal rules. In any case, any such attempt would defeat the spirit of applied mathematics which is concerned with living observational science quite as much as with formal mathematics. An endeavour to meet the difficulty is made in this book by giving guiding notes in various places. But the student has to build up his own experience. One moral to be drawn is that the ordinary student cannot succeed in this subject unless he works many of the set problems.

1·67. Many problems in mechanics require numerical answers. Measurements of the external world are never precise. Thus when, for example, a mass is stated to be 6·41 kilogrammes, it is understood that the measure is reliable to only three significant figures, and the answer to any calculation using this value is likewise reliable to at most three significant figures. Again, if π is taken as 22/7, it is invalid to presume four-figure accuracy in the answer.

1·68. It is not expected that all of the foregoing remarks will be appreciated on first reading. It is suggested that the reader turn back to this chapter later, as occasions warrant.

1·69. The following problems are designed to set the reader thinking on matters that have some general bearing on the theory to follow. It is not intended that much time should necessarily be spent on these problems. (On the other hand, carefully worded answers to many of the exercises set in subsequent chapters should be sought.)

EXAMPLES I

1. Is the statement 'x satisfies the equation $1 + \sqrt{x} = 2x$' equivalent to the statement 'x satisfies the equation $x = (2x - 1)^2$' ?

2. Why in elementary algebra does one write
$$(-a)(-b) = ab \,?$$

3. A student is required to find the area A of that part of the Earth's surface (assumed spherical, of radius a) which is visible from an aeroplane at height h above the surface. He obtains the result
$$A = \frac{2\pi a^3 h}{a + h}.$$
Why should he know at a glance that he must have made an error ? [Cf. § 3·21.]

4. Each of two pieces of thread, A, B, encircles the Earth once round the Equator. The shorter piece, A, is everywhere in contact with the Earth's surface. The longer piece, B, is outside A and is everywhere distant 1 ft from the surface. Given that the Earth is roughly spherical, find the difference in the lengths of A, B.

[Answer : About 6 ft.]

5. If we use four-figure logarithm tables in the usual way to evaluate
$$\frac{2 \cdot 854 \times 0 \cdot 3687 \times 88 \cdot 53}{0 \cdot 07811 \times 593 \cdot 4 \times 4 \cdot 057},$$
we should obtain 0·4949. If we use the processes of long multiplication and division, or a calculating machine, we should obtain 0·4954, correct to four significant figures. Why the discrepancy ?

6. If each of the numbers in the given fraction in Ex. 5 is given correct to four significant figures, are we entitled to infer that the result 0·4954 is correct to four significant figures ?

7. Would the weight of a given body as indicated on a given spring balance be the same on Mars as on the Earth ? Would the mass be the same ? [See § 3·31.]

8. At the instant when a tram with the brakes hard on comes to rest, a passenger usually jerks *backwards*. Can you explain why ?

9. An apple leaves a tree. What is the basis of your inference that it will fall towards the ground ?

10. Does the Earth revolve about the Sun, or the Sun about the Earth ? Or is the question unsatisfactory ?

11. Criticise the following arguments:
> (i) Henry VIII had six wives; therefore Henry IV had three.
> (ii) If one ship leaving Sydney can cross the Tasman Sea in three days, then three ships leaving Sydney can cross the Tasman Sea in one day.
> (iii) If one man can mow a lawn in 17 minutes, a million men can mow the lawn in about a thousandth of a second.

12. Criticise this sentence: Describe an experiment to prove Boyle's law, namely that the product of the pressure and the volume of a given mass of given gas is constant.
[It is possible to make three separate criticisms.]

13. A car travels from A to B at an average speed of 60 m.p.h., and back from B to A at 20 m.p.h. What is the average speed over the whole journey ?
[Answer: 30 m.p.h.]

14. All the pasture of a field could be eaten by 8 cows in 10 days, or by 6 cows in 15 days. Assuming that each cow eats at the same uniform rate and that the grass grows at a uniform rate, find how long the pasture would last 2 cows.
[Answer : For ever.]

15. A and B start walking down an escalator at the same instant. A takes two steps to each one of B, and reaches the bottom in 60 steps, while B takes 48 steps. What fraction of the distance has B covered when A reaches the bottom, and how many steps are showing in the escalator ? [Answer: Five-eighths; 80.]

16. A man is resting inside an artificial satellite which is in orbit circling the Earth at a constant distance R above the Earth's surface, where R is the Earth's radius. (*a*) Is the man weightless? (*b*) If not, what proportion of his weight at the Earth's surface has he lost? [Answer: (*a*) See § 17·9; (*b*) Three-quarters (see § 17·6).]

CHAPTER II

MOTION OF A POINT IN A STRAIGHT LINE

> " A point once told a line
> ' It's a belief of mine
> that it is merely a pretension
> to have such a thing as a dimension.' "
> —Reproduced by permission of the Proprietors of PUNCH.

In this book, the usual intuitive ideas of time and space are adopted without examination. Thus the postulates include those of elementary geometry.

The study of motion in time and space is called *Kinematics*.

The theory of the present chapter from § 2·2 onward is concerned with the motion of a *point* in a *straight line*. This is to be understood without further mention being necessarily made.

2·1. Units and measures. A first step in measuring is to specify a suitable unit; e.g. the second, hour, day, etc., are units in measuring time. The number of units in a measurement is called the *measure*. If an object is measured in different units, the measure is inversely proportional to the size of the unit; e.g., the measure of a length in yards is one-third of the measure in feet.

In classical mechanics, there are three entities—usually taken as time, length and mass (see § 3·2)—the units of which may be selected to be of any size. When a particular selection is made, a unit system is determined. The units of other entities, e.g. area, velocity, etc. (see later), are uniquely derived from the fundamental units of time, length and mass. The units, fundamental or derived, which apply to a particular unit system will be called 'theoretical units' for that system.

2·11. Units of time and length. Two unit systems, the 'f.p.s.' and the 'c.g.s.' have long been used in mechanics. Latterly, a third system, the 'm.k.s.' has been steadily replacing the 'c.g.s.'. In this book, all three systems will be used, with emphasis on the m.k.s. rather than on the c.g.s.

In all three systems, the theoretical unit of time is the second (denoted as sec or s). In the f.p.s., m.k.s. and c.g.s. systems, the theoretical units of length are the foot (ft), metre (m) and centimetre (cm), respectively.

Efforts are at present being made to encourage the use of SI (Système International) units generally in physical science. The m.k.s. units used in this book conform with and are included in the SI units.

It is a task of specialist physicists to specify the sizes of *standard* units of time and length to the highest precision that experimental science permits. The second, foot, metre, etc. are defined in terms of these standard specifications. In each country, there is a 'standards' laboratory concerned with securing the closest possible conformity in practice with the agreed international standard units.

Units such as the mile, kilometre, ounce, ton, minute, hour etc. are not theoretical units in any of the above three systems. All such units will be called 'practical units' in relation to the system concerned.

In the theoretical structure of mechanics, symbols in formulae stand for measures in some unit system. A particular unit system need be specified only when numerical applications are being made; but the selected unit system must be adhered to throughout the main course of the numerical calculations. *If units which are 'practical' in relation to the chosen unit system appear in the data of a problem, measures must be converted to measures in theoretical units before being fed into formulae.*

2·12. A convention. In ordinary algebra, symbols denote numbers. Correspondingly, in a phrase such as 'a length s ft', s denotes the measure of a length in feet. It is convenient, however, to refer sometimes to to 'a length s', meaning 'a length s *theoretical units* (in some theoretical system not necessarily specified)'. In problems where a particular unit system has to be specified, this convention has to be limited to usage of the type illustrated in Ex. (1) of § 2·51 and Ex. (1) of § 2·82; statements such as 'length $= 5$' are not tolerated because they fail to convey a precise meaning.

(Another convention is used in the *Stroud system*, where one can write, for example,

$$s = \tfrac{1}{2} \text{ mile} = 2{,}640 \text{ ft},$$

the s here denoting the length and not merely its measure. The Stroud system has some advantages.)

2·2. Position coordinate at time t. Consider a point moving in the straight line OX. Let P be its position on OX at time t; by 'at time t' we mean at the instant at which the elapsed time since the point occupied some specific initial position, A say, is t. Then its *position coordinate x*, relative to O, at time t is the distance from O to P with sign attached according to the sense of OP. The term *distance* is used when we are not concerned with the direction.

2·3. Velocity. Let Q be the position of the moving point at time $t + \Delta t$, and let $OQ = x + \Delta x$. The mean velocity relative to O over the time interval Δt is defined as $\Delta x/\Delta t$.

The phrase 'relative to the origin' is to be understood in the sequel in connection with all position coordinates, displacements, velocities and accelerations. The phrase will be written down explicitly when there might otherwise be uncertainty (see § 2·61) or ambiguity (see § 2·8) as to the particular origin involved. (See also the second paragraph of Chapter III.)

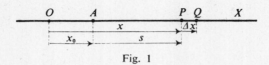

Fig. 1

Knowledge of the mean velocity over an interval Δt gives less information about the point's motion in this interval than would be the case if the interval were subdivided and the mean velocity determined for each sub-interval. Such subdivision may be continued indefinitely. This focuses attention on the important concept of *velocity at an instant*.

The velocity v at time t is defined as $\lim\limits_{\Delta t \to 0} \left(\dfrac{\Delta x}{\Delta t} \right)$, i.e., the *rate of change* of the position coordinate. Thus

$$v = \frac{dx}{dt} \quad (=\dot{x}), \tag{1}$$

where a dot placed on top of a symbol (as will be the case throughout this book) denotes the operation d/dt.

As with position coordinate, the direction as well as the magnitude of a velocity is significant; the measure of a velocity may be negative. The term *speed* is used when the direction is not being considered.

The theoretical f.p.s., m.k.s. and c.g.s. units of velocity are the foot per second (denoted as ft/sec or ft-sec^{-1}), metre per second (m/s), and centimetre per second (cm/sec).

It is useful to know that a speed of 15 m.p.h. (mi/h, or miles per hour) is equal to a speed of 22 ft/sec. This is easily checked.

2·4. Acceleration. The acceleration, f say, of the moving point, at any instant, is defined as the rate of change of velocity at that instant. Thus

$$f = \frac{dv}{dt} (= \dot{v}) = \frac{d^2x}{dt^2} (= \ddot{x}). \tag{2}$$

The theoretical f.p.s., m.k.s. and c.g.s. units are the foot per second per second (denoted as ft/sec² or ft-sec⁻²), the m/s² and the cm/sec², respectively.

Again, direction as well as magnitude is significant. A *retardation* of say α ft/sec² (α would usually be positive when the term 'retardation' is used) is an acceleration of $-\alpha$ ft/sec².

2·41. Acceleration expressed as a space-derivative. In (2), f is expressed as a differential coefficient with respect to t. On very many occasions it is useful to have f expressed as a derivative with respect to x. The formula is

$$\ddot{x} = \frac{d}{dx}(\tfrac{1}{2}\dot{x}^2) \quad \left(\text{or } f = \frac{d}{dx}(\tfrac{1}{2}v^2)\right). \tag{3}$$

The proof is:

$$\ddot{x} = \frac{d\dot{x}}{dt} = \frac{d\dot{x}}{dx}\frac{dx}{dt} = \dot{x}\frac{d\dot{x}}{dx} = \frac{d}{dx}(\tfrac{1}{2}\dot{x}^2).$$

2·5. Motion in a straight line ; uniform acceleration. Take circumstances as in § 2·2 (Fig. 1). Let x, v, f be as already defined; let x_0 and u be the initial position coordinate and velocity; at time t, let the *displacement* $AP = x - x_0 = s$. We are here given that f is *uniform*, i.e., is constant.

Summing up the data in symbolic form, we have the (very simple) differential equation

$$\ddot{x} = f, \text{ where } f \text{ is constant;} \tag{i}$$

together with the initial conditions

$$x = x_0 \text{ and } v \equiv \dot{x} = u, \text{ when } t = 0. \tag{ii}$$

Integrating* (i) with respect to t, we have

$$\dot{x} = ft + C,$$

where C is an integration constant. From (ii),

$$u = 0 + C.$$

Hence, eliminating C,

$$(\dot{x} =) \; v = u + ft. \tag{4}$$

Integrating (4) with respect to t, and proceeding similarly, we obtain

$$(x - x_0 =) \; s = ut + \tfrac{1}{2}ft^2. \tag{5}$$

Using (3) to express (i) in the form

$$\frac{d}{dx}(\tfrac{1}{2}v^2) = f, \quad \text{where } f \text{ is constant,}$$

---

* The reader who cannot as yet integrate is advised to accept this and similar results for the present, and check them later when his pure mathematics has 'caught up'. In most of the exercises set in the earlier chapters of this book the calculus is not needed.

and integrating now with respect to x, we obtain in a similar way
$$v^2 - u^2 (= 2f(x - x_0)) = 2fs. \tag{6}$$
(Note that (6) may be alternatively obtained by eliminating t between (4) and (5).)

The formulae (4), (5), (6) are of central importance in uniformly accelerated rectilinear motion. Two others, viz.,
$$s = \tfrac{1}{2}(u + v)t,$$
$$s = vt - \tfrac{1}{2}ft^2;$$
which are deduced by simple algebra from (4), (5), and (6), are useful in some problems.

In some contexts, it is convenient to interpret x and v as *general* values of the position coordinate and velocity; in others, as the *final* values for a particular motion. (And similarly with s.)

2·51. Examples.

(1) *A train running at 60 m.p.h. can be brought to rest by its brakes in 900 yards. Assuming that when the brakes are on there is always a constant retardation, find the speed of the train when it could be brought to rest in 100 yards.*

We first convert the data to give measures in f.p.s. units.

Thus in the first motion (the direction of motion being taken positive),
the initial velocity ('u'*) = 88 ft/sec ;
the displacement ('s'*) = 2700 ft.

In the second motion,
the initial velocity = V ft/sec, say;
the displacement = 300 ft.

Let the (constant) acceleration = f ft/sec² (f is of course negative).

Using '$v^2 = u^2 + 2fs$', we then have
(i) $0 = 88^2 + 5400f$;
(ii) $0 = V^2 + 600f$.

Hence $V^2 = 88^2/9$; and the positive value of V is relevant.
Thus the required speed = 88/3 ft/sec = 20 m.p.h.

(2) *A point is moving in a straight line with uniform acceleration. It is seen to cover a distance of* 15 m *in the third second of the motion and* 19 m *in the fifth second. How far does it go in the eighth second?*

Lemma. Distance in the nth unit of time (in usual notation)
$= un + \tfrac{1}{2}fn^2 - u(n-1) - \tfrac{1}{2}f(n-1)^2$, by (5),
$= u + f(n - \tfrac{1}{2})$.

Hence $15 = u + 5f/2,$
$19 = u + 9f/2,$
$s = u + 15f/2,$ where s m is the required distance.

These equations give $u = 10, f = 2, s = 25.$
Thus the required distance is 25 m.

* Units are understood; see §2·12.

(3) *A point starting at A moves in a straight line AX with initial velocity 1·0 m/s in the direction AX and a constant retardation of 0·2 m/s². (a) When is its velocity zero ? What is then its distance from A ? (b) What is its velocity when its distance in front of A is 0·9, 2·5, 3·0 m, respectively ? (c) When is it distant 2·1, 2·5, 3·0 m, respectively, in front of A, and when 3·9 m behind ?*

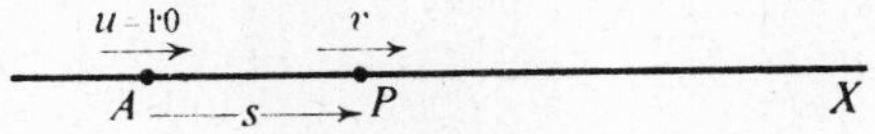

<pre>
          u = 1·0   v
           ——>   ——>
    ————————•———————————————•———————————————————
    A ——s——>P   X
</pre>

Fig. 2

Select the direction AX as the positive direction for all reference purposes.

Let u, v, f, s, t be as defined in § 2·5; s and v correspond to any time t; u is the initial value of v.

(a) We have $u = 1·0$, $f = -0·2$, $v = 0$, $s = ?$, $t = ?$

By (4), $0 = 1·0 - 0·2t$.
$\therefore t = 5$.
By (5), $s = 1·0 \times 5 + \frac{1}{2}(-0·2)5^2$
$= 2·5$.

Hence when the velocity of P is zero, the elapsed time is 5 sec, and the displacement is 2·5 m (in the positive direction, of course).

(b) We have
$u = 1·0$
$f = -0·2$
$s =$ (i) 0·9, (ii) 2·5, (iii) 3·0
$v = ?$

By (6),
$v^2 = 1·00 - 0·4(0·9, 2·5, 3·0)$
$= 0·64, 0, -0·2$;
i.e.,
$v =$ (i) $\pm 0·8$, (ii) 0,
(iii) $\pm\sqrt{(-0·2)}$.

In case (i), the required velocity is 0·8 m/s in either of the directions AX, XA.

In case (ii), the velocity is zero (given twice over), consistently with the result in (a).

In case (iii), the formal answer is not real. The interpretation is that the point is never 3·0 m in front of A. (In the light of (a) and (ii), it is evident that s cannot exceed 2·5.)

(c) We have
$u = 1·0$
$f = -0·2$
$s =$ (i) 2·1, (ii) 2·5, (iii) 3·0
  (iv) $-3·9$
$t = ?$

By (5),
2·1, 2·5, 3·0, $-3·9 = 1·0t - 0·1t^2$;
i.e.,
$t^2 - 10t + (21, 25, 30, -39) = 0$;
i.e.,
$t =$ (i) 3 or 7, (ii) 5,
(iii) $5 \pm \sqrt{(-5)}$,
(iv) 13 or -3.

In case (i), both times are relevant; after 3 sec, the point is moving in the direction AX, and after 7 sec, XA.

In case (ii), there is just one elapsed time, 5 sec, (given twice over, corresponding to repeated roots of the quadratic equation in t). This is because (see (b, ii), also (a)) the velocity is zero when $s = 2 \cdot 5$.

In case (iii), the answer is not real, for the same reason as in (b, iii).

In case (iv), the elapsed time is 13 sec. The value $t = -3$ is not relevant, as the data in the question give no information as to any motion prior to the time $t = 0$.

(The irrelevant answer shows that if the point had been moving with the given acceleration $-0 \cdot 2$ m/s prior to passing through A in the sense AX, then it would have been $3 \cdot 9$ m behind A at a time 3 sec before first reaching A.)

(4) *A point moving with constant acceleration from A to B in the straight line AB has speeds u, v at A, B respectively. (a) Find its speed at C the mid-point of AB. (b) Show that if the time from A to C is twice that from C to B, then $v = 7u$.*

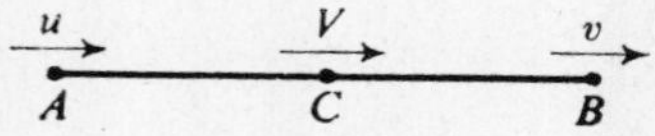

Fig. 3

Take the positive direction as AB.

Let $AB = s$; let the velocity at $C = V$.

(a) For the motion from A to C and from C to B, we have

$$\left.\begin{array}{l} \text{`}u\text{'} = u \\ \text{`}s\text{'} = \tfrac{1}{2}s \\ \text{`}v\text{'} = V \end{array}\right\} \qquad \left.\begin{array}{l} \text{`}u\text{'} = V \\ \text{`}s\text{'} = \tfrac{1}{2}s \\ \text{`}v\text{'} = v \end{array}\right\}$$

respectively, where 'u', 's', etc., correspond to symbols in § 2·5. (The use of inverted commas in this way avoids ambiguities.)

To use the equations in § 2·5, we need a fourth item in each case; we take 'f' rather than 't', because the acceleration, f say, is common to both motions.

Using '$v^2 = u^2 + 2fs$', we then have

$$V^2 = u^2 + 2fs/2,$$
$$v^2 = V^2 + 2fs/2.$$

Eliminating f, we have $V^2 = \tfrac{1}{2}(u^2 + v^2)$, giving
$$V = \pm \sqrt{\{\tfrac{1}{2}(u^2 + v^2)\}}.$$

Only the positive result is relevant.

(b) Let t be the time from C to B. Then we have

$$\left.\begin{array}{l} \text{`}u\text{'} = u \\ \text{`}s\text{'} = \tfrac{1}{2}s \\ \text{`}t\text{'} = 2t \\ \text{`}v\text{'} = V = \sqrt{\{\tfrac{1}{2}(u^2 + v^2)\}} \end{array}\right\} \qquad \left.\begin{array}{l} \text{`}u\text{'} = V = \sqrt{\{\tfrac{1}{2}(u^2 + v^2)\}} \\ \text{`}s\text{'} = \tfrac{1}{2}s \\ \text{`}t\text{'} = t \\ \text{`}v\text{'} = v \end{array}\right\}$$

Using '$s = \tfrac{1}{2}(u + v)t$', we then have

$$\tfrac{1}{2}s = \tfrac{1}{2}(u + V)2t, \qquad \tfrac{1}{2}s = \tfrac{1}{2}(V + v)t.$$

Eliminating t, we derive
$$v - 2u = V = \sqrt{\{\tfrac{1}{2}(u^2 + v^2)\}}. \tag{i}$$

Squaring would give
$$v^2 - 8uv + 7u^2 = 0,$$
yielding
$$v = 7u \text{ or } v = u. \tag{ii}$$

The result $v = u$ is invalid; it would imply that the time from A to C is equal to the time from C to B, which disagrees with the data. It arises from the step of squaring. The valid argument is that if (i) holds, then v/u has *one* of the values given by (ii).

The other result, $v = 7u$, fits all requirements.

S.P.I.C.E.* The above examples have been chosen to illustrate (while we are on easy problems) points which are important throughout the study of mechanics.

(α) In formulae such as (1), (2), (3), (4), (5), (6), and most others to follow, the symbols may represent negative as well as positive numbers.

(β) It is desirable at the outset in the solution of any problem to take one suitable direction as positive, and to relate *all* measures to this direction throughout the course of the solution.

(γ) The formal solution of the equations involved may yield more than one unique result. This may be because two (sometimes more) distinct results are actually compatible with the data, as in Ex. (3), (b, i) and (c, i). It may be that one or more of the results is irrelevant, as in Ex. (3), (c, iv), though compatible with the equations used. Again it may be that one of the results is invalid, as in Ex. (4), (b), due to a peculiarity of the algebraic process followed.

Further, it may be that the data are incapable of yielding a real answer, as in Ex. (3), (b, iii) and (c, iii).

These features illustrate that the *interpretation* of the formal mathematical results yielded is a matter of much significance; (cf. § 1·62).

(δ) The lemma in Ex. (2) is in some treatments given as 'bookwork'. We suggest that the student should know how to derive such auxiliary formulae when needed, rather than that he should try to memorise them (see § 5·3).

(ϵ) Ex. (1) is a case in which the formal theory is applied to an object, in this case a train, not included in the theory (see § 1·66). We in effect replace the train by a 'representative point'; it is left to the student's intuition to discern that the representation is suitable for the context. Similar use of intuition is needed in many problems to follow.

(ζ) Ex. (1) also illustrates the need for translating to measures in theoretical units before feeding into formulae. As a matter of courtesy, one usually translates back at the end and gives the answer in practical units corresponding to those in the data.

(η) In Ex. (1), the displacements, velocities and acceleration are understood to be relative to the Earth's surface.

* Special point(s) in choosing examples.

(θ) It will be noticed that the answer to Ex. (4) (*a*) is not the simple mean of *u* and *v*. Blind use of 'averages' leads to frequent errors.

2·6. Vertical motion under gravity.

2·61. Experimental background. Let P be any point of an ordinary body which is let fall from rest, relative to the Earth, in a vacuum in a limited region near the Earth's surface. Observational evidence indicates* that, relative to the Earth: (*a*) P moves approximately in a straight line; (*b*) the direction of the line, and the acceleration of P, are approximately constant and independent of the particular body. These effects are attributed to the gravitational attraction of the Earth, and the acceleration is called the 'acceleration due to gravity'.

2·62. Remarks. The inductive statement in § 2·61 is capable of much elaboration. Following are some of the main points:

(i) Refined experiments indicate that in general the path of P deviates slightly from a straight line and that the acceleration changes to some degree in both magnitude and direction during the motion. The smaller the dimensions of the region involved, compared with the Earth's radius, the less physically significant are these deviations.

(ii) The magnitude of the acceleration at points near the Earth's surface is found to be approximately

$$g_0(1 - 0.0053 \cos^2 \phi),$$

where ϕ is the *latitude*, i.e. the angular distance from the equator; and the measure of g_0 is 9·832 m/s², or 32·26 ft/sec².

Thus the magnitude ranges from about 9·780 m/s² at the equator to about 9·832 m/s² at the poles; i.e. from about 32·09 to about 32·26 ft/sec². (At Sydney, the value is 9·796 m/s² or 32·14 ft/sec².)

(iii) Through appreciable distances above the Earth's surface, the magnitude of the acceleration due to gravity varies approximately inversely as the square of the distance from the Earth's centre.

(iv) The *downward vertical* at any point near the Earth's surface is defined as the direction of the acceleration due to gravity at the point. The *horizontal* at the point is the plane at right angles to the vertical.

(v) The downward vertical is not precisely normal to the Earth's surface, nor does it in general pass through the Earth's centre. It is affected by (*a*) the Earth's rotation (see § 7·6, Ex. (5), (*b*)) the Earth's ellipticity of figure, (*c*) further irregularities in the Earth's internal and surface structure.

(vi) Further comment could be made on the words 'ordinary body' and 'vacuum'.

* The word 'indicates' is intended to convey a meaning somewhere between 'suggests' and 'shows'.

2·63. Working hypothesis. An artificial simplification will now be made by taking as a working hypothesis (see §1·42) that the acceleration due to gravity is constant in magnitude and direction in problems to be considered.

The magnitude of the acceleration due to gravity, commonly denoted by g, will, moreover, usually be taken to be either 32 ft/sec² or 9·80 m/s² (or 980 cm/sec²). It should be noted that, in the light of § 2·62 (ii), this simplification introduces a numerical error which may reach the order of 8 or 3 parts in 1,000 in the two cases, respectively.

In problems on motion through the atmosphere, we shall, further, unless the contrary is stated, ignore air-friction.

This mathematical model representation serves for the limited range of problems treated in this book. There exist important contexts where the model is seriously inadequate, and its limitations should not be forgotten. Similar remarks apply to various other mathematical models met later in the book.

2·64. Example. *A balloon has been ascending vertically from the ground at a constant speed for* 10 *sec. A stone is then let fall from the balloon and reaches the ground after a further 5 sec. Find the speed of the balloon and its height at the instant when the stone was let fall.*

Fig. 4

Take the positive direction vertically upwards.
Let the speed of the balloon = u ft/sec.
Let the required height = h ft.
Then considering the motion of the stone, we have

$\left.\begin{array}{l}\text{`}u\text{'} = u \\ \text{`}t\text{'} = 5 \\ \text{`}f\text{'} = -g = -32 \\ \text{`}s\text{'} = -h.\end{array}\right\}$ Hence, by '$s = ut + \tfrac{1}{2}ft^2$', we have
$-h = 5u - \tfrac{1}{2} \times 32 \times 5^2.$ (i)

Considering the motion of the balloon, we have
$$h = 10u. \tag{ii}$$

From (i), (ii), we derive $u = 26.7$, $h = 267$.
Thus the required speed and height are 26·7 ft/sec and 267 ft.

S.P.I.C.E. To illustrate (α) the use of the working hypothesis of § 2·63, (β) the handling of signs.

2·7. Motion in a straight line; variable acceleration. If the acceleration is not constant, the formulae (4), (5), (6), etc., of § 2·5 are not relevant. The *method* of § 2·5 is, however, relevant in so far as we often have to solve a differential equation subject to certain initial conditions. No additional theory of mechanics is needed, but some degree of skill in elementary calculus operations is called for.

(If the student lacks this skill as yet, he may ignore the following example until he reaches § 8·5. It is more important just now that he should appreciate the first sentence of § 2·7. Accelerations are constant in so many of the elementary problems set in mechanics that there is a risk of the student deifying the formulae (4), (5) and (6).)

2·71. Example. *A point P moves in a straight line with an acceleration whose direction is always towards a fixed point O in the line and whose magnitude is proportional to the distance OP. Initially the point is at rest at A, distant a from O. Find expressions giving (a) the speed of P in any position; (b) the position of P at any time.*

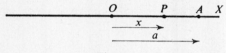

Fig. 5

Take the positive direction in the sense OA.
Let x be the position coordinate of P at time t.
Then (cf. § 2·5 (i), (ii)), we are given:
$$\ddot{x} = -n^2 x, \qquad \text{(i)}$$
where n is a real constant, together with the initial conditions
$$x = a \quad \text{and} \quad \dot{x} = 0, \quad \text{when } t = 0. \qquad \text{(ii)}$$
(The student should check that (i) holds when P is to the left as well as when P is to the right of O.)

Integrating (i) with respect to x (*Note*: x, not t), we have, taking (3) into account,
$$\tfrac{1}{2}\dot{x}^2 = -\tfrac{1}{2}n^2 x^2 + C,$$
where C is an integration constant. Substituting in this from (ii), we have
$$0 = -\tfrac{1}{2}n^2 a^2 + C,$$
and therefore
$$\dot{x}^2 = -n^2 x^2 + n^2 a^2. \qquad \text{(iii)}$$
Hence when the displacement is x, the speed is $n\sqrt{(a^2 - x^2)}$.
From (iii), we have
$$\pm (a^2 - x^2)^{-\frac{1}{2}} \, dx/dt = n.$$

Integrating with respect to t, we have
$$\mp \cos^{-1}(x/a) = nt + \epsilon,$$
i.e.
$$x = a \cos(nt + \epsilon),$$
where ϵ is an integration constant. By (ii), we take $\epsilon = 0$.
Hence the position at time t is given by
$$x = a \cos nt.$$

S.P.I.C.E. This problem is important in connection with simple harmonic motion—see Chapter VIII.

2·8. Change of reference origin. Suppose now that there are two reference origins O, O', where O' is in motion relative to O; and let P

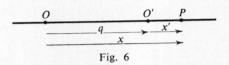

Fig. 6

be any other moving point. Let OO', OP, $O'P$ (with sign significance attached according to the order of the letters) be denoted by q, x, x', respectively; thus $x' = x - q$.

Differentiating with respect to the time t, we then have
$$v_{P \text{ rel } O'} = v_{P \text{ rel } O} - v_{O' \text{ rel } O}, \tag{7}$$
where $v_{P \text{ rel } O'}$ denotes the velocity of P relative to O', etc.

A similar formula holds for relative accelerations.

2·81. Applications. In a practical problem, there is often one reference origin which is regarded as being 'more basic' than others. For example, an origin O taken as a point attached to the Earth's surface would commonly be regarded as 'more basic' than one, O' say, attached to a body moving relatively to the Earth's surface. The solution of some problems is, however, sometimes expedited by changing from O to O' as reference origin, and formulae such as (7) enable this to be done. (In (7), the right-hand side is often written as '$v_P - v_{O'}$', the 'rel O' being understood.)

In a problem where the acceleration of P relative to O' is constant, we may use (4), (5), (6), with the interpretation that x, v and f now denote the position coordinate, velocity and acceleration of P relative to O'; (x_0, u are the initial values of the new x, v; t is still the elapsed time).

For the special case of two bodies moving under gravity in a vertical line, the acceleration of either one relative to the other $= g - g = 0$.

Thus there is the specially simple feature that the velocity of the one relative to the other is constant.

2·82. Examples. (1) *A body A is projected vertically upwards from a point O with speed* 128 *ft/sec. Five seconds later a body B is let fall from rest at O down a mineshaft. Find when and where A meets B.*

Take the positive direction vertically upwards.

Take the instant when *B* starts as the initial instant.

It is easy to show, using equations (5) and (4), that at the initial instant, the position coordinate and velocity of *A*, both taken relative to *O*, are 240 ft and -32 ft/sec, respectively.

Then, relative to *A*, we have:

Initial position coordinate, x_0* say, of $B = -240$ ft.

Position coordinate, x* say, of B when A, B meet is zero.

By (7), initial velocity, u* say, of $B = 0 - (-32)$ ft/sec
$= 32$ ft/sec.

Acceleration of B is zero.

If t sec is the elapsed time when A, B meet, then since '$x - x_0 = ut + \frac{1}{2}ft^2$', we have
$$0 - (-240) = 32t; \text{ i.e., } t = 7\cdot5. \tag{i}$$

Hence the two bodies meet 7·5 seconds after B starts.

Now considering the motion of B relative to O, we easily calculate, using (i), that the place where A, B meet is 900 feet below O.

(2) *A tram starts from rest with constant acceleration a. Simultaneously a man distant b behind the tram starts to chase the tram with constant speed V. Prove that he can catch the tram if* $V^2 \geqslant 2ab$.

Take as positive direction the direction of motion of the tram.

Relative to the tram:

Initial coordinate of man $(x_0) = -b$;
Final coordinate of man $(x) \quad = 0$;
Initial velocity of man $(u) \quad = V - 0 = V$;
Acceleration of man $(f) \quad = 0 - a = -a$.

Let t be the elapsed time if and when man catches tram.

Then since $\quad x - x_0 = ut + \frac{1}{2}ft^2$,
we have $\quad b = Vt - \frac{1}{2}at^2$,
yielding $\quad t = \{V \pm \sqrt{(V^2 - 2ab)}\}/a$.

The man will catch the tram if t is real, i.e. if $V^2 \geqslant 2ab$.

S.P.I.C.E. (α) The above examples are chosen to illustrate the method of change of origin, and also give further experience in the handling of signs.

(β) It may be noted that the examples could have been solved independently of the method of change of origin.

* Units are understood; see § 2·12.

E.g., in Ex. (1), let t sec be (as before) the time between the starting of B and the meeting of A, B; and let s ft be the height above O of the point where A, B meet. Considering the motions of A, B, separately, we can then write
$$s = 128(t+5) - \tfrac{1}{2} \times 32(t+5)^2,$$
$$s = 0 \times t - \tfrac{1}{2} \times 32t^2,$$
yielding $t = 7\cdot 5$ and $s = -900$, and hence the same answer as before.

(γ) The reader may be amused to interpret the two values yielded for t in Ex. (2).

(δ) In Ex. (1), it is tacitly assumed of course that the mineshaft is at least 900 ft deep.

2·9. Velocity-time graph. The graph of the velocity v of a moving point P against the time t has these two important properties:

(i) The gradient at any point of the graph, being dv/dt, represents (by equation (2)) the acceleration of P at the corresponding instant.

(ii) The area between a pair of ordinates, e.g. the shaded area of Fig. 7, being expressible in the form $\int_{t_1}^{t_2} v\, dt$, represents (by equation (1)) the displacement during the interval from t_1 to t_2.

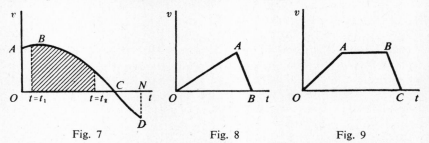

Fig. 7   Fig. 8   Fig. 9

In Fig. 7, notice these incidental features: between (the instants corresponding to) A and B, the acceleration is positive; at B the velocity is a maximum, the acceleration zero; between B and D, the acceleration is negative; at C the velocity is zero; between C and D the velocity is negative. The displacement corresponding to the whole curve $ABCD$ is given by the magnitude of the area of $OABC$ minus the magnitude of CDN.

Fig. 8 would correspond to the motion of a vehicle between two stops in which there is uniform acceleration, equal to $\tan \angle BOA$, for a time, followed immediately by uniform retardation, equal to $\tan \angle ABO$, until the vehicle again comes to rest. In Fig. 9, there is the further feature that a steady speed is maintained in between the accelerating and the retarding stages.

Problems in which the motion conforms to Fig. 8 or Fig. 9 (and some other particular types) are specially quickly solved by use of velocity-time graphs. The solutions are reduced to the solving of fairly simple pieces of geometry or trigonometry in place of more complicated algebra.

If the accelerations are (as in Fig. 7) not uniform the problem may be approximately solved by measurements on the graph. In this book, however, the set problems are mainly solvable by 'exact' mathematical processes.

2·91. Example. *A train T_1 passed a station S at 20 mi/h, maintained this speed for 3 miles, and was then uniformly retarded, coming to rest at a second station S', $3\frac{1}{2}$ miles from S. Prove that the time from S to S' was 12 minutes.*

A second train T_2 started from rest at S at the instant when T_1 passed through S and arrived at rest at S' simultaneously with T_1. T_2 was first uniformly accelerated, then ran at constant speed over one mile and finally was uniformly retarded. Find the constant speed.

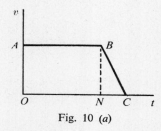
Fig. 10 (a)

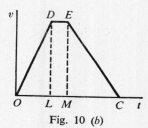
Fig. 10 (b)

In Fig. 10 (a), ABC is the v-t graph for T_1; BN is perpendicular to OC.

The data of the question (converted to measures in f.p.s. units) and the properties referred to in § 2·9 give:

$$OA = 88/3; \quad OA.ON = 3 \times 5280; \quad \tfrac{1}{2}OA.NC = 2640.$$

We are required to find OC.

We deduce:
$$ON = 9 \times 5280/88 = 540;$$
$$NC = 3 \times 5280/88 = 180.$$

Hence $OC = 720$.

Thus the time taken by T_1 to go from S to S' was 12 minutes.

In Fig. 10 (b), $ODEC$ is the v-t graph for T_2; DL, EM are perpendicular to OC.

We are now given:
$$OC = 720; \quad \text{area } LDEM = 5280; \quad \text{area } ODEC = 7 \times 5280/2.$$

We are required to find LD.

We have:
$$LM.LD = 5280,$$
$$\tfrac{1}{2}(OC + DE).LD = 7 \times 5280/2;$$
whence
$$OC.LD = 6 \times 5280.$$

Hence $LD = 6 \times 5280/720 = 44$.

Thus the steady speed reached by $T_2 = 44$ ft/sec $= 30$ mi/h.

S.P.I.C.E. (α) In this problem, it would have been quite safe, and more rapid in fact, to have worked in miles and hours instead of feet and seconds. The student is nevertheless advised to work in general in the f.p.s., m.k.s. or c.g.s. system for the present.

(β) The data do not fix the gradients of OD or EC; but these gradients are not needed in getting the answer.

(γ) The student should solve the problem independently by direct use of formulae such as (4), (5) and (6)—and thereby appreciate the amount of effort saved by using the v-t graph.

(δ) The motions of the two trains need not have been in straight lines. The student will discern in fact that much of the theory of this chapter can be applied to motion along a curved line, provided the symbols are appropriately interpreted, e.g. the 'x' in §§ 2·2, 2·3, 2·4, 2·5 as an arc-length.

EXAMPLES II

1. If the speed v and the position coordinate x of a point moving in a straight line are connected by the relation $v^2 = a + bx$, where a, b are constants, prove that the acceleration is constant and equal to $\frac{1}{2}b$.

2. Given that 1 knot is a speed of 1 sea-mile (6,080 feet) per hour, express in m/s² an acceleration of 20 knots per hour. [Answer: 0·00286 m/s².]

3. The speed of a train increases from 30 to 40 m.p.h. while the train covers a mile. Find the acceleration, assumed constant, and the time taken to cover the mile.

[Answers: 0·143 ft/sec²; 103 sec.]

4. A point moves from O in the straight line $X'OX$ with an initial velocity of 20 ft/sec in the positive direction, the acceleration being constant. After 20 sec the point is again at O. (a) Find the acceleration and the velocity at the instant of the return through O. (b) Find the position of the point when its velocity is zero. (c) Find when the displacement is 64 ft on the positive side of O, and when 44 ft on the negative side of O.

[Answers: (a) -2 ft/sec²; -20 ft/sec; (b) 100 ft from O; (c) after 4, and also after 16, sec; after 22 sec.]

5. At time t seconds after starting, the position coordinate x cm of a point moving in the straight line $X'OX$ is given by $x = 4 + 4t - 3t^2$. (a) Prove that the acceleration is constant, and find its value. (b) Find when the point is at O. (c) Find the position and velocity when the elapsed time is 5 seconds. (d) Find the velocity when (i) $x = 4$, (ii) $x = -11$.

Illustrate the results in (b), (c), (d) on a graph of x against t.

[Answers: (a) -6 cm/sec²; (b) after 2 sec; (c) -51 cm; -26 cm/sec; (d) (i) ± 4 cm/sec; (ii) -14 cm/sec.]

*6. At time t seconds after starting, the position coordinate x m of a point moving in the straight line $X'OX$ is given by $x = 2t^3 - t^2 - 4t + 3$. Find: (a) the initial

* An asterisk before the number of an example denotes that the example is a little harder or uses slightly more advanced pure mathematics than neighbouring examples.

position, velocity and acceleration; (b) the velocity and the acceleration when x is zero; (c) the position and velocity when the acceleration is zero.

Illustrate the results on a graph of x against t.

[Answers: (a) 3 m from O; -4 m/s; -2 m/s^2; (b) zero; 10 m/s^2; (c) 2·31 m from O; $-4·17$ m/s.]

7. A ball is thrown vertically upwards from a point A with initial speed V. Find the greatest height reached and the time that elapses before the ball returns to A.

[Answers: $\frac{1}{2}V^2/g$; $2V/g$.]

8. An object falling in a vertical line passes a window 6 ft high in 0·1 sec. Find to the nearest foot the distance above the window sill of the point from which the object was let fall. [Answer: 59 ft.]

9. A juggler throws up six balls, one after the other at equal intervals of time t, each to a height of 9 ft. The first ball returns to his hand at time t after the sixth was thrown up and is immediately thrown to the same height, and so on continually. Find the heights of the other balls at an instant when any one reaches the juggler's hand.

[Answer: 5, 8, 9, 8, 5 ft.]

10. A rough and ready form of photographic shutter is made of a vertical sheet of wood containing a horizontal slit. The wood is let fall freely from a position in which the bottom edge of the slit is 6 inches above the centre of the camera lens. Find the vertical breadth of the slit in order that it may take one-fiftieth of a second in passing the centre of the lens. [Answer: 1·43 in.]

*11. The depth of a well is to be measured by dropping in a stone and observing the time T which then elapses before the impact at the bottom is heard. Prove that the depth of the well is given by the smaller root of the equation

$$gx^2 - 2x(V^2 + gVT) + gV^2T^2 = 0,$$

where V is the (assumed constant) speed of sound in air.

Hence show, with the help of a binomial expansion, that if gT is small compared with V, the depth of the well is approximately $\frac{1}{2}gT^2 - \frac{1}{2}g^2T^3/V$.

If $T = 1·5$ sec and $V = 1,100$ ft/sec, find approximately the percentage error that would be made in estimating the depth if no allowance were made for the time occupied by the sound travelling up the well. [Answer: 4·4%.]

12. A stone P is let fall from rest under gravity from the point A. After time T a second stone Q is let fall also from rest at A. Find the distance between P and Q after a further time t. [Answer: $\frac{1}{2}gT(T + 2t)$.]

13. An object is projected vertically upwards from the point A with speed 49 m/s. Two seconds later, a second object is projected vertically upward from A with the same speed. Find when and where the two objects meet.

[Answers: 6 s after the first object was projected; 117·6 m above A.]

14. A car P starts from rest from a point A of a straight road, and moves with uniform acceleration 3 ft/sec^2. Five seconds later, another car Q starts from rest at A and chases P with uniform acceleration 4 ft/sec^2. Find: (a) the greatest distance of P in front of Q; (b) the time Q is in motion before it overtakes P.

[Answers: (a) 150 ft; (b) 32·3 sec.]

15. A motor car P is running in front of a second motor car Q, both having constant speeds of 45 m.p.h. At a certain instant P starts to brake with a constant retardation of 15 miles per hour per second; $\frac{1}{2}$ sec later, Q starts to brake with a constant retardation of $7\frac{1}{2}$ miles per hour per second. Find the minimum initial distance between the cars if Q does not bump P. [Answer: 132 ft.]

16. Illustrate the formulae $v = u + ft$, $s = ut + \frac{1}{2}ft^2$, $s = \frac{1}{2}(u + v)t$ on a velocity-time diagram.

17. A train running all the time at constant speed slips a carriage which is then uniformly retarded by the brakes, coming to rest in 600 yards. How far has the train travelled meanwhile? [Answer: 1,200 yards.]

18. The distance between a pair of stops on a tram-line is s. Between these stops the tram travels with uniform acceleration α till its highest speed is reached and then with uniform retardation β till it comes to rest. Prove that the time taken is $\sqrt{\{2s(\alpha + \beta)/\alpha\beta\}}$.

19. A lift runs between the ground and upper floors of a building, a distance of 180 ft. It is capable of moving with an acceleration of 2·4 ft/sec², a retardation of 4·8 ft/sec², or at uniform speed. The greatest permissible speed is 12 ft/sec. Find the least time in which the journey can be made, and find also the least time if the speed limitation is removed. [Answers: $18\frac{3}{4}$ sec; 15 sec.]

20. A train which normally runs at 60 m.p.h. over a certain section of the journey is checked by signal to 15 m.p.h. over a mile of the section in which the railroad is under repair. The train covers 1 mile while slackening speed for this purpose and $1\frac{1}{2}$ miles while recovering its normal speed. Assuming the retardation and acceleration to be constant, find the time lost due to the check. [Answer: $4\frac{1}{2}$ minutes.]

21. A motor-car has a maximum acceleration of 4 ft/sec² and a maximum retardation of 33 ft/sec². In a journey of 2 miles, starting and finishing at rest, the car first accelerates to a speed of 30 m.p.h., and maintains this speed until the end of the first mile; then accelerates to a speed of 45 m.p.h., and maintains this speed until the brakes are applied to bring it to rest. Prove that the least time of the journey is about 207 sec.

*22. If in the rectilinear motion of a point, the time t and position x satisfy the equation $t = ax^2 + bx + c$, where a, b, c are given constants, prove that: (a) the velocity in the position x is $(2ax + b)^{-1}$; (b) the acceleration is inversely proportional to the cube of the distance from a certain fixed point in the line of motion, and find the coordinate of the fixed point. [Answer: $-b/2a$.]

*23. The velocity v at time t of a point moving in a straight line is given by $v = at + bt^2$, where a, b are constants. Prove that the distance described when a time T has elapsed is the same as would have been described from rest in the time T if the acceleration had been constant and equal to the actual acceleration at the instant $t = \frac{1}{3}T$.

*24. The acceleration of a motor lorry which starts from rest is 4 ft/sec² initially and 1 ft/sec² after 10 sec. Assuming that the acceleration diminishes at a constant rate until it is zero, find: (a) the distance covered and the speed after 10 sec; (b) the elapsed time, the distance covered and the speed when the acceleration reaches zero.

[Answers: (a) 150 ft, 25 ft/sec; (b) 13·3 sec, 237 ft, 26·7 ft/sec.]

CHAPTER III

DYNAMICS OF A PARTICLE MOVING IN A STRAIGHT LINE

"The Newtonian laws of motion form the starting-point of most treatises in dynamics, and it seems to me that physical science, thus started, resembles the mighty genius of an Arabian tale emerging amid metaphysical exhalations from the bottle in which for long centuries it has been corked down. When the mists have quite cleared off we shall see more clearly its proportions, and there is a special need for a strong breeze to clear away our confused notions as to matter, mass and force. The writer is far from imagining that he can accomplish this clearance, but he is convinced that a firm basis for physics will be found only when scientists recognise that mechanism is no reality of the phenomenal world—that it is solely the mode by which we conceptually mimic the routine of our perceptions."

—KARL PEARSON, *The Grammar of Science*.

Some features of the behaviour of *matter* will now be examined. As with time and space, there will be no attempt at any ultimate explanation of matter. What will be done will be to set up a mathematical model theory—the 'Newtonian' or 'classical' model—designed to represent the features in question. The study of these features is called *Dynamics*.

To begin with, it will be postulated that there exists in the Universe some 'frame of reference' (set of coordinate axes) which is absolutely at rest. (A slightly weaker postulate is actually sufficient for Newtonian dynamics (see § 17·5); the postulate is superseded in Relativity Theory.) From now on, it is to be understood that displacements, velocities and accelerations are measured relative to such a frame, unless the contrary is stated. The adjective 'absolute' will be included explicitly on occasions when we wish to emphasise that a reference frame is taken to be absolutely at rest.

In Chapters III and IV (as in II), the model theory will be limited to the case of motion in a straight line; also, the directions of all directed quantities will be limited to the straight line of the motion. In this case, the absolute frame of reference is simply a point of the line which is taken to be absolutely at rest.

3·1. The concepts of force and mass.

3·11. Resultant force. Our 'perception' of matter arises in the first instance through our sense-experience of what we call 'force'. In dynamics, we introduce the technical term *force* in a more precise sense, but so as to correspond with our experience of 'force'.

Experience indicates that if an assigned 'body' of matter moves with different accelerations on different occasions, then it is actuated by different 'forces'; further, the greater the acceleration the 'greater' the 'force'. (This is subject to qualifications to be stated in § 3·13.)

In keeping with this experience, we introduce the concept of measurable *resultant force*, the measure, R say, being taken to be proportional to the measure, f say, of the acceleration for an assigned 'body'.

(Note that on the experience quoted, there would be no logical objection to taking R as proportional to some other power of f, e.g. f^3; but to do this would make the mathematical model unnecessarily cumbrous. It is a guiding principle in Science that where choice of postulates is permissible, the simplest postulates are always selected.)

3·12. Mass. Experience further indicates that an assigned 'force' when applied to different 'bodies' produces in general different accelerations; this indicates the existence of some inherent quality of a 'body', which determines its degree of response to an applied 'force'.

This experience is met by introducing the concept of (measurable) *mass*, the measure, m say, being taken to be inversely proportional to f when the resultant force is assigned.

3·13. Particle. The 'acceleration of a body', referred to in §§ 3·11, 3·12, lacks definite meaning for a 'body' of finite extent. So a further step in the construction of the mathematical model is to introduce the *particle*, having mass concentrated at a point and having no extent beyond this point. We now adopt as postulates the quantitative statements in §§ 3·11, 3·12, with the word 'body' replaced by *particle*.

It is further postulated that the mass of any particle remains constant as the time changes, and is never negative. (In some special problems, particle masses may be taken to be zero—see e.g. § 4·6.)

Note that a *force* has no relevance except in so far as it acts on a particle.

In applications of the theory to problems we shall, until the end of Chapter IX, treat bodies as single particles, except where the contrary is stated, e.g. in § 4·9.

3·14. Momentum. The *momentum* of a particle is defined as an entity having the direction of the velocity (v) and measure equal to the product mv.

The *mass-acceleration* of a particle (sometimes called the *effective force* on the particle) is similarly defined, the direction being that of the acceleration (f), and the measure equal to mf.

3·15. Fundamental equation of particle dynamics. By §§ 3·11, 3·12, we have $f \propto R$ when m is assigned, and $f \propto 1/m$ when R is assigned. Hence, by the property of joint variation in algebra, $f \propto (R/m)$ in general. The final constant of proportionality is, again for the sake of simplicity, selected as unity; thus

$$R = mf. \qquad (8a)$$

By (2), we may alternatively write

$$R = \frac{d}{dt}(mv). \qquad (8b)$$

Thus * the resultant force on a particle is equal to * the mass-acceleration, i.e. to the rate of change of momentum. These relations presuppose the use of a standard reference direction in the measuring of R, f, mv.

It is emphasised that f and v here denote *absolute* acceleration and velocity.

3·16. Addition of forces. Since two or more 'forces' may act simultaneously on the one 'body', it is necessary to consider the relation between individual *forces* on a particle and the resultant force. For this, an additional postulate is needed.

(Cf. an analogous case : From a knowledge of the outputs of two men working separately and no further knowledge, it is not possible to deduce the output when they work together.)

In the light of relevant experimental results, it is postulated that if $F_1, F_2, \ldots$ are separate forces acting on a particle, then the resultant force R is given by

$$R = F_1 + F_2 + \ldots; \qquad (9)$$

it is understood that each of the measures $R, F_1, F_2, \ldots$ is positive or negative according to the direction in which the corresponding force acts on the particle.

This postulate embodies what sometimes has been called 'the physical independence of forces'.

The simplicity of the addition rule (9) is in part due to R having been taken as simply proportional to f in § 3·11.

3·17. Equilibrium. If the resultant force R is zero, the particle is said to be in *equilibrium*; sometimes it is then said that the set of forces $F_1, F_2, \ldots$ (§ 3·16) is a *set in equilibrium*.

The subject of *Statics* is concerned with equilibrium problems. The theory of Statics is evidently included in the theory of Dynamics, but special techniques are developed (see Chapter XV).

* The words 'the measure of' are to be understood here and in similar places.

3·18. Units. Theoretical units of mass are the *pound* (lb), *kilogramme* (kg), and *gramme* (g) in the f.p.s., m.k.s. and c.g.s. systems, respectively. These units are specified in terms of the matter in certain standard pieces of metal (cf. § 2·11).

By (8a), the unit of force is then derived as the force which would produce the unit of acceleration in a particle of unit mass. In the three systems, respectively, the theoretical units of force are called the *poundal* (pdl), *newton* (N) and *dyne* (dyn), respectively.

The theoretical units of momentum are the lb-ft/sec, kg-m/s and g-cm/sec.

The units of mass-acceleration are the same as those of force.

3·19. Remarks.

(i) The theme in §§ 3·11 to 3·15 is substantially that envisaged in the first two of Newton's 'laws of motion', enunciated in Newton's '*Principia*' in 1687:

I. *Corpus omne perseverare in statu suo quiescendi vel movendi uniformiter in directum, nisi quatenus a viribus impressis cogitur statum illum mutare.*

II. *Mutationem motus proportionalem esse vi motrici impressae, et fieri secundum lineam rectam qua vis illa imprimitur.*

The law I associates force with acceleration; and II may be regarded as embodying (8a) or (8b). (The Latin version is left untranslated since it relates to a somewhat restricted context.)

(ii) Newton's 'third law' will be referred to in Chapter IV.

(iii) We introduce the technical term *light*, meaning 'of negligible mass'. The term *heavy* when used means that the mass is not to be neglected.

(iv) The reader should check that one newton is equal to 10^5 dynes. One of the reasons that the m.k.s. system is superseding the c.g.s. is that the dyne and some other c.g.s. units are inconveniently small in many contexts. Another reason is that the m.k.s. system is specially serviceable in Electromagnetism.

3·1(10). Examples. (1) *A certain constant (resultant) force, acting on a (particle of) mass 20 lb for 5 sec, generates in it a speed of 50 ft/sec. Find: (a) the force; (b) the acceleration this force would produce in a mass of 1 ton; (c) the distance through which the force would move in generating from rest a speed of 30 m.p.h. in the mass in (b).*

(a) Since the resultant force, R pdl say, is constant, we may by (8b) write

$$Rt = mv - mu,$$

where t, m, u and v denote the measures in f.p.s. units of the elapsed time, mass, initial and final velocities.

Thus
$$R \times 5 = 20 \times 50 - 20 \times 0.$$

Hence the force is 200 pdl.

(b) The second mass = 1 ton = 2240 lb.
Hence, by (8a), the required acceleration, f ft/sec² say, is given from
$$200 = 2240f,$$
and is therefore 5/56 ft/sec².

(c) Since the acceleration is constant for the motion, formulae in § 2·5 may be used. We use '$v^2 = u^2 + 2fs$'.
Since 30 m.p.h. = 44 ft/sec, we have
$$44^2 = 0 + 10s/56.$$
The value of s gives the required distance, i.e. 10,800 ft (approx.).

(2) *Find in dynes the constant (resultant) force which will in 5 seconds move a mass of 1 kilogramme through a distance of 6 metres from rest.*

Since the resultant force R is constant, it follows by (8a) that the acceleration is constant. Hence we can use '$s = ut + \tfrac{1}{2}ft^2$' to determine the acceleration.

In m.k.s. units, we have $s = 6, u = 0, t = 5$.
Hence $6 = \tfrac{1}{2}f \times 5^2$, giving $f = 0.48$ (m/s²).
Hence, by (8a), $R = 1 \times 0.48$.
Thus the required force = 0·48 N = 48,000 dyn.

(3) *A body of mass m moving initially with a speed of u is acted on by a resistance force (of direction against the motion) whose magnitude at any instant is kv, where v is the speed at that instant and k is constant. Find the speed when a distance s has been covered.*

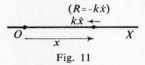

Fig. 11

In this problem, the resultant force is not constant, and hence the acceleration is not constant. So we cannot use the formulae of § 2·5. We have to rely on (8b) and whatever skill we may have in the calculus.
(If this skill is still lacking, the student may refrain from reading the solution below just now, but should convince himself of the invalidity of using a formula such as '$v^2 = u^2 + 2fs$' in such a problem.)

Take the direction of motion as positive.
Let x be the position coordinate at time t referred to any fixed origin O in the line of motion.
Then, using (8a), we have
$$-k\dot{x} = R = m\ddot{x}; \tag{i}$$
$$\dot{x} = u \quad \text{and} \quad x = x_0 \text{ say}, \quad \text{when } t = 0. \tag{ii}$$
Integrating (i) with respect to t, we have
$$-kx = m\dot{x} + C,$$
where C is constant. By (ii), we have
$$-kx_0 = mu + C. \tag{iii}$$

Hence $-k(x - x_0) = m(\dot{x} - u)$.

Hence when the distance covered, viz. $x - x_0$, is equal to s, the speed $\dot{x}$ is equal to $u - ks/m$.

S.P.I.C.E. (α) To give experience in the use of the fundamental equation (8a) or (8b). Either form provides a key to relating the effects of force to the kinematical features discussed in Chapter II.

(β) The student should note the importance of the preliminary argument in Ex. (2) needed to justify the use of '$s = ut + \frac{1}{2}ft^2$'.

(γ) Ex. (3) is included to show that there are problems in which the use of '$s = ut + \frac{1}{2}ft^2$', etc., would be invalid.

(δ) It is evident from the solution of Ex. (3) that the given distance s must not exceed mu/k; otherwise there would be no solution possible.

3·2. Dimensions of a dynamical entity.
The units of time, length and mass are fundamental in the sense that the units of all other dynamical entities are derivable from them. This is the case for example with area, velocity, acceleration, force, and will be the case with all other dynamical entities to be met.

(In kinematics, mass has no relevance and there are just the two fundamental units of length and time; in ordinary geometry, there is just the one fundamental unit, viz., length.)

We use [L], [T], [M] to denote the theoretical units of length, time and mass in any system of units. The theoretical unit of velocity is then denoted by [L]/[T]. This statement is expressed in the form [Velocity] = [LT^{-1}]; and similarly with all other cases. Thus we have:

$$[\text{Area}] = [L^2].$$
$$[\text{Volume}] = [L^3].$$
$$[\text{Velocity}] = [LT^{-1}].$$
$$[\text{Acceleration}] = [LT^{-2}].$$
$$[\text{Momentum}] = [MLT^{-1}].$$
$$[\text{Mass-acceleration}] = [\text{Force}] = [MLT^{-2}].$$

Such equations are said to express the dynamical *dimensions* of the entities appearing on the left-hand sides.

Correspondingly (see e.g. §§ 2·3, 2·4), it is sometimes convenient to denote the unit of velocity in the m.k.s. system as 1 m-s^{-1}; and similarly in other cases. For example, 1 pdl = 1 lb-ft-sec^{-2}.

3·21. Dimension checks.
It is evident from the way in which the theory is set up that in every theoretical formula, the dimensions of each term are the same. The student should check this for all the formulae so far produced.

The property just stated gives a swift means of detecting certain algebraic errors in solving problems.

For example, in Ex. (3) of § 3·1(10), it follows from (i) that $[k] = [MT^{-1}]$. Hence if in place of (iii) we had obtained say $-kx_0 = mu^2 + C$, a dimension check would show that we must have made an algebraic slip.

It is good economy to form the habit of making dimension checks wherever possible; the average student could in this way, with little extra effort, greatly improve the reliability of his work.

Dimension checks of course cannot be made after numerical measures have been substituted into a formula (unless a special allowance is made).

3·22. Change of units. If we wish to change from measures in one unit system to measures in another, we can do this readily using dimension theory and the theorem (already stated in § 2·1) that the measure of a given quantity is inversely proportional to the size of the unit. The following example illustrates this.

3·221. Example. *Given that* 1 *ft* = 30·48 *cm and* 1 *lb* = 0·4536 *kg, find the number of newtons in* 1 *pdl, and show how to convert a measure in newtons to a measure in poundals.*

Let $[L_1]$, $[T_1]$, $[M_1]$ denote the fundamental units of the m.k.s. system; $[L_2]$, $[T_2]$, $[M_2]$, of the f.p.s. system.

We are given:
$$[L_2] = 0\cdot 3048[L_1],$$
$$[T_2] = [T_1],$$
$$[M_2] = 0\cdot 4536\,[M_1].$$

Hence
$$[M_2 L_2 T_2^{-2}] = 0\cdot 4536 \times 0\cdot 3048\,[M_1 L_1 T_1^{-2}]$$
$$= 0\cdot 13826\,[M_1 L_1 T_1^{-2}].$$

Thus 1 pdl = 0·13826 N.

Hence a measure in pdl = $0\cdot 13826^{-1}$ of the measure in newtons
= 7·233 × the measure in newtons.

For example, a force of 5 N = 5 × 7·233 pdl
= 36·16 pdl.

3·23.

Dimension theory can further be used in a positive way to help in advancing knowledge. For example, in Aerodynamics, dimension theory is used in designing, and in interpreting the results of, experiments on small-scale models in wind-tunnels.

3·3. Weight. We now make the artificially simplified postulate that the Earth's surface is in a state of absolute rest. This will henceforth be assumed to be the case except when the contrary is stated.

In § 2·6, we discussed the acceleration of a freely falling body in a region near the Earth's surface. The resultant force causing this acceleration is called the *weight* of the body, and acts vertically downwards. By (8*a*), its magnitude, W say, is given by

$$W = mg, \qquad (10)$$

g being as in § 2·6.

It is evident that W, like g, depends on the latitude of the particular region and on other circumstances (as outlined in § 2·62). On the other hand, the mass m of a given body is taken as fixed independently of the locality.

The units and dimensions of weight are those of force, *not* of mass; the m.k.s., c.g.s. and f.p.s. units of weight are the newton, dyne and poundal, respectively.

3·31. Remarks.

(i) It happens that in practice the masses of different bodies are commonly compared by comparing their weights. This practice is reliable only to the extent that g may be assumed constant. There can be significant error in taking the ratio of the masses of two bodies in different localities to be equal to the ratio of their weights in these respective localities.

(ii) The weight of a body is commonly estimated by using some mechanism such as a spring balance. The force exerted on the body by the spring is, by (8*a*), equal and opposite to the body's weight if the body is at rest. The extension of the spring is assumed correlated (in one-to-one correspondence*) with the force exerted by the spring and hence with the weight of the body.

An ordinary spring balance carries a pointer and scale on which measures of weight are read; the marks on the scale are the result of a 'calibration' using bodies whose weights have, directly or indirectly, been previously estimated in relation to the standard pieces of metal referred to in § 3·18.

If a body being 'weighed' on a spring balance has an acceleration (e.g. inside a balloon ascending with acceleration), the force exerted by the spring is, by (8*a*), not equal and opposite to the body's weight. The force as read on the scale in such circumstances is sometimes referred to as the 'apparent weight'.

(iii) There is little precision in the use of the words 'mass' and 'weight' in ordinary speech. But the student must appreciate the clear-cut distinction between the dynamical terms *mass* and *weight*, and especially the dimensional difference.

(iv) On account of (i), 'practical' units of force called the kilogramme weight (kg.wt. or Kg), the gramme weight (g.wt.) and the pound weight (lb.wt.) are in use. One Kg is the weight of a body whose mass is 1 kg, and is therefore the same as g N, where g is about 9·80; 1 g.wt. $\approx$ 980 dyn; 1 lb.wt. $\approx$ 32 pdl. These practical units vary, as g does, with the particular locality, etc.

* I.e., corresponding to any one particular extension there is just one value of the force, and vice versa. Other cases of 1–1 correspondence are innate in §§ 3·11, 3·12.

In solving problems in dynamics, the student should invariably translate from measures in these (and all other—see § 2·11) practical units into measures in the theoretical units.

Furthermore, to write (for example) 'lb'—in place of 'lb.wt.'—as a unit of force, is an error on the theory here presented.

(v) In some text-books, a unit of force is used which is g times the f.p.s. unit of force, and called a 'pound'. The corresponding unit of mass is called a 'slug', and is g times as great as our pound.

(vi) If account is taken of the motion of the Earth's surface, (10) does not follow from (8a), for f in (8a) denotes absolute acceleration, whereas g denotes an acceleration relative to the Earth's surface. The equation (10) is usually retained, however, and the weight is re-defined as having the direction of the 'acceleration due to gravity' and magnitude given by mg. Thus unless the simplification referred to in the first paragraph of § 3·3 is made, the weight of a particle near the Earth's surface is not in general quite equal to the Earth's gravitational attraction on it. (See also § 17·9.)

3·4. Examples.

The following remarks are relevant to examples commonly set at this stage.

(i) Trains and the like will be treated as particles until further notice.

(ii) In some problems in Chapters III and IV, there is a deviation from the stipulation in the third paragraph of Chapter III in that there are some forces acting outside the line of motion. This is the case in Ex. (1) below, where the weight of the train and the supporting force by the rails are at right angles to the direction of motion. The student is asked to accept our statement that in this and similar cases that appear in Chapters III and IV, such forces may be ignored. He will be able to substantiate this when he has reached § 6·41.

(iii) The force exerted on a body by a *smooth* surface (see § 6·6) is at right angles to the surface. The circumstances of (ii) apply here also, and such forces can be ignored in Chapters III, IV.

(iv) Frictional forces (see § 6·6 again) are forces which act on a body always against the direction of the body's motion.

(v) Reference will often be made to the 'force exerted by the engine on a train'. It will be sufficient in elementary problems to regard this as a simple force acting forwards at the front of the train. Actually (this will be better appreciated after Chapter IV is read), the external forward force is exerted on the train by the rails as a result of certain complicated internal happenings inside the engine. (Cf. the solution of Ex. (3) of § 4·72.)

(1) *The engine of a train of total mass* 300 *tons moving horizontally in a straight line exerts a constant forwards force of* 4 *tons wt., and frictional resistances amount to* 10 *lb.wt. per ton of the train's mass. Find the acceleration.*

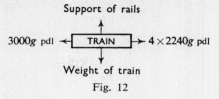

Fig. 12

Take the direction of motion of the train as positive.
Resultant force (R, say) on train $= (4 \times 2240g - 3000g)$ pdl.
Mass of train (m) $\qquad = 300 \times 2240$ lb.
Let the acceleration of the train $= f$ ft/sec².
Since $R = mf$, we then have
$$5960g = 672{,}000f.$$
Thus the required acceleration $= 0 \cdot 28$ ft/sec².

(2) *Investigate the force between a man's feet and the floor of a lift for various representative cases.*

Fig. 13

Let m denote the mass and W the weight of the man; and let F denote the upward force exerted by the lift on the man in any particular circumstances.

Fig. 13 shows the forces acting on the man. (In the figure, the circle representing the man is inserted at some distance from the floor of the lift in order to show clearly the forces on the man. Such procedure is common, and in fact desirable, in solving problems in mechanics.)

The resultant upward force on the man is seen to be $F - mg$.

(*a*) *Lift at rest.*
 By (8*a*), $\qquad F - mg = m \times 0.$
 So $\qquad\qquad\qquad F = mg = W.$

(*b*) *Lift moving with uniform speed.*
 By (8*a*), $\qquad F - mg = m \times 0.$
 So $\qquad\qquad\qquad F = mg = W$, again.

(*c*) *Lift moving with upward acceleration f.*
 By (8*a*), $\qquad F - mg = mf.$
 So $\qquad\qquad\qquad F = m(g + f).$
 Thus the man's 'apparent weight' is $(g + f)/g$ times his 'true weight' W.

(d) Lift moving with downward acceleration f, where $f < g$.
  By (8a), $F - mg = m(-f)$.
  So $F = m(g - f)$.   (i)
  The man's 'apparent weight' is $(g - f)/g$ times W.
(e) As (d), except that $f = g$.
  By (i), $F = 0$.   The floor exerts no support.
(f) As (d), except that $f > g$.
  (i) would now yield F negative. This would not accord with the physical circumstances unless there was some means of attachment between the man and the floor. Failing this, the floor would separate from the man, who would move with downward acceleration g (until he hit the roof).

(3) *A mass of* 10 kg *falls freely from rest through* 10 *metres and then comes to rest again after penetrating* 0·2 m *of sand. Find in* Kg *the resistance of the sand, assumed constant.* (See Fig. 17, p. 49.)

Using '$v^2 = u^2 + 2fs$', we see that the speed, V m/s say, of the mass just before entering the sand is given by
$$V^2 = 2g \times 10, \quad \text{(where } g = 9·80\text{)}. \tag{i}$$
During the penetration of the sand, let F newtons be the (upward) resistance force of the sand on the mass.
  Then the resultant downward force on the mass
$$= (mg - F) \text{ newtons, where } m = 10.$$
  Since this resultant force is constant, the downward acceleration, f m/s² say, is constant. Hence '$v^2 = u^2 + 2fs$' is again relevant, and gives
$$0 = V^2 + 2f \times 0·2. \tag{ii}$$
By (i) and (ii), $f = -50g$.
Since $mg - F = mf$, we then have
$$10g - F = -500g.$$
Hence the required resistance force $= 510g$ newtons
$$= 510 \text{ Kg}.$$

(4) *A mass of* 200 lb *is suspended at the lower end of a light vertical rope and is being hauled up vertically. Initially the mass is at rest and the pull on the rope is* 300 lb.wt. *The pull diminishes uniformly at the rate of* 6 lb.wt. *per each foot through which the mass is raised. Find the speed of the mass when it has been raised* 20 ft.

At time t sec, let x ft be the height of the mass above its initial position. The pull of the rope (F pdl say) is given by
$$F = (300 - 6x)g;$$
for this gives $F = 300g$ initially, and $dF/dx = -6g$, as required by the data.

The resultant upward force on the mass $= (F - 200g)$ pdl
$$= (100 - 6x)g \text{ pdl}.$$
We note that this resultant force, and hence also the acceleration, is not constant.

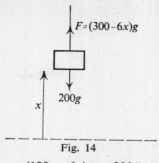

Fig. 14

By (8a), we have $(100 - 6x)g = 200\ddot{x}$.
Hence, using (3),

$$100 \frac{dv^2}{dx} = (100 - 6x)g,$$

where v ft/sec is the speed in the position x ft.

Let V ft/sec be the required speed, i.e. when $x = 20$. Then, integrating with respect to x over the range $x = 0$ to $x = 20$, we have

$$100 \left[v^2 \right]_0^V = g \left[100x - 3x^2 \right]_0^{20};$$

i.e., $V^2 = 8g$.

Hence the required speed is 16 ft/sec.

S.P.I.C.E. (α) It is well for the student to form the habit of showing in a figure, as in Figs. 12, 13, 14, *all* the forces on the body considered.

(β) In Ex. (1), note (ii) of § 3·4 is illustrated.

(γ) In connection with Ex. (2), note that the presence of a uniform velocity of the lift has no bearing on the force exerted between the floor and the man. Notice also that in the solution of this problem the standard direction (upwards) was kept unchanged throughout, even although the lift had sometimes a positive downward acceleration.

(δ) In Ex. (3), it is necessary to justify the use of '$v^2 = u^2 + 2fs$'.

(ϵ) Ex. (4) is a case where the resultant force and acceleration are variable.

3·5. Principle of momentum.

3·51. Impulse. The impulse of a force F acting on a particle from time $t = t'$ to time $t = t''$ is defined to be $\int_{t'}^{t''} F\,dt$. We shall write $\int_0^t F\,dt$ to denote the impulse of F from time $t = 0$ to any general time t.

In particular, if (but only if) F is constant, the impulse of F during the time t is equal to Ft.

3·52. Principle of momentum.
By (8b), we have
$$\int_0^t R\, dt = mv - mu, \tag{8c}$$
where *mu, mv* are the initial and final momenta of the particle. Thus the increase in momentum of a particle is equal to the impulse of the *resultant* force (both measured of course in the same sense).

Either of (8b), (8c) expresses the *principle of momentum* for a single particle moving in a straight line.

3·53. Units and dimensions.
The units and dimensions of impulse are evidently the same as those of momentum (§ 3·14).

(The student should check that the left- and right-hand sides of (8c) have the same dimensions.)

The f.p.s., m.k.s. and c.g.s. units of impulse may be given the alternative names of pdl-sec, N-s, dyn-sec, respectively. Thus, e.g., 1 pdl-sec = 1 lb-ft/sec.

3·54. Instantaneous impulse.
The instantaneous impulse of a force F at any time t_0 is defined to be $\lim_{\tau \to 0} \int_{t_0}^{t_0+\tau} F\, dt$. For ordinary forces, this limit is equal to zero.

But there exist some forces so great as to produce significant momentum changes while acting during times too short to be appreciated. This experience is incorporated in the mathematical model by postulating the concept of *impulsive force*, for which the above limit is finite and not zero.

(E.g., if $F = J/\tau$, where J is a (finite) constant, the value of the limit is J.)

The principle of momentum in the form (8c), and not the equation (8a), is appealed to in estimating the effects of impulsive forces.

As already implied, the instantaneous impulse of any non-impulsive force (e.g. weight) is zero.

Hence, by (8c), *the impulse of the resultant impulsive force acting on a particle at any instant is equal to the jump at that instant in the momentum of the particle*.

3·541.
Let σ be the change in the position of a particle during the time τ, where $\tau \to 0$, of action of an impulsive force. By the definition of mean velocity (§ 2·3), $\sigma = V\tau$, where V may be taken as intermediate between the prior and subsequent values of the velocity of the particle. Since V is finite, $\sigma \to 0$.

Thus there is no jump in the position coordinate of a particle during the action of an impulsive force. This is an important simplifying factor in the solving of problems involving impulsive forces.

It should be noted, however, that σ cannot be ignored in connection with the work (see § 3·61) of an impulsive force.

3·55. Example. *A marble of mass 4 oz is let fall from rest through 9 feet on to a pavement, and it rebounds to a height of 4 feet. Find the impulse exerted by the pavement on the marble.*

Take the positive direction as vertically upwards.

The velocity of the marble just before striking the ground is, using equation (6), found to be -24 ft/sec; and, just after, $+16$ ft/sec.

Hence the resultant upward impulse on the marble at the instant of impact $= \frac{1}{4} \times 16 - \frac{1}{4} \times (-24)$ lb-ft/sec
$= 10$ lb-ft/sec.

This instantaneous impulse is the impulse exerted by the pavement, since the instantaneous impulse of the only other force present, viz. the weight of the marble, is zero.

3·6. Work.

The theory already given is sufficient for solving problems on the dynamics of a particle (in the circumstances set down on p. 28). But by introducing further concepts, of which *work* is the first, the processes of solution can be greatly facilitated in many cases. Further, the new concepts are of far-reaching importance in applied mathematics and science generally.

Let P be a particle whose position coordinate changes over an interval of time from an initial value x_0 to any general value x. Let F (measured in the same direction as x) be a particular one of the forces acting on P; F may be variable. Then the work done *by* F during the motion is defined to be

$$\int_{x_0}^{x} F \, dx. \tag{11}$$

We sometimes say that the work done *against* F during the motion is

$$-\int_{x_0}^{x} F \, dx.$$

In the *particular case* where F is constant, the work done by F is $F(x - x_0) = Fs$, where s is the particle's displacement in the direction of F.

3·61. Work of impulsive force.

Corresponding to (11), the work done by an impulsive force acting on a particle when in the position x_0 is given by $\lim_{\sigma \to 0} \int_{x_0}^{x_0 + \sigma} F \, dx$, where σ is as in § 3·541.

It can be deduced, using § 3·54, that this work (which may be positive or negative) *is in general not zero*. Mistakes in problems are often made by neglecting to allow for the work of impulsive forces.

3·62. Work of resultant force.

The work done by the resultant force R on a particle during any displacement is equal to the algebraic sum of the works of all the forces, $F_1, F_2, \ldots$, acting on the particle. For, the work done by R

$$= \int_{x_0}^{x} R \, dx$$

$$= \int_{x_0}^{x} (F_1 + F_2 + \ldots) dx$$

$$= \int_{x_0}^{x} F_1 \, dx + \int_{x_0}^{x} F_2 \, dx + \ldots$$

3·63. Units and dimensions. The m.k.s., c.g.s. and f.p.s. units of work are the N-m or *joule* (denoted as J), the dyn-cm or *erg*, and the ft-pdl. One $J = 10^7$ ergs.

The dimensions of work are $[ML^2T^{-2}]$.

Practical units include the Kg-m, equal to g J, where $g \approx 9.80$; the g.wt.-cm, equal to g ergs, where $g \approx 980$; and the ft-lb.wt., equal to g ft-pdl, where $g \approx 32$. It is unsatisfactory to write 'ft-lb', meaning 'ft-lb.wt.' The *calorie*, approximately equal to 4·18 joules, is frequently used in thermodynamics.

3·64. Examples. (1) *A gun of mass* 200 kg *is drawn* 50 *metres along horizontal ground. The magnitude of the resistance due to the roughness of the ground is one tenth of the weight of the gun. Find in joules the work done against the resistance.*

The resistance force $= 0.1 \times 200g$ newtons, of direction opposite to the displacement.

This force is constant.

Hence the work done by it $= -20g \times 50$ J
$= -1000g$ J.

Thus the work done against the resistance force $= 9.80 \times 10^3$ joules.

(2) *Find the work done by the pull on the rope during the displacement of the mass in Ex.* (4) *of* § 3·4.

In this problem the force F is variable.

The work done by $F = \int_0^{20} F \, dx$ ft-pdl

$$= \int_0^{20} (300 - 6x)g \, dx \text{ ft-pdl}$$

$$= g \left[300x - 3x^2 \right]_0^{20} \text{ ft-pdl}$$

$$= 4800g \text{ ft-pdl}$$

$$= 4800 \text{ ft-lb.wt.}$$

S.P.I.C.E. (α) To give exercise on various units and on questions of sign.

(β) To emphasise that the formula Fs gives the work done by F only if F is constant.

3·7. Power. The *power* of a force F at any instant is defined as the rate at which F is doing work (on some particle) *at that instant.*

Between times t, $t + \delta t$, the work done by F is approximately $F\delta x$, where δx is the displacement (measured in the same direction as the force). Hence, at time t, the power of F is $\lim\limits_{\delta t \to 0} \left(\dfrac{F\,\delta x}{\delta t}\right)$; i.e., $F\,dx/dt$. Thus

$$\text{Power of } F = Fv, \qquad (12)$$

where v is the velocity of the particle on which F is acting. The power is variable if either F or v is variable (unless $F \propto v^{-1}$).

(The formula Fs/t, sometimes given for power, is limited to the narrow special case in which both F and v are constant, and is often given undue prominence.)

3·71. Units and dimensions. The m.k.s., c.g.s. and f.p.s. units of power are the J/s or watt (denoted as W), the erg/sec, and the ft-pdl/sec. The dimensions are $[ML^2T^{-3}]$.

Practical units include the g.wt.-cm/sec, equal to (about) 980 erg/sec; the ft-lb.wt./sec (*not* ft-lb/sec), equal to 32 ft-pdl/sec; the *horse-power* (h.p.) equal to 550 ft-lb.wt./sec or $550g$ ft-pdl/sec, where $g \approx 32$.

3·72. Examples. (1) *Given that* $1\,ft = 30\cdot48\,cm$, $1\,lb = 453\cdot6\,grammes$, *and* $g = 32\cdot2\,ft/sec^2$, *show that* $1\,h.p. = 746\,watts.$

With notation as in the example of § 3·221.

$$[M_2 L_2{}^2 T_2{}^{-3}] = 0\cdot4536 \times (0\cdot3048)^2 [M_1 L_1{}^2 T_1{}^{-3}]$$
$$= 4\cdot214 \times 10^{-2} [M_1 L_1{}^2 T_1{}^{-3}]$$

Hence 1 ft-pdl/sec $= 4\cdot214 \times 10^{-2}$ watts.

Hence 1 h.p. $= 550 \times 32\cdot2 \times 4\cdot214 \times 10^{-2}$ watts
$= 746$ watts.

(2) *An engine raises in 3 minutes at uniform speed a mass of 12 cwt vertically through a height of 300 feet. Find the horse-power in operation.*

Since the velocity is constant,

$$F - mg = 0,$$

where F pdl is the upward force exerted by the engine on the mass, and $m = 12 \times 112$ (lb).

The upward velocity (v ft/sec) $= 300/180$ ft/sec $= 5/3$ ft/sec.

Hence the power needed $= Fv$ ft-pdl/sec

$$= (12 \times 112 \times g \times 5)/3 \text{ ft-pdl/sec}$$
$$= \dfrac{12 \times 112 \times 5g}{3 \times 550g} \text{ h.p.}$$
$$= 4\cdot07 \text{ h.p.}$$

(3) *A train of total mass* 120 *tons starts from rest along a level track and moves with uniform acceleration. After* 3 *minutes its speed is* 30 *m.p.h. There is a resistance of* 10 *lb.wt. per ton. Find in h.p. the rate at which the engine must be capable of working.*

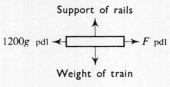

Fig. 15

The acceleration, f ft/sec^2 say, is given constant. Hence, by (8a), the resultant force $(F - 1200g)$ pdl, where F pdl is the force exerted by the engine at any instant, is constant; hence F is constant.

The velocity, v ft/sec say, varies, its maximum being the final value of 44 ft/sec. Hence, by (12), F being constant, the needed power is $44F$ ft-pdl/sec.

It remains to determine F.

Since the acceleration is constant, we may use '$v = u + ft$'. By the data, this gives $f = 11/45$.

Then, by (8a), $F - 1200g = 120 \times 2240 \times (11/45)$,

which gives $F = 1{\cdot}04 \times 10^5$.

Hence the power needed $= 44 \times 1{\cdot}04 \times 10^5$ ft-pdl/sec

$$= \frac{44 \times 1{\cdot}04 \times 10^5}{550g} \text{ h.p.}$$

$= 260$ h.p.

S.P.I.C.E. (α) The formula (12) is relevant to both Exs. (2) and (3), but the formula 'Fs/t' could not be validly used in Ex. (3).

(β) In Ex. (3), it should be noted that the power is continually changing. The engine must be capable of the *maximum* power involved in the given motion.

3·8. Energy. Energy is defined as work capable of being done in various special sets of circumstances, some of which will now be introduced.

The units and dimensions of energy are those of work.

3·81. Kinetic Energy. The *kinetic energy* (K.E.) of a particle at an instant is defined as the work that would be done against any resultant force which reduced the particle's speed to zero.

3·811. Formula for kinetic energy. At time $t = t_0$, let x_0 and v_0 be the values of the position coordinate x and the velocity v of a particle of mass m. Let R (measured in the same direction as x) be the value at any later time t of a resultant force under which the particle comes to rest, say at time $t = t_1$ in the position $x = x_1$. (R will of course be negative during at least part of the motion if $x_1 > x_0$.)

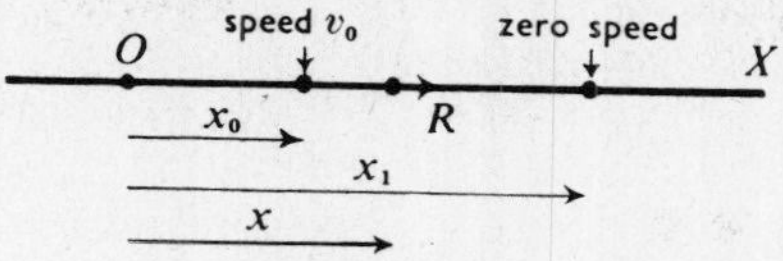

```
        speed v₀      zero speed
    O      ↓              ↓         X
    •——————•——————————————•—————————
           x₀       R
           ——————————→
                    x₁
           ——————————————→
        x
    ——————→
```

Fig. 16

By (11) and the definition of kinetic energy, the K.E. T_0 of the particle in the position x_0 is given by

$$T_0 = -\int_{x_0}^{x_1} R\,dx$$

$$= -\int_{t_0}^{t_1} m\frac{dv}{dt}v\,dt \qquad \text{(by (8a) and (1))}$$

$$= \int_{t_1}^{t_0} \frac{d}{dt}(\tfrac{1}{2}mv^2)\,dt$$

$$= \left[\tfrac{1}{2}mv^2\right]_0^{v_0}$$

$$= \tfrac{1}{2}mv_0^2.$$

Thus when the speed of a particle of mass m is v,

$$\text{K.E. of particle} = \tfrac{1}{2}mv^2. \tag{13}$$

Note that the formula (13) is derived independently of particular values of R; this justifies the use of the word 'any' in the definition of kinetic energy in § 3·81.

3·82. Principle of energy. By the definition of kinetic energy, the kinetic energy of a particle moving with a given speed is also equal to the work which would be done *by* a resultant force under which the particle *acquires* the given speed from rest.

It follows that during any motion of a particle:

$$\left\{\begin{array}{c}\textbf{\textit{Work done on particle by}}\\\textbf{\textit{resultant force}}\end{array}\right\} = \left\{\begin{array}{c}\textbf{\textit{Increase in K.E.}}\\\textbf{\textit{of particle.}}\end{array}\right. \tag{14a}$$

This is the *principle of energy* for a single particle.

For the present one-dimensional case, we may express the principle in a formula as follows. Let R denote the resultant force; x_0, u the initial position and velocity; and x, v any later position and velocity. Then

$$\int_{x_0}^{x} R\,dx = \tfrac{1}{2}mv^2 - \tfrac{1}{2}mu^2.$$

3·821. Example. *A particle of mass* 0·5 *kg is moving at the rate of* 1 *m/s. What constant force will bring it to rest in* (a) 4 *sec*, (b) 4 *cm* ?

Take the direction of motion as positive.

Let R_1, R_2 newtons denote the required forces in the two cases. (R_1, R_2 will of course be negative.)

(a) Since R_1 is constant, we have by the principle of momentum, (8c),
$$R_1 t = mv - mu,$$
where $t = 4$, $m = 0·5$, $u = 1$, $v = 0$, in m.k.s. units.

Hence $R_1 = -0·125$.

The required force is 0·125 newtons, against the direction of motion.

(b) Since R_2 is constant, we have by the principle of energy, (14a),
$$R_2 x = \tfrac{1}{2}mv^2 - \tfrac{1}{2}mu^2,$$
where $x = 0·04$, $m = 0·5$, $u = 1$, $v = 0$.

Hence $R_2 = -6·25$.

The required force is 6·25 newtons, against the direction of motion.

S.P.I.C.E. To illustrate how the principles of momentum and energy may be used as alternatives to the form '$R = mf$'. Either of the parts (a) or (b) of the Ex. could of course be solved using '$R = mf$', together with an equation from § 2·5.

3·83. Conservative forces. Consider motions of a particle, under one or a number of forces, over various closed paths. (A closed path is a path in which the starting and finishing points coincide. A one-dimensional example is given in the first sentence of § 3·832.)

A *field of force* is a region throughout which the force concerned is specified as a function of position. If the net work done by the force on a particle is zero whenever the particle moves through a closed path, the field is said to be *conservative*.

Examples of conservative fields are those associated with constant gravity (§ 3·832) and with the tension of a perfectly elastic string (§ 3·92).

Many forces are not conservative, e.g. friction; by the statement in Note (iv) of § 3.4, the work done by a friction force on a body is always negative.

3·831. Potential energy. Let x denote the position coordinate of a particle P, and F a force, belonging to a conservative field, acting on P. The particle P is said to have *potential energy*, in relation to the field of F and any standard position, x_0 say, equal to the work which would be done by F on P if P were displaced from x to x_0. That is,

$$\left. \begin{array}{l} \textbf{P.E. (referred to the point } x_0\textbf{)} \\ \textbf{\textit{associated with the conservative force F}} \end{array} \right\} = \int_x^{x_0} F\, dx = \int_{x_0}^x (-F)\, dx. \quad (15)$$

The standard position x_0 may be taken at any convenient point in the line of motion, since only *changes* in potential energy are involved in applications.

The formulae in (15) may also be interpreted as denoting the increase in potential energy (associated with the field) when the particle moves from x_0 to x.

It must be appreciated that the term 'potential energy' cannot be associated with non-conservative forces.

When use is made of the potential energy concept in relation to one or more of the forces present, the other forces will be called *extraneous* forces.

3·832. Potential energy associated with weight. If a particle of mass m moves in a vertical line from any point A to any point B and then back to A, it is easily verified that the net work done by its weight is zero. The field of gravity (here treated as constant), which gives rise to weight, is therefore conservative.

By (15), the corresponding potential energy is

$$mg(x - x_0), \qquad (16)$$

where $x - x_0$ is the height above a standard position x_0. (The positive direction being vertically upward, F of (15) is here equal to $-mg$.)

3·833. Addition of potential energies. It is possible for more than one conservative field to be simultaneously present in the same region. The corresponding total potential energy is the algebraic sum of the individual potential energies.

This follows since, by §3·62, the work done on a particle by the resultant of a set of forces is equal to the sum of the works done by the individual forces.

3·834. Alternative form of the principle of energy. By equations (14a) and (15), the increase in the kinetic energy of a particle moving from the position x_0 to the position x is equal to the sum of the works done by the extraneous forces, minus the increase in potential energy. Thus

Increase in $(K.E. + P.E.) =$ Work done by extraneous forces.

$$(14b)$$

3·835. Conservation of energy. For the *particular case* in which the extraneous forces do no work, the right-hand side of (14b) is zero, and the *mechanical energy*, i.e. kinetic plus potential energy, of the particle remains constant.

This is simply illustrated in the case of a particle moving in a vertical line under its weight and no other force. Let v_1, v_2 be the speeds in any two positions at heights x_1, x_2 above some level taken as standard. Then by equation (6), the positive direction being taken vertically upwards, we have
$$v_1^2 - v_2^2 = -2g(x_1 - x_2).$$
Hence $\tfrac{1}{2}mv_1^2 + mgx_1 = \tfrac{1}{2}mv_2^2 + mgx_2.$

The left-hand side is the mechanical energy in the position x_1; the right-hand, in x_2.

3·836. Remarks.

(i) It needs to be emphasised that mechanical energy is conserved only in special artificially simplified cases. In Nature, there are always additional forces, such as friction, which cause some dissipation of mechanical energy into other forms (see § 4·54).

(ii) The work done on a body by the force at a contact with a *smooth* fixed surface is zero. This follows from theory to come later (§ 6·834).

(iii) Impulsive forces are in general responsible for sudden changes in mechanical energy. In some *special* cases, e.g. in a *perfectly* elastic impact of a body against a fixed wall (see § 4·81 (iii)), the mechanical energy of the body, is however, not changed.

3·84. Examples. (1) *Solve Ex. (3) of § 3·4, using the principle of energy.*

Let A be the initial position of the mass, B the point where it enters the sand, and C the point where it comes to rest (Fig. 17).

During the motion from A to B, the only force acting is the weight, $10g$ newtons. Hence the mechanical energy is conserved between A and B.

Hence, taking B as the standard position for estimating the P.E., we have:
Mechanical energy at B = Mechanical energy at A
$= 10g \times 10$ J,
the K.E. at A being zero.

During the motion from B to C, an additional force, due to the resistance of the sand, acts. This force, F newtons upward say, is extraneous.

Hence, by (14b), the K.E. being zero at C, we have:
$-10g \times 0·2$ J $=$ Mechanical energy at C
$=$ Mechanical energy at B + Work done by F
$= (100g - 0·2F)$ J,

using (11), F being given constant.
Hence $0·2F = 102g,$
giving again that the required resistance force is $510g$ newtons, or 510 Kg.

NOTE.—The problem could have been solved still more rapidly by applying (14b) directly to the whole motion from A to C. The student should ascertain that this would immediately give:

P.E. at C − P.E. at A = Work done by F between B and C,

and should verify that this equation leads to the answer.

The last equation could also of course be written:

P.E. at A − P.E. at C = Work done *against* F between B and C.

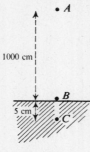

Fig. 17

(2) *An engine pumps water continuously through a hose. If the speed with which the water passes through the nozzle is v, prove that the rate at which kinetic energy is being imparted to the water is proportional to v^3.*

Let λ be the mass per unit length of the jet where the water leaves the nozzle.

During a small elapsed time δt, we have, to sufficient accuracy:

Mass of water ejected $= \lambda(v\delta t)$;

K.E. of this water $= \tfrac{1}{2}(\lambda v \delta t)v^2$.

Hence the rate at which K.E. is being imparted $= \lim\limits_{\delta t \to 0} \left(\dfrac{\tfrac{1}{2}\lambda v^3 \delta t}{\delta t} \right)$

$= \tfrac{1}{2}\lambda v^3.$

NOTE.—(i) This result explains the efficacy of a hose in controlling an unruly crowd.

(ii) A tacit appeal to experience has been made in assuming that the theory of particle dynamics can be reasonably applied to the motion of a fluid in the circumstances of the problem.

(iii) If v is variable, the mass of water ejected in time δt is only approximately equal to $\lambda v \delta t$. But since the error will be of the order of $(\delta t)^2$, its effect disappears when the limit is later taken. This explains the presence of the words *to sufficient accuracy* in the course of the solution; similar use of this phrase will often be made in the sequel.

(3) *A waterfall 30 yards high is fed by a stream 40 yards wide and 5 feet deep approaching the top at $7\frac{1}{2}$ m.p.h. Find in h.p. the rate at which work could be done by the waterfall, assuming no energy dissipation. (Assume also that the mass of 1 cubic foot of water $= 62.5$ lb.)*

(NOTE.—We shall here simplify by supposing the water to enter the fall vertically with the speed of $7\frac{1}{2}$ m.p.h., thus neglecting the curvature of the paths of water particles. Actually, theory of Chapter IX shows that allowance for this curvature would not affect the answer.)

During a small elapsed time δt seconds, we then have:

Mass of water entering fall $= (120 \times 5 \times 11\delta t) \times 62.5$ lb.

P.E. of this mass (taking the standard level at the base of the fall)
$$= (600 \times 11 \times 62.5\delta t) \times g \times 90 \text{ ft-pdl}.$$
K.E. of this mass $\quad = \frac{1}{2}(600 \times 11 \times 62.5\delta t) \times 11^2$ ft-pdl.
Sum of this P.E. and K.E. $= (6600 \times 62.5\delta t)(90g + 60.5)$ ft-pdl.

The last expression gives the work which could be done by the waterfall in δt seconds, ignoring dissipation of energy. (It is assumed that the water is reduced to rest at the base of the fall by the time the work is done.)

Hence the required power $= (6600 \times 62.5)(90g + 60.5)$ ft-pdl/sec

$$= \frac{(6600 \times 62.5)(90g + 60.5)}{550g} \text{ h.p.}$$

$$\approx 6.9 \times 10^4 \text{ h.p.}$$

S.P.I.C.E. (α) The first form of the solution in Ex. (1) illustrates a case of conserved mechanical energy (in the motion from *A* to *B*).

(β) Both forms of the solution in Ex. (1) show the use of the principle of energy when there is an extraneous force. The student is advised to follow out thoroughly details of sign in the solution.

(γ) The solution in Ex. (1) should be compared with the solution in Ex. (3) of § 3·4.

(δ) Ex. (2) is a case in which it is most expedient to consider circumstances in a small time δt, and later take a limit. The student should relate the remark in Note (iii) of Ex. (2) to his work in the calculus.

(ϵ) Ex. (3) would from a theoretical standpoint be better introduced after the theory of Chapter IX. We have put it here partly because problems of this difficulty are commonly introduced to students at this stage, and partly to illustrate the extent of gaps which may lie between the model theory and some of the problems which a student may be called upon to treat in practice.

In the set exercises in this book, however, the student will be expected to make suppositions of the type set down in the Note above only to the extent that the text or worked examples have given reasonably sufficient clues for such purposes.

3.9. Hooke's law.

We shall now consider certain limited aspects of the dynamical behaviour of elastic strings and springs.

We shall neglect here all possible complications due to the mass and weight of the string or spring. This has the effect of limiting our consideration either to light strings or springs, or to certain very special cases in which the system moves in a horizontal plane.

Let a ($=AB$) be the *natural length*, i.e. the length of the string or spring when at rest under no applied forces, and let $a + x$ ($=AP$ say) be the length on any other occasion. Then x (sometimes x/a) is called the *extension*.

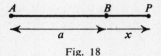

Fig. 18

Suppose a particle is attached at the end P. Then according to *Hooke's law*, the particle is subject to a force towards A, of magnitude F say, given by

$$F = kx, \qquad (17a)$$

where k, called the string or spring 'constant', is independent of F and x. Use is sometimes made of the *modulus of elasticity* λ, equal to ka, so that (17a) may also be written as

$$F = \lambda x/a. \qquad (17b)$$

Hooke's law is a *mathematical model* based on observations of the elastic behaviour of wires. The law agrees with the observations very closely in many cases, but by no means all.

A material is said to be *perfectly elastic* when the Hooke's law model applies. It is said to be *inelastic* (this is another model representation) when x is always zero for finite F. If a material is not inelastic and does not obey Hooke's law, it is *imperfectly elastic*. (All real materials are imperfectly elastic to greater or less degree.)

It should be noted that, in the more complete theory of perfect elasticity, a single modulus of elasticity is in general insufficient to describe the relation between stress and strain.

In (17a) or (17b), F is the *tension* of the string or spring. (Further details on tension will be given in § 4.6.)

With a string, F and x must be positive or zero; but they may also be negative with a spring.

The dimensions of λ are those of force, $[MLT^{-2}]$; and of k, force per unit distance, $[MT^{-2}]$.

3·91. Work during change of extension. Let the end A be fixed. Then by (11), if the extension changes from x_0 to x say, the work W done against the tension F (F here acts oppositely to the direction of increase of x) is given by

$$W = \int_{x_0}^{x} F\,dx = \int_{x_0}^{x} kx\,dx = \tfrac{1}{2}kx^2 - \tfrac{1}{2}kx_0^2.$$

This work is equal to the product of the increase in extension and the mean of the initial and final tensions; for the last expression is equal to

$$(x - x_0)(kx_0 + kx)/2.$$

3·92. Conservation of energy. By § 3·91, if the extension changes from any value x_1 to any other value x_2 and then back to x_1, the work done by the tension

$$= -(\tfrac{1}{2}kx_2^2 - \tfrac{1}{2}kx_1^2) - (\tfrac{1}{2}kx_1^2 - \tfrac{1}{2}kx_2^2),$$

which is zero. Hence the tension of a perfectly elastic string or spring is conservative.

Hence there exists an associated potential energy. Putting $x_0 = 0$ in § 3·91, we see that this potential energy is equal to

$$\tfrac{1}{2}kx^2, \tag{18}$$

referred to the unstretched position as standard.

3·93. Force-displacement graph. The graph of the relation (17) between F and x is a straight line through the origin. The foregoing results can therefore be simply derived by reference to a graph such as in Fig. 19. For example, the work done against the tension when the extension changes from x_1 to x_2 is represented by the shaded area in Fig. 19.

The student should verify graphically the statement in small print in § 3·91.

3·94. Examples. (1) *A perfectly elastic horizontal india-rubber cord has lengths of 46, 54 inches under tensions of 3, 6 lb.wt., respectively. Find (a) the modulus of elasticity; (b) the work done by the resultant of applied forces under which the cord is stretched from rest in its natural length to rest at a length of 60 inches.*

Let a ft be the natural length of the cord, and k pdl/ft its elastic constant. By (17a), the data give

$$3g = k\left(\frac{46}{12} - a\right), \qquad 6g = k\left(\frac{54}{12} - a\right),$$

yielding $\quad k = 4\cdot 5g, \quad a = 38/12.$

The modulus of elasticity (λ) $= ka$ pdl
$\qquad\qquad\qquad\qquad\quad = 4\cdot 5 \times (38/12)$ lb.wt.
$\qquad\qquad\qquad\qquad\quad = 14\cdot 25$ lb.wt.

By (14b), the required work = increase in P.E., since the K.E. is zero both initially and finally.

Hence, by § 3·92, this work $= \tfrac{1}{2}k \left(\dfrac{60}{12} - a\right)^2$ ft-pdl

$= 2 \cdot 25 \times (22/12)^2$ ft-lb.wt.

$= 7 \cdot 56$ ft-lb.wt.

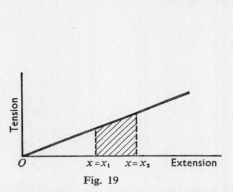

Fig. 19

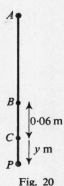

Fig. 20

(2) *A mass of $0 \cdot 012$ kg is suspended at the lower end of a light perfectly elastic vertical string whose upper end is fixed, the extension being 6 cm when the mass is in its equilibrium position C. The mass is then pulled up 2 cm above C and set in motion with an initial upward velocity of 40 cm/sec. Find (a) the potential energy when the mass is in any position; (b) the highest position of the mass; (c) its speed when 2 cm below C; (d) its position when the speed is greatest.* (See Fig. 20.)

We shall use m.k.s. units. As an exercise, the student may be interested to re-write the solution using c.g.s. units.

Let A be the fixed point of suspension, AB the natural length of the string, k N/m its elastic constant. Let P be any general position of the mass, distant y m below the equilibrium position C.

Let F newtons denote the tension when the mass is at P; and F_C the particular value of F when the mass is at C.

Since the mass is in equilibrium at C, (8a) gives $0 \cdot 012g - F_C = 0$.

Since, by (17a), $F_C = 0 \cdot 06k$, we have $k = 0 \cdot 2g$, where $g = 9 \cdot 80$. \hfill (i)

(a) We shall throughout refer potential energies to C as standard position.

When the mass is at P (provided the string is not slack, i.e., provided P is not above B), we have:

Elastic P.E. $= \{\tfrac{1}{2}k(0 \cdot 06 + y)^2 - \tfrac{1}{2}k \times 0 \cdot 06^2\}$ J;

Gravitational P.E. $= -0 \cdot 012gy$ J.

Hence, by § 3·833 and (i), (provided $y \geqslant -0 \cdot 06$),

total P.E. $= \tfrac{1}{2}ky^2$ J \hfill (ii)

$= 0 \cdot 1gy^2$ J. \hfill (iii)

(b) All the forces involved are conservative.
Hence, P.E. + K.E. = constant. (iv)
We are given that initially $y = -0.02$, $\dot{y} = -0.4$.
By (iii), initial P.E. $= 0.00004 \times 9.80$ J.
Also, initial K.E. $= \frac{1}{2} \times 0.012 \times 0.4^2$ J.
∴ initial (P.E. + K.E.) $= 1.380 \times 10^{-4} \times 9.80$ J. (v)
Let the highest position be given by $y = A$.
Since the K.E. is zero in this position, it would follow from (iii), (iv) and (v) (provided (iii) is relevant), that
$$0.980 A^2 = 1.380 \times 10^{-4} \times 9.80,$$
yielding $A = \pm 0.0371$.
The positive result would give the lowest position of the mass, and is therefore not relevant.
The negative result gives the highest position as 3·71 cm above C, which is the answer required, provided (iii) has been validly used. This is the case since the position found is below B.

(c) By (iii) and (iv), we have
$$0.1 g y^2 + \tfrac{1}{2} \times 0.012 \dot{y}^2 = \text{constant}. \quad \text{(vi)}$$
Hence $\dot{y}^2$ is determined when y^2 is given.
Hence $\dot{y}^2$ is the same when $y = 0.02$ as when $y = -0.02$.
Hence the required speed is 40 cm/sec.

(d) By (vi), $\dot{y}^2$ is greatest when y^2 is least, i.e. when $y = 0$. Hence C is the position of the mass when its speed is greatest.

S.P.I.C.E. (α) Ex. (1) was chosen as an easy exercise on the use of the energy principle with perfectly elastic strings, and as a preliminary to the slightly more complicated Ex. (2).

(β) Ex. (2) was chosen primarily as one in which two different types of potential energy, gravitational and elastic, are involved.

(γ) Attention is drawn to the necessity of considering the possibility of the string becoming slack in Ex. (2).

(δ) In problems such as Ex. (2), the practice is sometimes followed of measuring the 'extension' from the equilibrium position C instead of from B. We notice by (ii) that the formula $\tfrac{1}{2}ky^2$, where y is as in Fig. 20, gives the *total* potential energy (elastic plus gravitational).

(ε) It may be noted that the mass in Ex. (2) moves in simple harmonic motion. The course of the above solution should later be compared with the theory given in Chapter VIII.

EXAMPLES III

1. A constant force acting on a mass of 1·5 kg while the mass moves in a straight line over a distance of 10 metres, increases the speed from 0·2 m/s to 1 m/s. Find the force and the speed it would generate in a minute in a mass of 10 kg initially at rest. [Answers: 0·072 newton; 0·432 m/s.]

Exs. III]   **DYNAMICS OF A SINGLE PARTICLE**   55

2. A train passing a point A was observed two minutes later passing a point B one mile distant at a speed of 60 m.p.h. Assuming constant acceleration, find the ratio of the resultant force on the train to the weight of the train.
[Answer: 0·023.]

3. A train of mass 400 tons, moving along a horizontal track with constant acceleration, has speeds of 20 and 40 m.p.h. at the beginning and end of a two-minute interval. If the force of the engine is 5 tons wt., find the resistance in lb.wt. per ton of the train's mass.
[Answer: 10·9 lb.wt. per ton.]

4. A balloon whose total mass is M is falling with downward acceleration f_1. Neglecting frictional resistance due to the air, find how much ballast should be ejected from the balloon in order that it should then have an upward acceleration f_2.
[Answer: $M(f_1 + f_2)/(f_2 + g)$.]

5. A balloon has an upward acceleration of 1 m/s². In the balloon, a spring balance indicates the apparent weight of a body as 1 Kg. What is the body's weight?
[Answer: 0·907 Kg.]

6. A man falls down a lift well on to the top of a lift which is descending with uniform speed V, and is distant h below the man at the instant when he starts to fall. Prove that the impulse of the shock he receives is the same as he would receive if he fell through a distance $h + V^2/2g$ on to a fixed platform.

7. A mass of 100 lb, after falling from rest for 10 sec, penetrates a bog which brings the mass to rest in 0·1 sec. Find: (a) the resistance of the bog, assumed constant, in lb.wt.; (b) the impulse of this resistance during the motion, in lb-ft/sec; (c) the distance the mass penetrates into the bog.
[Answers†: (a) 10,100 lb.wt.; (b) 32,300 lb-ft/sec; (c) 16 ft.]

*8. A body of mass m moving initially with speed u is acted on by a resistance force of magnitude kv^2, where k is constant and v is the speed at any instant. Find the speed when a distance s has been covered.
[Answer: $ue^{-ks/m}$ (where $e = 2·718$ . . .).]

*9. The magnitude of the acceleration due to gravity at a point outside the Earth at a distance x from the Earth's centre is (very nearly) μ/x^2, where μ is constant. Neglecting atmospheric resistance, prove that if an object is projected normally outwards with speed V from a point of the Earth's surface, its speed $\dot{x}$ in any position will be given by $V^2 - \dot{x}^2 = 2gR^2(R^{-1} - x^{-1})$, where R is the Earth's radius, and g is the magnitude of the acceleration at the Earth's surface. Hence (taking the Earth's radius as 3,960 miles) prove that if the speed of projection exceeds 7 miles per second, the object will escape the Earth's influence.

10. A ball of mass 100 g is struck by a club, the contact lasting 0·01 sec; and is thereby given a speed of 10 m/s. Assuming that, during the contact, the force increases at a constant rate from zero to a maximum and then decreases at a constant rate from this maximum to zero, find: (a) the impulse of the blow; (b) the maximum force on the ball.
[Answers: (a) 1 kg-m/s; (b) 200 newtons.]

† In these (and many other) answers, figures after the first three are without significance.

11. A jet of water has mass-per-unit-length λ and speed v at the place where it impinges against a wall at right angles. Neglecting any rebound of the water from the wall, find the magnitude of the force exerted on the wall. (This is also the magnitude of the force exerted on the wall—see § 4·2.) [Answer: λv^2.]

*12. A mass of 1 cwt. is suspended at the lower end of a light vertical rope, and is being hauled up vertically. The pull on the rope is given by $(112 + 3\cdot5x)$ lb.wt., where x ft is the distance of the mass above its initial position of rest. Prove that the measure in ft/sec² of its acceleration at any instant is equal to the measure in ft/sec of its velocity at that instant.

13. A venetian blind has 20 bars, each of mass 12 oz and $\frac{1}{8}$ inch thick. When the blind is down the distance between the tops of successive members is $2\frac{5}{8}$ inches. Neglecting friction, find the work done in drawing up the blind, the highest bar being raised 3 inches during this process. [Answer: 33·4 ft-lb.wt.]

*14. Find the work done by the pull of the rope in Ex. 12 in raising the mass through x ft. Verify that the work done by the *resultant* force during the process is equal to the increase in the kinetic energy of the mass.

[Answer to first part: $(112x + 1\cdot75x^2)g$ ft-pdl.]

15. In order to drive a steamer 120 ft long at a steady speed of 10 knots, a power of 180 h.p. has to be communicated to the propellers. Assuming the resistance to be proportional to the square of the speed and to the area of the wetted surface, find the steady speed of a steamer of similar shape 200 ft long when the communicated power is 4,000 h.p. [Answer: 20 knots.]

16. It has been stated that the human heart expends enough energy in 24 hours to raise a 150-pound man from the bottom to the top of the Empire State Building, New York (of height 1,250 ft). Find the corresponding mean value of the horse-power of the human heart. [Answer: About 0·004 h.p.]

17. Assuming no dissipation of energy, find the horse-power needed to raise steadily from rest 500 lb of water per minute through 20 ft, the water leaving the top of the pipe with a speed of 16 ft/sec. [Answer: 0·364 h.p.]

18. A mass m lb falls from rest under gravity through h ft and acquires kinetic energy equal to αmh J. Find α, given that 1 ft = 30·48 cm, 1 lb = 453·6 g.

[Answer: 1·35.]

19. An object of mass 5 lb falls vertically from rest through 16 ft on to a fixed horizontal plane. If one-fifth of its energy is dissipated at the impact, find: (*a*) the height to which the ball rebounds; (*b*) the impulse at the impact.

[Answer: (*a*) 12·8 ft; (*b*) 303 pdl-sec.]

20. A body of mass 1 kg falls from rest through the air through 218 m, and acquires a speed of 50 m/s. Find: (*a*) the loss in potential energy; (*b*) the work done against air-friction (not here neglected). [Answers: (*a*) $2\cdot14 \times 10^3$ J; (*b*) $0\cdot89 \times 10^3$ J.]

21. An engine is required to generate uniformly a speed of 15 m.p.h. on the level in a train of mass 100 tons in two minutes after starting. The resistance to motion is 10 lb.wt. per ton. Find the horse-power needed.

[Answer: 91·3 h.p.]

22. A mass of M cwt is moved from rest under a constant driving force of P tons wt. and a constant retarding force of Q tons wt. Show that the horse-power of the driving force after t minutes is approximately $4{,}890P(P - Q)gt/M$.

23. The engine of a car of mass M cwt works at the constant rate of H horse-power for t minutes, and the car as a result acquires a speed of V miles per hour, starting from rest. Neglecting all forces other than that due to the engine, find the value of MV^2/Ht.
[Answer: 8,770.]

24. A train of total mass 400 tons starts from rest on a level track. The engine exerts a constant pull of 10 tons wt. until the speed is 15 m.p.h., and thereafter the engine works at a constant rate. The resistance throughout the motion is 2 tons wt. (a) Find the elapsed time when the speed of 15 m.p.h. is acquired; (b) show that the speed can never exceed 75 m.p.h.
[Answer: (a) 34·4 sec.]

25. A train running on the level at a steady speed of 40 m.p.h. slips a carriage of mass 20 tons, and the ultimate steady speed of the train then becomes 45 m.p.h., the engine continuing to work at the same rate. The resistances due to friction amount to 15 lb.wt. per ton of the moving mass. Find the horse-power at which the engine is working, and the mass of the train.
[Answers: 288 h.p.; 180 tons.]

26. The speed of a body of mass m kg, initially at rest, is v m/sec after t sec have elapsed. (a) If the resultant force is constant, find its magnitude and also the maximum value of its power (during the interval in question). (b) If the power of the resultant force is constant, find this power and the maximum value of the resultant force.
[Answers: (a) mv/t N; mv^2/t W; (b) $\tfrac{1}{2}mv^2/t$ W; initially the resultant force must be indefinitely great.]

*27. The resultant force on a train starting from rest is P, where P is constant, until a certain speed V is acquired; subsequently, the power of the resultant force has the constant value PV. When the speed is v, where $v > V$, prove that the elapsed time is $M(V^2 + v^2)/2PV$, and that the distance from the start is $M(\tfrac{1}{2}V^3 + v^3)/3PV$, where M is the mass of the train.

28. A light perfectly elastic string of modulus 0·16 Kg has a length of 2·4 metres under a tension of 20 g.wt. Find: (a) its natural length; (b) the work in joules done when the string's length is increased from 2·3 to 3·1 metres.
[Answers: (a) 2·13 m; (b) 0·333 J.]

29. A mass of 2 lb is suspended at the lower end of a light vertical perfectly elastic string whose upper end is fixed, and in the equilibrium position the length of the string is 40 in. When the mass is then pulled up through 3 in, the tension in the string is 8 oz.wt. Find the magnitude of the greatest (vertical) velocity which can now be given to the mass without the string ever becoming slack.
[Answer: 2·16 ft/sec.]

30. A force of Q lb.wt. produces a compression of b inches in a buffer spring. A mass of M lb impinges on the free end with a speed of V ft/sec. Assuming that no energy is dissipated, find in lb.wt. the maximum thrust in the buffer.
[Answer: $\sqrt{(12QMV^2/bg)}$ lb.wt.]

CHAPTER IV

DYNAMICS OF A SYSTEM OF PARTICLES MOVING IN A STRAIGHT LINE

"It hurts me as much as it hurts you."
—The action, reaction principle.

4·1. System of particles. So far, only single particles have been considered. We now consider dynamical systems which may consist of one or more particles.

The set of positions occupied at any instant by the particles of a system gives the *configuration* at that instant.

It is postulated that when two particles of masses m_1, m_2 are joined together, the result is a particle of mass $m_1 + m_2$. This corresponds to many experimental observations (ignoring certain results in modern physics). Thus masses are taken to be simply additive.

(The need for a postulate is illustrated by noting that if two chemically interacting liquids are mixed, the resulting volume is not in general equal to the sum of the two volumes; cf. also § 3·16.)

The mass of any system of particles is defined as the sum of the masses of the constituent particles.

In Chapter IV, the theory continues to be restricted to the one-dimensional case.

Much of the theory is set up for the case in which the number of particles of a system is finite. This theory will, however, sometimes be applied to cases (e.g. in § 4·9) where the number of particles is indefinitely great; the question of the justification of this will be deferred to § 11·2.

4·11. Centre of mass. Let $x_1, x_2, \ldots$ be the position coordinates of a system of particles referred to an origin O, and let $m_1, m_2, \ldots$ be their masses. The *centre of mass* (C.M.) of the system is defined to be the point, G say, whose coordinate, $\bar{x}$, is given by

$$\bar{x} = \frac{m_1 x_1 + m_2 x_2 + \ldots}{m_1 + m_2 + \ldots}. \tag{19a}$$

The position of G as thus given is independent of the particular origin O taken. To show this, take another origin O', and let $x_1', x_2', \ldots$ be the coordinates of the particles referred to O'. Let G' be the centre of mass obtained on applying (19a) with O' as origin. Thus

$$O'G' = (m_1 x_1' + m_2 x_2' + \ldots)/(m_1 + m_2 + \ldots).$$

Since $x_1' = x_1 - OO'$, etc., it is easy to show that

$$O'G' = \bar{x} - OO' = OG - OO' = O'G.$$

Hence G and G' are one and the same point.

The expression $(m_1 x_1 + m_2 x_2 + \ldots)$ will often be denoted as $\Sigma(mx)$; and similarly in other cases. Thus

$$\bar{x} = \frac{\Sigma(mx)}{\Sigma m}. \tag{19b}$$

If $\bar{v}$ and $\bar{f}$ are the velocity and acceleration of G, differentiation of (19b) gives

$$\bar{v} = \frac{\Sigma(mv)}{\Sigma m}, \qquad \bar{f} = \frac{\Sigma(mf)}{\Sigma m}, \tag{20}$$

where v, f are the velocity, acceleration of a typical particle of the system.

4·12. Example. *If a system of particles moves under gravity (and no other forces) in a vertical line, the centre of mass moves with downward acceleration g.*

For, by (20), the downward acceleration of the C.M.

$$= \frac{m_1 g + m_2 g + \ldots}{m_1 + m_2 + \ldots} = g.$$

It may be noted that this result is also an immediate consequence of the statement in the second paragraph of § 4·33 (to follow).

4·13. Momentum. The momentum of the system is defined to be $\Sigma(mv)$. By (20), the momentum is also equal to $M\bar{v}$, where $M = \Sigma m$, the mass of the system.

4·2. Action and reaction forces. Consider two bodies P_1, P_2 (treated, as usual, as particles) of masses m_1, m_2, rigidly attached together, and let one of them be acted on by a force F as shown in Fig. 21. (The bodies are drawn separated in Fig. 21 in order to show the forces clearly.)

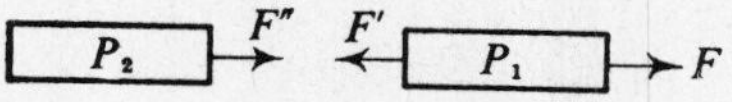

Fig. 21

Let F' be the force on P_1 due to the presence of P_2, and F'' the force on P_2 due to P_1, as in Fig. 21; F', F'' are commonly referred to as the 'action' and 'reaction' between the two bodies. Let f be the acceleration (in the direction of F) of P_1 and hence also of P_2. Then

$$F - F' = m_1 f; \qquad F'' = m_2 f.$$

Also, considering the combined body, we have by § 4·1 and (8a)

$$F = (m_1 + m_2) f.$$

From these three equations it follows that $F' = F''$, i.e. that the action and reaction are equal and opposite in the case taken.

In other cases, for example if P_1 and P_2 are not in contact and move with different accelerations (suppose e.g. P_1 and P_2 were magnets attracting each other and F a further force applied to P_1), it is not possible to *deduce* that the action and reaction forces are equal and opposite. But in the light of experimental observations this is postulated to be the case. (It can be shown that this new postulate absorbs and supersedes the additivity postulate of § 4·1.)

The postulate is equivalent to Newton's 'third law of motion': *Actioni contrariam semper et aequalem esse reactionem: sive corporum duorum actiones in se mutuo semper esse aequales et in partes contrarias dirigi.*

Action, reaction forces to which the postulate is relevant include forces arising from impacts between bodies, and from explosions, e.g. when a shot is fired from a gun; tension and thrust forces due to light string or rod attachments between particles (see § 4·6 for details); forces due to electric or magnetic attractions or repulsions; etc.

For a system consisting of more than two particles, the postulate is taken to be relevant to each pair of particles.

4·3. Principle of momentum.

4·31. Internal and external forces. When a system of particles is *considered as a whole*, the action, reaction forces between pairs of the particles are called *internal forces*. All other forces acting on members of the system (e.g. the force F in Fig. 21, weight forces, etc.) are called *external forces*.

In problems, it is often convenient to apply formulae to parts of systems which do not include all the particles present. A force may be external for one part of a system, but internal for another. For example, in Fig. 21, F' is an external force so far as P_1 alone is concerned, but is an internal force for the system consisting of both P_1 and P_2.

It needs to be emphasised that the terms 'internal' and 'external' depend on the particular part of a system being considered. In solving problems, it should be stated which part of a system is being considered, and a separate diagram should normally show *all* the external forces on that part.

4·32. Principle of momentum. Let P_1, P_2, P_3, ... be a system of particles of masses m_1, m_2, m_3, ... Let F_1 be the resultant of those forces which act on P_1 and which are external for the whole system;

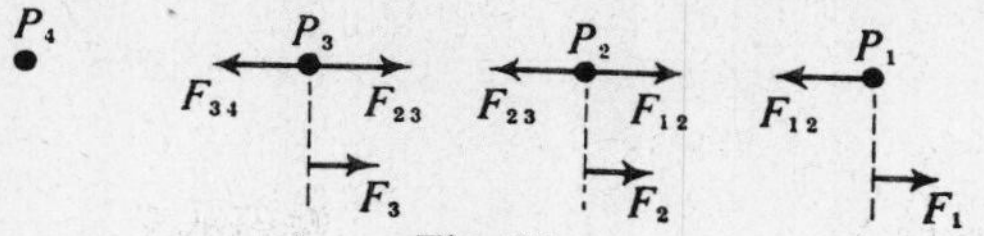

```
      P₄         P₃            P₂           P₁
       •     ←——•         ←——•         ←——•
           F₃₄  |   F₂₃   F₂₃  |  F₁₂   F₁₂  |
                |→            |→           |→
                 F₃            F₂           F₁
```
Fig. 22

similarly F_2, F_3, etc. Let F_{12} be the magnitude of the action, reaction forces between P_1, P_2; similarly F_{23}, F_{13}, etc. (the forces F_{13} are not shown in Fig. 22). Let $f_1, f_2, \ldots$ be the accelerations of $P_1, P_2, \ldots$ Then, by (8a),

$$F_1 + (-F_{12} - F_{13} \ldots) = m_1 f_1,$$
$$F_2 + (F_{12} - F_{23} - \ldots) = m_2 f_2,$$
$$F_3 + (F_{13} + F_{23} - \ldots) = m_3 f_3,$$
$$\ldots \ldots \ldots \ldots \ldots \ldots \ldots \ldots$$

It is evident that the sum of the whole set of terms in brackets is zero. Hence, on adding these equations, we have

$$R = \Sigma(mf), \tag{21a}$$

where R is the resultant of all the *external* forces on the whole system. The earlier equation (8a) is a particular case of (21a).

By (21a), each m being postulated (§ 3·13) to be constant,

$$R = \frac{d}{dt}\{\Sigma(mv)\}; \tag{21b}$$

or, integrating with respect to t,

$$\int_0^t R\, dt = \Sigma(mv) - \Sigma(mu), \tag{21c}$$

where mu, mv are the initial and final momenta of a typical particle.

Each of (21b), (21c) embodies the principle of momentum for a system of particles (moving in a straight line). By (21b), the resultant *external* force at any instant is equal to the rate of change of the momentum of the system at that instant. (21c) connects the impulse of the resultant external force with the change in momentum over any interval of time.

4·33. Motion of centre of mass. Let M, $= \Sigma m$, be the mass of the system. Then, by (20) and (21a), (21b), (21c), we may write

$$R = M\bar{f}, \tag{21d}$$

or
$$R = \frac{d}{dt}(M\bar{v}), \tag{21e}$$

or
$$\int_0^t R\, dt = M\bar{v} - M\bar{u}, \tag{21f}$$

where $\bar{u}$, $\bar{v}$ are the initial and final velocities of the centre of mass.

Any one of (21*d*), (21*e*), (21*f*) shows that the motion of the centre of mass of the system is the same as that of an *equivalent particle* of mass M acted on by all the external forces present. (Cf. § 4·12). This is a highly useful way of looking at the principle of momentum.

4·34. Conservation of momentum. If *in particular* the resultant force R on the system is zero at any instant, it follows from (21*b*) that the rate of change of momentum is zero at that instant.

So long as this holds, the momentum of the system is conserved, and by (21*e*) the velocity of the centre of mass is constant.

4·35. Influence of impulsive forces. If impulsive forces are present at any instant, their influence is estimated (as in § 3·54) in terms of their instantaneous impulses.

On the basis of (21*c*), (21*f*), it follows that *internal* impulsive forces, e.g. those arising at an impact between a pair of particles, have no effect on the momentum of the whole system (or on the velocity of its centre of mass).

At any instant, any jump that there may be in the momentum of the system is equal to the resultant of the impulses of the external impulsive forces acting at that instant. (It has been seen (§ 3·54) that the instantaneous impulse of any non-impulsive force is zero.)

4·4. Examples. (1) *Two bodies moving in opposite directions with speeds of* 8 *cm/sec and* 10 *cm/sec impinge and then adhering move with a speed of* 3 *cm/sec in the direction of the first body's original motion. Compare the masses of the two bodies.*

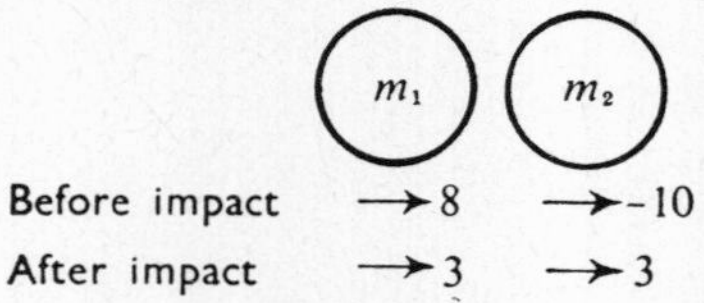

$$
\begin{array}{lcc}
& m_1 & m_2 \\
\text{Before impact} & \rightarrow 8 & \rightarrow -10 \\
\text{After impact} & \rightarrow 3 & \rightarrow 3
\end{array}
$$

Fig. 23

Let m_1, m_2 be the two masses.
Take the positive direction as that of the velocity of m_1 before the impact.
The data of the question are then incorporated in Fig. 23.
There are no external impulses. Hence by the principle of momentum
$$8m_1 - 10m_2 = 3(m_1 + m_2).$$
Hence $\qquad\qquad m_1 : m_2 = 13 : 5.$

(2) *A shot of mass 2 lb is projected horizontally from a gun of mass 5 cwt, the speed relative to the barrel being 2000 ft/sec. Find the initial speed of recoil of the gun, assuming there is no initial hindrance to the recoil.*

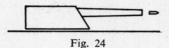

Fig. 24

Take the direction in which the barrel points as positive.

Let the initial velocities of the shot and gun just after firing be u, U ft/sec, respectively. (U must of course be negative.)

During the change in motion examined, there are no external horizontal impulses on the system consisting of the gun and shot. (Vertical forces may here be ignored; see Note (ii) of § 3·4.) Hence, by § 4·34, the momenta of the system before and after the firing are equal.

Thus $\qquad\qquad\qquad 0 = 2u + 5 \times 112 U.$ (i)

But $\qquad\qquad$ 2000 ft/sec = velocity of shot relative to barrel just after firing

$\qquad\qquad\qquad\qquad = (u - U)$ ft/sec. (ii)

From (i) and (ii), we deduce that $U = -2000/281 = -7·12$.

Thus the gun starts to recoil with a speed of 7·12 ft/sec (backwards of course).

(3) *A train consists of a railway engine of mass M tons coupled to two trucks each of mass m tons. Initially the train is at rest and each coupling is slack to the extent of α feet. A constant force of P tons weight is then applied to the engine to drive the train forwards. Neglecting frictional resistances, find (a) the time that elapses before the second truck starts to move; (b) the speed with which this truck starts to move.*

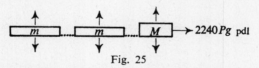

Fig. 25

As in previous similar cases, vertical forces are ignored.

The direction of the motion will be taken as positive.

We shall consider four separate stages of the motion:

(i) The engine moves α feet. During this stage of the motion:

$\qquad$ Resultant force on engine $= 2240 Pg$ pdl.

$\qquad$ Moving mass $\qquad\qquad = 2240 M$ lb.

$\qquad \therefore$ Acceleration $\qquad\qquad = Pg/M$ ft/sec^2 (which is constant).

$\qquad \therefore$ Velocity acquired (using '$v^2 = u^2 + 2fs$')

$\qquad\qquad\qquad\qquad\qquad = \sqrt{(2Pg\alpha/M)}$ ft/sec;

and time taken (using '$s = ut + \tfrac{1}{2}ft^2$')

$\qquad\qquad\qquad\qquad\qquad = \sqrt{(2\alpha M/Pg)}$ sec.

(ii) The first truck is jerked into motion.
By § 4·35, the momentum of the system (S_1 say) consisting of the engine and first truck is unchanged.
Let their common velocity after the impulse $= v_1$ ft/sec.
Then $2240(M+m)v_1 = 2240M\sqrt{(2Pg\alpha/M)}$, giving v_1.

(iii) The system S_1 moves α feet, and may now be treated as a single particle.
Its acceleration $= Pg/(M+m)$ ft/sec² (which is constant).
Velocity acquired (v_2 say)
$$= \sqrt{\{v_1^2 + 2Pg\alpha/(M+m)\}} \text{ ft/sec}$$
$$= \sqrt{\{2Pg\alpha(2M+m)\}/(M+m)} \text{ ft/sec.}$$
Time during this stage (using '$s = \tfrac{1}{2}(u+v)t$')
$$= 2\alpha/(v_1+v_2) \text{ sec}$$
$$= \frac{2\alpha(M+m)}{\sqrt{(2Pg\alpha)}\{\sqrt{M}+\sqrt{(2M+m)}\}} \text{ sec}$$
$$= \sqrt{\left\{\frac{2\alpha(2M+m)}{Pg}\right\}} - \sqrt{\left\{\frac{2\alpha M}{Pg}\right\}} \text{ sec,}$$
the last formula being obtained as a result of rationalising the denominator in the line above.

The answer (a) is now given as the sum of the two times found in (i) and (iii), i.e. $\sqrt{\{2\alpha(2M+m)/Pg\}}$ sec.

(iv) The second truck is jerked into motion.
Let the velocity of the train be v_3 ft/sec just after this impact.
Then $(M+m+m)v_3 = (M+m)v_2$.
From this and the result for v_2 in (iii), we derive the answer (b) as v_3 ft/sec, where
$$v_3 = \sqrt{\{2Pg\alpha(2M+m)\}}/(M+2m).$$

NOTE.—Using the velocity results found, we could have derived the answer (a) more rapidly as follows:
Let t sec be the required total time.
Then $2{,}240Pgt$ lb-ft/sec $=$ Impulse of resultant external force *on whole system* during the time t
$=$ Increase in momentum of whole system during this time
$= 2{,}240(M+2m)v_3$ lb-ft/sec
$= 2{,}240\sqrt{\{2Pg\alpha(2M+m)\}}$ lb-ft/sec.

This gives $t = \sqrt{\{2\alpha(2M+m)/Pg\}}$, as previously obtained.

S.P.I.C.E. (α) Ex. (1) was chosen to show how to handle signs in using the principle of momentum. A diagram of the type of Fig. 23 is very useful in such cases.

(β) Ex. (2) was chosen in order to emphasise that, in using the principle of momentum, it is the 'absolute' velocities that are involved in forming the momenta of the particles of the system. If the data give a relative velocity, this must be connected with the 'absolute' velocities as in equation (ii) of Ex. (2).

(γ) Ex. (3) was given as a problem involving several steps. The alternative method given in the note at the end is particularly instructive. The student should watch for the possibilities of making a general step of this character.

(δ) The student might note that in part (iii) of Ex. (3) the mathematical analysis was facilitated by rationalising a denominator.

4·5. Energy.

4·51. Kinetic Energy. The kinetic energy of a system of particles is defined to be $\Sigma(\tfrac{1}{2}mv^2)$, i.e. the sum of the kinetic energies of the individual particles.

4·52. Potential energy. If conservative fields are present, the corresponding potential energy of the system is defined as the sum of the potential energies of the individual particles.

In problems in this book, conservative forces are usually external. When *internal* conservative forces are present, the corresponding potential energy is calculated in practice for each *pair* of particles involved. (See e.g. § 4·61.)

4·521. In the particular case of particles in a vertical line, the potential energy associated with the weights is, by § 3·832, equal to $\Sigma(mgx)$, where x is the height of a typical particle above the reference level; and hence, by (19b), g being taken constant, is equal to $Mg\bar{x}$.

Thus the potential energy is the same as that of an equivalent particle of mass equal to the mass of the system and situated at the centre of mass.

4·53. Principle of energy. During any interval of time, the work done by the resultant force on each particle of a system is, by (14a), equal to its gain in kinetic energy. This resultant force is the resultant of all the forces on the particle concerned, *including* forces which are internal for the whole system.

Adding up for all the particles gives the principle of energy for a system of particles, viz.:

$$\left. \begin{array}{c} \textit{Sum of works by ALL forces} \\ \textit{(internal and external)} \end{array} \right\} = \left\{ \begin{array}{c} \textit{Increase in K.E.} \\ \textit{of system.} \end{array} \right. \quad (22a)$$

An alternative form of the principle of energy, for use when some of the acting forces are conservative, is:

$$\left. \begin{array}{c} \textit{Sum of works by extraneous} \\ \textit{forces (internal and external)} \end{array} \right\} = \left\{ \begin{array}{c} \textit{Increase in (K.E.} + \\ \textit{P.E.) of system.} \end{array} \right. \quad (22b)$$

In the *particular case* in which the net work of the extraneous forces is zero, the mechanical energy of the system is conserved.

4·54. Remarks. (i) *In using the principle of energy, we cannot in general ignore internal forces.* This is in sharp contrast to what is permissible in using the principle of momentum.

(ii) In wider contexts than the present, reference is made to a *principle of conservation of energy*. But the term 'energy' as thus used includes not only mechanical energy but a variety of other forms—heat energy, electrical energy, chemical energy, etc. According to this principle, any change in the mechanical energy of an isolated system is a conversion of mechanical energy into some other form of energy.

For example, for a system consisting of two rough bodies in contact and sliding relatively to each other, the net work of the action, reaction forces is negative. Corresponding to (22b), we should then say that mechanical energy is being *dissipated*; i.e., mechanical energy is being converted into other forms of energy (mainly heat energy as a rule).

(iii) The term *efficiency* is used in mechanics when energy is being delivered from one source to another, and denotes the ratio of the undissipated energy to the original energy. The term is elaborated in § 15·9.

(iv) At any instant when there are sudden changes of velocity in parts of a system, there is in general a sudden change in the mechanical energy. (An exception is the model case of *perfectly* elastic impacts.)

E.g., consider a system consisting of two bodies of equal mass, moving with equal speeds in opposite directions, which collide and then adhere. The K.E. of the system is *suddenly reduced* from a finite value to zero. (The momentum of the system of course remains unchanged; it is zero both initially and finally.)

On the other hand, an explosion can cause a *sudden increase* in mechanical energy (usually at the expense of chemical energy).

In both these illustrative cases, there is net work done on the system by internal impulsive forces. In the first case, the net work is negative; in the second, positive.

(v) Because of (iv), it is as a rule more expedient to use the principle of momentum than the principle of energy when impulsive forces act.

4·55. Examples. (1) *A bullet of mass m travelling horizontally with speed v strikes a thick target of mass M and comes to rest relative to the target after penetrating a distance α into the target. The target is free to move horizontally in the line of motion of the bullet. Find the acceleration of the target during the process of penetration, assuming the resistance of the target to be a constant force.*

Take the origin O at the position of the nose of the bullet when it first meets the target; take the positive direction in the direction of motion of the bullet. (See Fig. 26, p. 67.)

Let V be the ultimate common velocity of the bullet and target (treated as a pair of particles).

Then considering the system consisting of the bullet and target, we have by the principle of momentum

$$(m + M)V = mv. \tag{i}$$

During the penetration the system loses K.E.

$$= \tfrac{1}{2}mv^2 - \tfrac{1}{2}(m + M)V^2$$
$$= \tfrac{1}{2}mMv^2/(m + M) \quad \text{(using (i))}. \tag{ii}$$

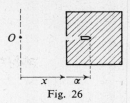

Fig. 26

Let the displacement of the target during the penetration be x, and let R be the magnitude of the resistance. (R is non-impulsive, since the bullet penetrates a finite distance into the target during the action of R.)

Then, work done on target $= Rx$;

work done on bullet $= -R(x + \alpha)$.

Hence, by the principle of energy,
$$Rx - R(x + \alpha) = -\tfrac{1}{2}mMv^2/(m + M);$$
i.e.,
$$R = \tfrac{1}{2}mMv^2/(m + M)\alpha.$$

By (8a), it follows that the acceleration of the target during the penetration $= \tfrac{1}{2}mv^2/(m + M)\alpha$.

(2) *A hammer of mass M moving with speed v strikes a nail of mass m which is initially at rest and just penetrating a fixed block of wood. Find the ratio of the energy lost at the instant of impact to the energy of the hammer just before the impact, assuming that the nail penetrates a finite distance further into the wood as a result of the blow.*

Since the nail penetrates a finite distance further into the wood, the resistance exerted by the wood on the nail is non-impulsive.

Hence the only impulsive forces acting on the hammer and nail consist of the action, reaction between them at the instant, t_0 say, when the hammer meets the nail.

Since these forces are internal for the system consisting of the hammer and nail, the momentum of this system just after t_0 is equal to that just before, viz. Mv.

The loss of K.E. at the instant t_0 is then found, by algebra similar to that used in deriving equation (ii) of Ex. (1), to be $\tfrac{1}{2}mMv^2/(m + M)$.

The K.E. of the hammer before the impact $= \tfrac{1}{2}Mv^2$.

Hence the required ratio $= m/(m + M)$.

S.P.I.C.E. (α) In Ex. (1), the principles of momentum and energy were both used in getting the answer. The problem could have been solved, but less rapidly, independently of the principle of energy. The student may do this as an exercise.

(β) In Ex. (2), a careful discussion of impulsive and non-impulsive forces was necessary.

(γ) It is interesting to note that if the mass of the hammer in Ex. (2) is great compared with that of the nail, the fraction of energy lost at the instant of impact is small. But if the hammer and nail are of comparable mass, the process is rather inefficient.

4·6. Tension and thrust. In this and the following sections of the chapter, the main concern will be with applications of theory set down in §§ 4·1-4·5.

In § 3·9, the tension of a string was considered, but only in so far as the associated force on a particle attached at one end was concerned. Further detail will now be given.

As a model representation of a string, we take a system of particles lined up in fixed order, adjacent particles being indefinitely close to each other. If the string is 'light', each of the particles is taken to have zero mass.

Between any two points of attachment of a light string, the internal forces between all pairs of adjacent particles of the string are the same. To show this, take P_1, P_2, P_3 in Fig. 22 to represent adjacent particles of a string. By (8a), $F_{12} - F_{23} = 0$, since the mass of P_2 is zero. Hence $F_{12} = F_{23}$; similarly, $F_{23} = F_{34}$; etc. On account of this series of equalities, we can speak of the *tension* of a light string as the pull exerted on any particle of it by an adjacent particle, or as the pull exerted by the string on a particle attached (as in § 3·9) at either end. (These remarks all apply whether the string is moving or at rest.)

If the string is not light, the tension may vary from point to point; but we shall not consider such cases in dynamical problems in this book.

Some later problems will involve a pair of particles connected by a light rod. The above discussion may be applied in full to the rod, with the further feature that the directions of the internal forces may sometimes be opposite to those indicated in Fig. 22. In the latter case we speak of the *thrust* of the rod; a thrust (in the present context—see also § 16·3) is thus a negative tension. A rod may be in thrust or in tension; in the former case it is sometimes called a *strut*, in the latter, a *tie*. A string can be only in (positive) tension, or else slack.

In the model representation of a rod, the distances between its particles will be taken to be invariable; i.e., rods will be taken to be *rigid* (see § 13·1). (Real rods are of course deformable to greater or less degree; investigations of deformation belong to the theory of elasticity.)

Strings will be taken to be *inelastic* (i.e. unstretchable) or *perfectly elastic* (thus satisfying Hooke's law). (This leaves out of account many types of imperfectly elastic behaviour that occur in practice.)

Some of the later problems of this chapter will involve strings which are bent round smooth curved surfaces, and thus do not lie entirely in straight lines. In such cases, the strings will be taken to be perfectly *flexible*; without defining this term in detail, we state here that it enables us to take the tension of the string to be unaffected by the bending.

4·61. Energy considerations. (i) The discussion given in § 3·92 on the potential energy associated with a perfectly elastic string can be made more general.

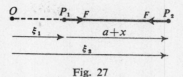

Fig. 27

Consider a system consisting of two particles P_1, P_2 joined by a perfectly elastic light string of natural length a, and let the system be in motion in a straight line. At time t, let the coordinates of P_1, P_2, referred to an origin O, be ξ_1, ξ_2, and let $P_1 P_2 = \xi_2 - \xi_1 = a + x$. Then the tension F at time t is kx, where k is the string constant.

The sum of the works done by the tension forces on P_1, P_2 during the ensuing small time δt is, to sufficient accuracy, $F\delta \xi_1 + (-F)\delta \xi_2$; i.e., $= -F\delta(\xi_2 - \xi_1) = -F\delta x$.

Hence the sum of the works done by these forces when the length of $P_1 P_2$ changes from a to $a + x$

$$= -\int_0^x kx\, dx = -\tfrac{1}{2}kx^2.$$

This is the net work done by all the tension forces present, since the works of the two tension forces on any particle of the string itself are equal in magnitude and opposite in sign and so cancel each other.

It can now be inferred, as in § 3·92, that the whole system has elastic potential energy equal to $\tfrac{1}{2}kx^2$.

(ii) Suppose two particles P_1, P_2 of masses m_1, m_2 are joined by a light string, the system being initially at rest, say on a smooth horizontal table, and the string slack: and suppose that P_1 is then projected away from P_2 with speed v.

If the string is inelastic, it may be shown, by argument similar to that in Exs. in § 4·55, that there is a loss of kinetic energy of amount $\tfrac{1}{2}m_1 m_2 v^2/(m_1 + m_2)$ at the instant when the string becomes taut. The energy thus dissipated is equal to the work done against the impulsive tensions acting on P_1, P_2.

If on the other hand the string is perfectly elastic, the energy principle (with use of the formula $\tfrac{1}{2}kx^2$ for elastic potential energy) can often be used with advantage.

(iii) If two particles joined by a light inelastic string are in motion in a line, and there are no sudden changes of motion during a given interval of time, the total work of the tension forces during this time is zero. In these circumstances, the tension forces may be ignored in applying the energy equation to the system.

This can be shown to apply also in the case of a flexible string bent round a *smooth* surface.

4·62. Examples. (1) *Particles P, Q, each of mass 2 lb, are joined by a perfectly elastic light string, and the system can move in the line of the string on a smooth horizontal table. Initially P and Q are at rest together, and Q is then projected at a speed of 15 ft/sec away from P. Find the velocities of P and Q just before they next meet.*

Take the direction of Q's initial motion as positive.

Let u, v ft/sec be the required velocities of P, Q.

Since the momentum of the system is conserved, we have
$$2u + 2v = 2\times 15. \tag{i}$$
Since the mechanical energy is conserved, we have by § 4·61 (ii)
$$\tfrac{1}{2}\times 2u^2 + \tfrac{1}{2}\times 2v^2 = \tfrac{1}{2}\times 2\times 15^2, \tag{ii}$$
the elastic P.E. being zero both initially and finally.

(i) and (ii) yield $u = 0, v = 15$, or $u = 15, v = 0$.

The latter result gives the solution.

(2) *Particles P, Q of masses 0·2, 0·3 kg are joined by a light inelastic string of length 1 m. Initially P and Q are at rest together and Q is then projected vertically upwards with a speed 5 m/s. Find the speed of P when returning through its initial position.*

Take the positive direction vertically upwards.

Just before the string becomes taut, the velocity of Q
$$= \sqrt{(5^2 - 2\times 9{\cdot}80\times 1)} \text{ m/s}$$
$$= 2{\cdot}32 \text{ m/s}.$$
Just after the string is taut, P and Q have a velocity of u cm/sec, where
$$(0{\cdot}2 + 0{\cdot}3)u = 0{\cdot}3 \times 2{\cdot}32;$$
i.e., $\qquad\qquad u = 1{\cdot}39.$

Thereafter, the velocities of P and Q are always equal, and equal to the velocity of the C.M. The C.M. moves with acceleration $-g$.

Hence when P returns through its initial position, its velocity is 1·39 m/s downwards.

S.P.I.C.E. (α) To give further experience in the use of the principle of momentum.

(β) To show when and when not the principle of energy may be well used in string problems.

(γ) In Ex. (1), it was not necessary to calculate the elastic potential energy.

4·7. Problems involving pulleys. In this section all strings will be assumed to be light and inelastic, and the grooves of all pulleys smooth.

4·71. Atwood's machine. A simple version of this machine consists of a string and masses m_1, m_2 fitted round a fixed pulley as shown in Fig. 28, and moving under gravity, the portions of the string not in contact with the pulley being vertical.

We take the positive direction vertically downwards, and let x_1, x_2 be the position coordinates, measured from some fixed level, of m_1, m_2 at any time t.

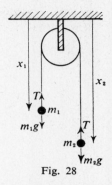

Fig. 28

Then $x_1 + x_2$ is constant during the motion, since it differs from the length of the string by a constant. Hence, differentiating twice with respect to t, we have

$$f_1 + f_2 = 0, \qquad \text{(i)}$$

where f_1, f_2 are the (downward) accelerations of m_1, m_2. (The result (i) is of course obvious, but the method here indicated is required in more difficult problems.)

Let T be the tension of the string at the instant in question. (If the pulley were rough, it may be shown that the tension would in general be different on the two sides of the pulley; we here use the result stated in the last paragraph on page 68.)

Considering separately the motions of m_1, m_2, we then have

$$m_1 g - T = m_1 f_1, \qquad \text{(ii)}$$
$$m_2 g - T = m_2 f_2. \qquad \text{(iii)}$$

From (i), (ii), (iii), we can find f_1, f_2, T when m_1, m_2 are given. The results are:

$$f_1 = \frac{m_1 - m_2}{m_1 + m_2} g; \quad f_2 = \frac{m_2 - m_1}{m_1 + m_2} g; \quad T = \frac{2 m_1 m_2}{m_1 + m_2} g. \qquad \text{(iv)}$$

(The student should remember how to derive and solve (i), (ii) and (iii), rather than try to memorise the formulae (iv).)

It will be seen in § 13·65 Ex. (4) how (iv) needs amending when the string is rough and the pulley rotates with the string.

Atwood used this machine in the eighteenth century to illustrate Newton's laws of motion. The machine was once used to determine values of g.

4·72. Further problems. (1) *In the system shown in Fig. 29, the pulleys A, B, C are each of mass 1 lb, D and E are fixed, and between pulleys each portion of the string is vertical. Circumstances hold as stated in the first sentence of § 4·7. Find the tension of the string, and the accelerations of A, B, C.*

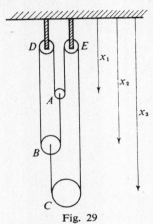

Fig. 29

Take the positive direction vertically downwards.

Let x_1, x_2, x_3 ft be the position coordinates, measured from a fixed level, of the centres of A, B, C at time t sec; and let f_1, f_2, f_3 ft/sec² be the corresponding accelerations.

Following the string from the end at the centre of A to the end at the centre of B, we have

$$(x_2 - x_1) + x_2 + x_1 + x_1 + x_3 + (x_3 - x_2) = \text{constant},$$

since the left-hand side differs from the (constant) length of the string by the sum of a number of constant lengths. Thus

$$x_1 + x_2 + 2x_3 = \text{constant}.$$

Differentiating twice with respect to t, we then have

$$f_1 + f_2 + 2f_3 = 0. \qquad \text{(i)}$$

Let T pdl be the tension of the string. (There is just one string and hence just one tension in this problem.) Considering separately the motions of A, B, C, we have (the mass being 1 lb in each case):

$$T + g - 2T = f_1, \qquad \text{(ii)}$$
$$T + g - 2T = f_2, \qquad \text{(iii)}$$
$$g - 2T = f_3. \qquad \text{(iv)}$$

Substituting from (ii), (iii), (iv) into (i) we obtain

$$(g - T) + (g - T) + 2(g - 2T) = 0.$$
$$\therefore T = 2g/3.$$

Hence by (ii), (iii), (iv),

$$f_1 = f_2 = g/3; \quad f_3 = -g/3.$$

Thus the tension is $2g/3$ pdl, the accelerations of A and B are both $g/3$ ft/sec^2 downwards, and the acceleration of C is $g/3$ ft/sec^2 upwards.

(2) *A mass M is in motion on a smooth horizontal table. On each side it is attached to a light inelastic string passing over a smooth pulley at an edge of the table. The other ends of the strings are attached respectively to masses m_1, m_2 which move vertically. Find the accelerations of M, m_1, m_2 and the tensions of the strings.*

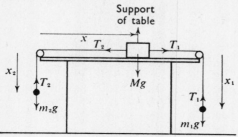

Fig. 30

Let the position coordinates of M, m_1, m_2, be given by x, x_1, x_2 as shown in Fig. 30, and let f, f_1, f_2 be the corresponding accelerations.
Let T_1, T_2 be the tensions of the two strings.
Since the lengths of the strings are fixed, we have

$$x + x_2 = \text{constant}; \qquad x_1 + x_2 = \text{constant}.$$

Hence $\qquad f + f_2 = 0; \qquad\qquad f_1 + f_2 = 0. \qquad\qquad \text{(i), (ii)}$

Considering the motions of M, m_1, m_2, we have:

$$T_1 - T_2 = Mf\,; \qquad\qquad\qquad\qquad\qquad \text{(iii)}$$
$$m_1 g - T_1 = m_1 f_1\,; \qquad\qquad\qquad\qquad \text{(iv)}$$
$$m_2 g - T_2 = m_2 f_2. \qquad\qquad\qquad\qquad \text{(v)}$$

From (i), (ii), (iii), (iv), (v), we obtain

$$T_1 = \frac{m_1(M + 2m_2)g}{M + m_1 + m_2}\,; \qquad T_2 = \frac{m_2(M + 2m_1)g}{M + m_1 + m_2}\,; \qquad \text{(vi)}$$

$$f = f_1 = -f_2 = \frac{m_1 - m_2}{M + m_1 + m_2} g. \qquad\qquad \text{(vii)}$$

The results (vi) and (vii) give the answers, signs being interpreted in relation to the directions of arrows in Fig. 30.

(3) *A man of mass M is holding on to a light flexible inelastic rope which passes over a smooth fixed pulley and has a counterpoise of mass M attached on the other side at the same level as the man. Initially the whole system is at rest. Show that, with only the vertical pull of one side of the rope available, the man cannot depart from the same level as the counterpoise.*

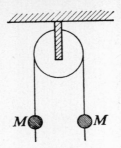

Fig. 31

(i) Suppose the man starts to climb up the rope.

He does this holding the rope and bringing to bear forces internal to himself. These forces influence the tension of the rope, but being internal, do not enter the equation of motion of (the centre of mass of) the man.

Let T be the tension of the string at any instant, and f the upward acceleration of the man. Then
$$T - Mg = Mf.$$
Let the upward acceleration of the counterpoise be f'.

Then for the motion of the counterpoise, we have
$$T - Mg = Mf'.$$
Hence $f = f'$, establishing the result for the case (i).

(ii) If the man moves down the rope, it can be similarly shown that the accelerations of the man and counterpoise are equal.

(iii) If the man lets go of the rope, he moves with acceleration g downwards. But T is then zero. Hence the counterpoise also moves with downward acceleration g.

(4) *A light inelastic string AB passes over a smooth fixed pulley and has masses 5, 4 oz at its ends A, B. A second light inelastic string BC, of length 2 ft, is attached at B to the 4 oz mass and at C to a mass of 3 oz. Initially the system is at rest with the string AB taut, the portions of this string not in contact with the pulley being vertical, and the mass at C resting on a support 1 ft vertically below B. Find the speed of the mass at C just after it is jerked into motion.*

Until the string BC becomes taut, A and B will move with accelerations found by § 4·71 to be of magnitude $g/9$ ft/sec².

The speed, v ft/sec say, acquired just before BC is taut, i.e. when a distance 1 ft is covered, is $\sqrt{(2g/9)}$ ft/sec, i.e. 8/3 ft/sec.

Let J lb-ft/sec be the impulsive tension in the string AB at the instant when the 3 oz mass is jerked into motion, and let V ft/sec be the common speed of the three masses just afterwards.

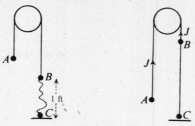

Fig. 32

Then considering the mass at A, we have
$$-J = (5/16)(V - v); \tag{i}$$
and, considering the system BC (the impulsive tension in BC is *internal* for this system),
$$J = \left(\frac{4}{16} + \frac{3}{16}\right)V - \frac{4}{16}v. \tag{ii}$$
Eliminating J from (i) and (ii), we have
$$\left(\frac{5}{16} + \frac{4}{16} + \frac{3}{16}\right)V = \left(\frac{5}{16} + \frac{4}{16}\right)v. \tag{iii}$$
Since $v = 8/3$, the required speed is found to be 2 ft/sec.

S.P.I.C.E. (α) In Exs. (1) and (2) it is necessary to develop certain kinematic formulae connecting the displacements and accelerations of different parts of the system. These examples illustrate the technique needed. Once again, the handling of signs should be noted.

(β) If, in Ex. (2), M is put equal to zero, the equations (vi), (vii) reduce to the formulae (iv) of § 4·71. The student should see why this is so, and note further that this gives a useful check on the answer to Ex. (2). The student should be on the look-out for such checks.

(γ) As a corollary to Ex. (3), if two men of equal mass start from the same level on a race to the top up the ropes on the two sides of a fixed pulley, the result must be a dead-heat. If the men's masses are unequal, the lighter man must necessarily win. (In practice there may of course be complications due to the mass of the rope, etc.)

(δ) Ex. (4) is a problem involving impulsive tension in a string round a pulley. When a little experience is gained in working problems of this type, an equation such as (iii) in the solution may be written down directly, without introducing the symbol J explicitly.

4·8. Elastic impacts. Let m_1, m_2 be the masses of two spherical bodies P_1, P_2 which collide. Let u_1, u_2 be their velocities just before, and v_1, v_2 just after, the collision.

By the principle of momentum applied to the whole system,
$$m_1v_1 + m_2v_2 = m_1u_1 + m_2u_2. \tag{i}$$
If m_1, m_2, u_1, u_2, are given, we cannot by (i) alone determine v_1 and v_2. For this purpose, it is necessary to introduce a further postulate based (as usual) on experimental evidence. This is Newton's 'law of impact', according to which
$$v_2 - v_1 = -e(u_2 - u_1), \tag{23}$$
where e, called the 'coefficient of restitution', is independent of particular values of the velocities. According to the postulate, the relative velocity of P_2 to P_1 after the impact is equal to $(-e)$ times the relative velocity before impact. It is also postulated that $0 \leqslant e \leqslant 1$.

By (i) and (23), v_1 and v_2 can be determined if m_1, m_2, e, u_1 and u_2 are given.

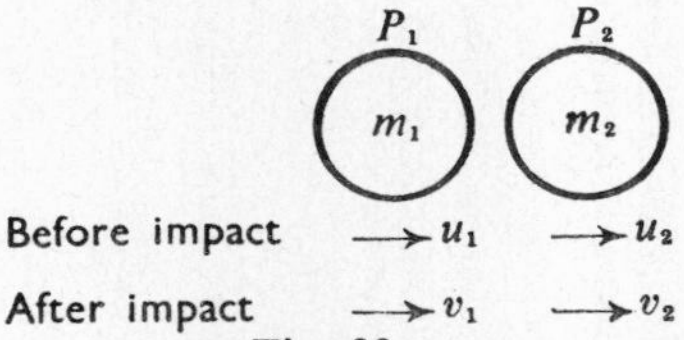

Fig. 33

4·81. Remarks.

(i) From the two equations in § 4·8, it can be shown by elementary algebra that there is a sudden loss of kinetic energy at the impact equal to
$$\tfrac{1}{2}\frac{m_1m_2}{m_1+m_2}(1-e^2)(u_1-u_2)^2.$$
(The student may if he wishes do this now as an exercise; in § 12·612, Ex. (3), a more rapid method will be indicated.)

(ii) If $e = 0$, it follows by (23) that the velocities of P_1, P_2 are equal after the impact. This would be the case if the bodies adhered on colliding, and is sometimes referred to as an 'inelastic impact'.

The case $e = 0$ corresponds dynamically to the case where an inelastic string joining two particles suddenly becomes taut, with no rebound.

(iii) If $e = 1$, the loss of kinetic energy is zero. In this, but only in this, case is the energy conserved at the impact, and the elasticity is *perfect*.

(iv) The case of an elastic impact of a body against a fixed wall is included in (23). If P_2 is the wall, we have $u_2 = v_2 = 0$, and hence, by (23), $v_1 = -eu_1$. (We cannot use the momentum equation since m_2 is now, in effect, infinite.)

(v) It is sometimes postulated that the process of impact consists of two stages, in the first of which the relative velocity is reduced to zero, while in the second stage the final velocities v_1, v_2 are acquired.

If I', I'' are the magnitudes of the action-reaction impulses during the two stages, respectively, and U is the common velocity of P_1, P_2 at the end of the first stage, we have

$$-I' = m_1(U-u_1); \qquad I' = m_2(U-u_2);$$
$$-I'' = m_1(v_1-U); \qquad I'' = m_2(v_2-U).$$

From these we deduce

$$\frac{I''}{I'} = \frac{U-v_1}{u_1-U} = \frac{v_2-U}{U-u_2} = \frac{v_2-v_1}{u_1-u_2}.$$

Hence, by (23), $\qquad I'' = eI'.$

This procedure can be useful in more complicated problems on impact.

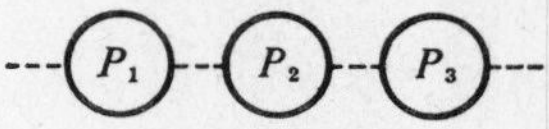

Fig. 34

(vi) As an instance of a case where more than two bodies are involved in an impact suppose (see Fig. 34) that P_1 impinges on P_2 which is at rest and in contact with P_3, also at rest. It is customary to regard P_2, P_3 as initially separated by an indefinitely small distance. The equations of § 4·8 are then applied (*a*) to the impact between P_1, P_2, (*b*) to the ensuing impact between P_2, P_3, (*c*) to the further possible ensuing impacts, e.g. between P_1, P_2 again.

4·82. Examples. (1) *Two balls whose masses are in the ratio* 1 : 2 *and which are initially moving in opposite directions with speeds in the ratio* 2 : 1, *impinge directly. The coefficient of restitution is* 5/6. *Prove that after the impact, each ball retires with* 5/6 *of its initial speed.*

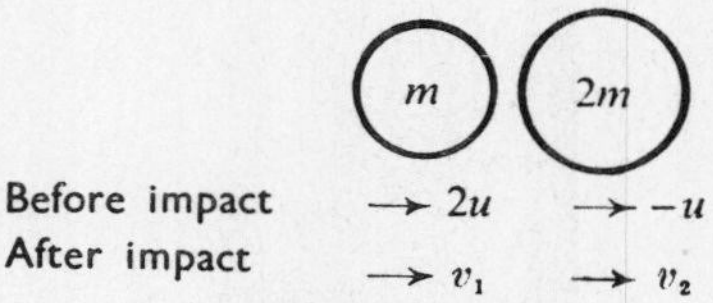

Fig. 35

Let m, $2m$ be the masses of the balls.
Take the positive direction as that of the velocity of m before impact.
Let $2u$ be the initial velocity of m, and hence $-u$ that of $2m$.
Let v_1, v_2 be the velocities after impact, as shown in Fig. 35.
There are no external impulses. So, by the principle of momentum,

$$mv_1 + 2mv_2 = m.2u + 2m(-u),$$

i.e. $\qquad v_1 + 2v_2 = 0.$ \hfill (i)

By Newton's law of impact,

$$v_2 - v_1 = -\tfrac{5}{6}(-u-2u),$$

i.e. $\qquad v_2 - v_1 = 5u/2.$ \hfill (ii)

From (i), (ii), we deduce that

$$v_1 = -\tfrac{5}{6}(2u); \quad v_2 = \tfrac{5}{6}u.$$

These results give the answer.

(2) *The masses of three perfectly elastic balls A, B, C are 2, 1, 2 kg, respectively. Initially B, C are at rest and in contact, and A is moving toward B in the line of B, C at a speed of 1 m/s. Find the final velocities of A, B, C.*

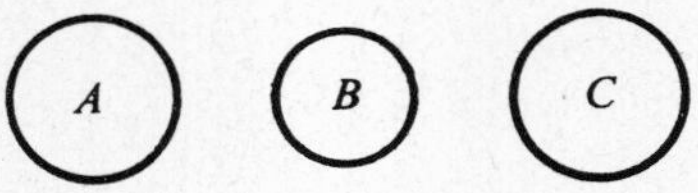

	A	B	C
Initially	$\to 1$	$\to 0$	$\to 0$
After impact (i)	$\to u_1$	$\to v_1$	$\to 0$
After impact (ii)	$\to u_2(=u_1)$	$\to v_2$	$\to w_2$
After impact (iii)	$\to u_3$	$\to v_3$	$\to w_3(=w_2)$

Fig. 36

Take the initial direction of motion of A as positive.

We take B, C to be initially slightly separated, and consider (i) the impact between A, B, (ii) the ensuing impact between B, C, (iii) the subsequent impact between A, B; (it will be verified that the impact (iii) occurs).

Symbols for the velocities after these impacts are indicated in Fig. 36. Of course $u_2 = u_1$, and $w_3 = w_2$.

For the impact (i), we have
$$v_1 - u_1 = -(0-1),$$
$$v_1 + 2u_1 = 2,$$
yielding $\quad u_1 = \tfrac{1}{3}, \quad v_1 = \tfrac{4}{3}.$

For (ii), we have
$$w_2 - v_2 = -(0 - v_1),$$
$$2w_2 + v_2 = v_1,$$
yielding $\quad v_2 = -\tfrac{1}{3}v_1 = -\tfrac{4}{9}, \quad w_2 = \tfrac{2}{3}v_1 = \tfrac{8}{9}.$

The impact (iii) will occur since $u_2(=u_1) > v_2$. We then have
$$v_3 - u_3 = -(v_2 - u_2) = \tfrac{7}{9},$$
$$v_3 + 2u_3 = v_2 + 2u_2 = \tfrac{2}{9},$$
yielding $\quad u_3 = -\tfrac{5}{27}, \quad v_3 = \tfrac{16}{27}.$

There are no further impacts, since $w_3(= w_2) > v_3 > u_3$.

Hence the final velocities of A, B, C are $-\tfrac{5}{27}, \tfrac{16}{27}, \tfrac{8}{9}$ m/s, respectively.

S.P.I.C.E. (α) Both examples show the care needed in handling signs with the use of equation (23).

(β) In Ex. (2), a special examination is needed of the number of impacts between pairs of bodies that have to be taken into account.

(γ) Since the elasticity is 'perfect' in Ex. (2), the answer may be checked by showing that the final kinetic energy is equal to the initial kinetic energy, viz. 1 joule.

4·9. Problems where the mass changes. We are assuming in this book (see § 3·13) that all bodies have constant mass except where we state the contrary. We now consider a class of problems in which the mass of a body is changing as time goes on (e.g. a falling hailstone, or a rocket expelling gas).

We take first the case in which the mass of a body is increasing continuously as time goes on. At time t, let the values of the mass and forward velocity be m, v; and of the resultant external force, R. At time $t + \delta t$, let the values be $m + \delta m$, $v + \delta v$, $R + \delta R$, respectively.

We shall regard m and δm as the masses of a pair of distinct particles. Suppose that, before joining m, δm had a forward velocity v'.

Then by (21c), we have, to sufficient accuracy,
$$R\delta t = m(v + \delta v - v) + \delta m(v + \delta v - v').$$

Dividing through by δt and letting $\delta t \to 0$, we then have that at time t,
$$R = m\frac{dv}{dt} + \frac{dm}{dt}(v - v')$$
$$= \frac{d}{dt}(mv) - v'\frac{dm}{dt}.$$

These formulae, suitably interpreted, also apply if the body is continuously losing mass. The rate of loss of mass is $-dm/dt$, and v' now denotes the forward velocity of an element of mass just after leaving the body. The student should carry out the demonstration in this case as a separate exercise.

4·91. Examples. (1) *A heavy thin flexible string AB, whose mass per unit length, λ, is constant, is initially at rest in a straight line on a smooth plane horizontal surface. A hand then moves from rest at A with constant acceleration f towards B, collecting the string with it as it goes. Find the force exerted by the hand on the string after time t has elapsed, the hand then being between A and B.*

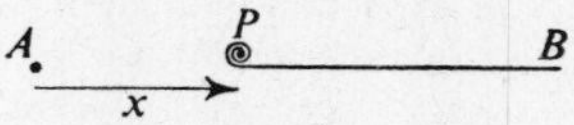

Fig. 37

Let P be the position and v the velocity of the hand at time t, and let $AP = x$.

Then $v = ft$; $x = \tfrac{1}{2}ft^2$.

At time t, the mass m of the string in motion $= \lambda x = \tfrac{1}{2}\lambda ft^2$.

Since the string is initially at rest, the required force, F say, is by § 4·9 given by
$$F = d(mv)/dt$$
$$= d(\tfrac{1}{2}\lambda ft^2 . ft)/dt$$
$$= \tfrac{3}{2}\lambda f^2 t^2.$$

(2) *Material is expelled from the tail of a rocket whose total initial mass is* 28,000 *lb at the rate of* 280 *lb/sec, the speed of expulsion relative to the rocket being* 6,600 *ft/sec. This rate of expulsion is maintained for one minute. Find the initial acceleration and the speed attained by the rocket if directed vertically upwards.*

Take the positive direction vertically upwards.
Let m lb be the mass, and f ft/sec^2 the acceleration after t sec.
We then have
$$m = 280(100 - t). \tag{i}$$
Also, by the formula in § 4·9,
$$-mg = mf - 280 \times 6600. \tag{ii}$$
By (ii), the initial acceleration
$$= (280 \times 6600 - 28000g)/28000 \text{ ft/sec}^2$$
$$= 34 \text{ ft/sec}^2.$$
By (i) and (ii),
$$\frac{dv}{dt} = f = -g + \frac{6600}{100 - t}.$$
Integrating with respect to t from $t = 0$ to $t = 60$, we have that the speed attained after one minute
$$= \left\{ -60g - 6600 \left[\log_e(100 - t) \right]_0^{60} \right\} \text{ ft/sec}$$
$$= (-60g + 6600 \log_e 2\cdot 5) \text{ ft/sec}$$
$$= 4100 \text{ ft/sec, approximately.}$$

S.P.I.C.E. (α) Both Exs. (1) and (2) show that the equation '$R = mf$' does not hold if the mass m is changing. (Immediate use of '$R = mf$' in Ex. (1) would give the false result $\frac{1}{2}\lambda f^2 t^2$.)

(β) The data in Ex. (2) correspond roughly to features of the V2 rocket used by the Germans at the close of World War II. The actual speed reached by the rocket was somewhat higher than that in our answer because we have simplified the data. In an actual rocket the speed of expulsion of matter does not remain constant. A number of other details, including variation in the value of g, atmospheric effects, etc., need to be taken into account in deriving a more accurate answer.

(γ) In Ex. (1), the student should check that the dimensions of the final expression obtained are those of force.

EXAMPLES IV

1. Prove that the centre of mass of two particles, m at P and M at Q, divides PQ inversely in the ratio of m to M.

2. At a certain instant, a particle P is projected vertically upward with speed V from a point A, and a particle Q of the same mass is let fall from rest at A down a mineshaft. Find the velocity of the centre of mass of P and Q after the lapse of time t.

[Answer: $\frac{1}{2}V - gt$, vertically upwards.]

Exs. IV] DYNAMICS OF A SYSTEM OF PARTICLES   81

3. A train travelling at 45 m.p.h. picks up 300 cubic feet of water at a uniform rate from a trough ½ mile long. The mass of 1 cubic foot of water is 62·5 lb. Neglecting the work done through raising the water vertically, find the extra horse-power that must operate if the train maintains a steady speed of 45 m.p.h. [Answer: 116 h.p.]

Show that one-half of this power is dissipated.

4. A gun of mass 10 tons fires a shot of 2 cwt horizontally with a velocity relative to the muzzle of 1,000 ft/sec. Find the initial speed of recoil of the gun, assuming that there is no initial hindrance to the recoil. [Answer: 9·9 ft/sec.]

5. A projectile of mass 100 lb is fired horizontally from a gun of mass 8 tons with a speed of 2,500 ft/sec. The recoil of the gun is stopped in 18 inches. Find in tons weight the force (assumed constant) on the recoil cylinders.
[Answer: 16·2 tons wt.]

6. Bullets, each of mass 2 oz, leave the muzzle of a quick-firing gun of mass 100 lb horizontally with speeds of 1,500 ft/sec. The recoil mechanism of the gun is so arranged that the velocity of the gun is reversed in direction but unchanged in magnitude at each instant when a bullet leaves the gun. Find the kinetic energy of the gun at such an instant.

If the bullets are being fired at a steady rate of 200 per minute, find the distance of recoil of the gun, assuming that a constant force F lb.wt. is applied to prevent the gun from having a progressive backward motion.

Also find F, (a) applying the principle of energy to the gun; (b) applying the principle of momentum to the system consisting of the gun and a bullet.
[Answers: 1·37 ft-lb.wt.; 0·84 in; 19·5 lb.wt.]

7. A moving shell is broken by an internal explosion into two portions of masses m_1, m_2. The mechanical energy released by the explosion is W. Assuming that the original shell and the two portions all move in the same straight line, prove that the relative speed of the two portions is $\sqrt{\{2(m_1+m_2)W/m_1m_2\}}$.

8. A 50-ton locomotive moving at 10 m.p.h. collides with a truck at rest, of mass 10 tons. After the impact the locomotive and truck move on together. Find: (a) the velocity just after the impact; (b) the impulse between the locomotive and the truck at the impact; (c) the loss in energy at the instant of impact.
[Answers: (a) 12·2 ft/sec; (b) 2·74 × 10^5 lb-ft/sec; (c) 2·0 × 10^6 ft-pdl.]

9. A pile-driver of mass 2 tons falls vertically through 10 ft on to the top of a pile of mass 8 cwt, and as a result the pile penetrates 1 ft into the ground. Find in tons wt. the resistance of the ground, assumed constant. [Answer: 19·1 tons wt.]

10. A hammer of mass M moving with speed V strikes horizontally a nail of mass m, and drives it a distance d into a fixed block of wood. Find the resistance of the wood, assumed constant. [Answer: $M^2V^2/\{2(M+m)d\}$.]

11. An electric crane, using electrical energy at the rate of 45 kilowatts, is raising a mass of 6 tons vertically at a uniform speed of 3 feet per 2 seconds. Find the efficiency.
[Answer: 60·8%.]

12. A bullet of mass m is fired horizontally with speed u into a box of sand of mass M, which is initially at rest on a smooth fixed horizontal table, and penetrates a distance b into the box before coming to rest relative to the box. If F is the resistance to penetration, assumed constant, prove that

$$Fb = \tfrac{1}{2}mMu^2/(M+m),$$

and that during the process of penetration the box moves through a distance $mb/(M+m)$.

13. A ten-stone man ascends a mountain at the rate of 1,400 feet vertically per hour. Find in h.p. his mean rate of work, assuming that he dissipates energy at a mean rate of 10 ft-lb.wt./sec. [Answer: 0.12 h.p.]

14. Particles P, Q of masses 4, 6 oz, are joined by a light inelastic string of length 4 ft. Initially P and Q are held at rest together at A, and there is a fixed table distant 3 yards below A. P is now let fall, and Q is released just before the string becomes taut. Find where Q next meets P, assuming a coefficient of restitution 0·5 between P and the table. [Answer: 1·02 ft above the table.]

*15. Particles P, Q, of masses 2, 3 kg, are joined by a perfectly elastic light string, the system being free to move in a straight line on a smooth horizontal table. Initially P and Q are close together, and Q is then projected at a speed of 1 m/s away from P. Later, when P strikes Q, the particles adhere. Find (a) their common velocity, (b) the impulse between P and Q when they meet. [Answers: 60 cm/sec; 1·2 N-s.]

16. A light inelastic string has masses m, m' lb attached at its ends, and passes over a smooth fixed pulley. The string can bear a maximum tension equal to four-ninths of the sum of the weights of the masses. Prove that the minimum acceleration of the system is $g/3$ ft/sec². Find the values of m and m' when the acceleration has this minimum value, given that the tension is then 4 lb.wt. [Answers: 3 lb, 6 lb.]

17. A and B are two objects of masses 1, 3 lb on a smooth horizontal table joined by a light taut inelastic string. B, which is nearer the edge of the table, is joined by a second light inelastic string to an object C of mass 2 lb hanging over the edge of the table and free to move vertically. If the system is left free to move, find the tensions of the strings. [Answers: $g/3$ pdl; $4g/3$ pdl.]

18. Two scale-pans, each of mass 10 oz, are connected by a light inelastic string which passes over a smooth fixed pulley. A mass of 4 oz is placed in one pan, and 8 oz in the other. Find the forces between these masses and the pans when the system is released, and also the resultant supporting force exerted by the pulley.
[Answers: 9, 14, 63 pdl.]

19. Two masses, 3 and 5 lb, are tied to the ends of a light inelastic string 13 ft long which passes over a small smooth horizontal peg fixed at a height of 8 ft above a horizontal table, the 5 lb mass lying on the table, and the 3 lb mass being held close to the peg. If the 3 lb mass is allowed to fall, show that it will not reach the table. Find the greatest height reached by the 5 lb mass, and the time that elapses between its first leaving the table and next arriving at the table.

[Answers: 2 ft $9\tfrac{3}{4}$ in; 1·68 sec.]

Exs. IV] DYNAMICS OF A SYSTEM OF PARTICLES 83

20. A light inelastic string has masses m, m' g attached at its ends and passes over a smooth fixed pulley. The system moves from rest for 10 seconds, when the lighter mass m' picks up from rest a third mass. After 10 more seconds, the system is in the same configuration as when the third mass was picked up. Find: (a) the third mass; (b) the speed of the system after the 20 seconds.

[Answers: (a) $3(m - m')$ g; (b) $5g(m - m')/(2m - m')$ cm/sec.]

21. Masses P and Q of 4, 5 kg are attached to the ends of the string in an Atwood's machine. A rider R of mass 2 kg carried by P can be removed by a ring through which P passes freely. The system is released from rest when P is at a point A, above the ring. After P has passed through the ring, P comes to rest at B, where $AB = 98$ cm. Find the speed when R strikes the ring.

After coming (instantaneously) to rest at B, P ascends, picks up R and then comes to rest at C, above the ring. Find the depth of C below A.

[Answers: 0·98 m/s; 17·8 cm.]

22. A smooth ring is threaded on a light inelastic string whose ends are passed over two smooth fixed pulleys close together, the ring hanging between the two pulleys, while masses of 2 lb, 5 lb are attached at the ends of the string. The system is released from rest with the various portions of the string hanging vertically, and it is found that the ring remains at rest. Prove that its mass is 40/7 lb.

If the ring were tied to the string and not free to slide relatively to the string, prove that its acceleration would be $9g/89$ ft/sec².

*23. Over a smooth pulley B of mass 1 g passes a light inelastic string having masses 6, 9 g at its ends. Over a smooth fixed pulley A passes a light inelastic string having a mass of 15 g at one end and attached to the axis of the pulley B at the other end. If the system moves so that the parts of the strings not in contact with the pulleys move vertically, find the acceleration of the 15 g mass. [Answer: 12·9 cm/sec².]

24. A ball A impinges directly on a ball B at rest. The velocity of A is reduced by two-fifths at the impact, and the coefficient of restitution is 0·6. Find: (a) the ratio of the masses of A and B; (b) the ratio of the velocities after impact.

[Answers: 3 : 1; 1 : 2.]

25. Find the conditions that the velocities of two spheres should be interchanged at a direct impact.

[Answer: The masses must be equal and the elasticity must be perfect.]

26. A smooth sphere moves on a horizontal table and strikes an identical sphere lying at rest on the table at a distance d from a vertical cushion, the impact being along the line of centres and normal to the cushion. If e is the coefficient of restitution for all impacts involved, prove that the next impact between the spheres will take place at a distance $2de^2/(1 + e^2)$ from the cushion.

Interpret the result when $e = 1$. [Answer: The second sphere in this case is finally at rest in its initial position.]

*27. The masses of three perfectly elastic spheres A, B and C are in the ratio $7 : 1 : 7$. Their centres are and remain in a straight line during their motion, and B is between A and C. Initially A and C are at rest and B is given a velocity along the line of centres towards A. Show that B strikes A twice and C once, and that its final velocity is $1/64$ of its initial velocity.

Verify that the initial and final kinetic energies of the system are the same.

*28. A uniform chain of mass M and length $2a$ has particles of masses m, m', where $m > m'$, attached at its ends. The chain is placed with equal lengths a hanging vertically over a small smooth horizontal fixed peg, and is slightly displaced from rest. Find the speed of either mass when the masses have moved a distance a.

[Answer: $\{(M + 2m - 2m')ga/(M + m + m')\}^{\frac{1}{2}}$.]

29. A uniform heavy thin flexible string AB of mass λ per unit length is lying, in a straight line, on a smooth horizontal table. A hand then moves from rest at A towards B, collecting the string with it as it goes. At time t after the start, the acceleration of the hand is kt, where k is constant. Prove that the force then exerted by the hand is $\frac{5}{12} \lambda k^2 t^4$.

*30. A spherical raindrop, whose radius is initially 0.05 in, falls from rest at a height of 1 mile, and during the fall its radius grows by precipitation of moisture (initially at rest) at the rate of 10^{-4} in/sec. Neglecting resistances, find: (a) the radius of the drop when it reaches the ground; (b) the excess of the time of the fall over the time that would have been taken if the drop had not grown.

[Answers: (a) 0.05185 in; (b) 0.33 sec.]

CHAPTER V

INTERLUDE

"Surely some revelation is at hand."
—W. B. YEATS.

So far, the treatment has been entirely one-dimensional. In problems, the motion and the directions of the forces have been in the one straight line (apart from inconsequential deviations, e.g. those referred to in § 3·4, Note (ii)).

It is well that we should now do a little stock-taking before stepping out into space of two and three dimensions, as we shall do in Chapter VI.

5·1. Review of earlier theory. In Chapters II, III, IV, rather a large number of pages have been used up in the course of explaining a theory whose essential structure is quite small. Many of the pages have been concerned with giving guidance in fitting the theory to practical problems and with issuing cautions against ways in which the theory is commonly abused. The student should make sure at this stage that the forest is not obscured by the trees.

Following are the main features of the theory to date:

(i) Introduction of the concepts of displacement, velocity and acceleration; and relative displacement, velocity and acceleration. (Chapter II.)

(ii) Introduction of the concepts of force, mass and particle; resultant force, and the question of addition of forces; the fundamental equation '$R = mf$'. (Chapter III.)

(iii) Definitions of momentum of a particle, impulse of a force; impulsive forces; the principle of momentum for a particle. (Chapter III.)

(iv) Definitions of the work and power of a force acting on a particle, and of the energy of a particle; work of impulsive force; work of resultant force; kinetic energy of a particle; principle of energy for a particle. (Chapter III.)

(v) Definitions of a conservative field of force and the associated potential energy of a particle; extraneous forces; form of energy principle when work is done by conservative forces; formulae for potential energy in two particular cases. (Chapter III.)

(vi) The additivity of mass, and systems of particles; definitions of the mass, centre of mass, momentum, kinetic energy, potential energy (when relevant), of a system; internal and external forces, and Newton's action-reaction law; the principle of momentum for a system; the principle of energy for a system. (Chapter IV.)

(vii) Discussion of units and dimensions. (Chapters II and III.)

In the process of assimilating Chapters III and IV, the student should appreciate the degree of analogy between relations for a single particle and for a system of particles.

5·2. Applications. In order to apply the theory, a number of auxiliary details have been set up, including:

(i) Special formulae for a point moving with constant acceleration. (Chapter II.)

(ii) Illustrations of the method needed when the acceleration is variable. (Chapters II, III.)

(iii) Consideration of simple effects of the Earth's gravitational attraction; the terms vertical and horizontal; weight; potential energy associated with weight. (Chapters II and III.)

(iv) Reference to the particular case of forces in equilibrium; statics. (Chapter III.)

(v) Tension, thrust; Hooke's law; potential energy associated with perfectly elastic string. (Chapters III, IV.)

(vi) Newton's law of impact. (Chapter IV.)

(vii) Treatment of cases of varying mass. (Chapter IV.)

(viii) Use of velocity-time and force-displacement graphs in special problems. (Chapters II and III.)

(ix) Remarks on practical units. (Chapters II and III.)

5·3. Formulae. To be successful in this subject, the student must think clearly; he will not be able to do this if his memory is over-burdened. He should aim to achieve a command of the principles of the subject rather than to remember a host of formulae.

All the more important formulae so far obtained have been numbered in arabic figures (from (1) to (23)).

In Chapter II, the formulae (1), (2), (4), (5), (6) will hardly fail to be remembered. The student should either memorise (3) and (7), or know how to derive them quickly.

The fundamental equation (8a) and the kinetic energy and gravitational potential energy formulae (13) and (16) in Chapter III will hardly fail to be remembered. All the other formulae up to (15) are easily deducible from earlier ones, or embody principles and definitions referred to in § 5·1. If any one of the Hooke's law formulae (17a), (17b), (18) is remembered, the other two are easily inferred.

In Chapter IV, (19b), (21b), (22a, b) and (23) should be remembered. All remaining formulae can then be readily derived. The student is advised to know the method of derivation of the formulae in § 4·9.

If he follows these remarks, a student should be little troubled with memorising formulae on the work so far covered.

5·4. The shape of things to come. Dynamical problems in two and three dimensions are now to be considered, but in Chapters VI, VII, VIII, IX, only single particles will be involved. The basic theory is contained in Chapter VI, while Chapters VII, VIII, IX are mainly concerned with special classes of problems.

Another interlude chapter (X) will be introduced before systems of particles in more than one dimension are treated.

Lest the student feel intimidated by the reference to 'three dimensions', we hasten to state that in practically all the problems set in this book the motion and the relevant forces will be essentially *coplanar*, i.e. in one plane. We include only such three-dimensional theory as is no more difficult to set up than the corresponding two-dimensional theory; the context will make it clear when results hold for three- as well as for two-dimensional space.

In § 6·1 on *vector analysis*, there is no dynamical theory. The purpose of § 6·1 is to establish a few results on the algebra of vectors which enable the dynamical theory from § 6·2 onwards to be set up with the minimum of trouble. Moreover, the use of vectors reveals the close resemblances between the one-dimensional formulae of Chapters II, III and the formulae for more than one dimension in Chapter VI.

It will be seen that many of the postulates and definitions in earlier chapters carry over without reservation to space of more than one dimension. In other cases, modifications are needed and must be appreciated. *It will be seen that very often a formula for one dimension can be adapted to the case of three dimensions simply by giving certain of the symbols vectorial characters*; this is incidentally a useful aid to memory.

5.5. The examples below are miscellaneous ones on the topics of Chapters II to IV, and are for the purpose of giving additional practice, especially to brighter students. Some of these examples are quite difficult.

It is suggested that, before venturing further, the student should be able to set out solutions to problems with the same attention to logical detail that has been given to the setting out of worked problems in this book. It is suggested that he should also be starting to appreciate some of the remarks in Chapter I.

EXAMPLES V

1. Write amended statements in place of the following statements which are all either wrong or in need of some qualification:

(a) If a point moves through a distance s ft in a straight line in t sec, the initial and final velocities being u, v ft/sec, respectively, then $s = \frac{1}{2}(u + v)t$.

(b) The kinetic energy of a particle whose mass is m lb and speed v ft/sec is $\frac{1}{2}mv^2$ ft-lb.

(c) If a force F pdl acts during t sec over a distance of s ft, the power is Fs/gt ft-lb.wt./sec.

(d) If a force F acts on a particle of mass m for time t, then $Ft = mv - mu$, where u, v are the initial, final velocities of the particle.

(e) The dimensions of impulse are $[MT]$.

(f) When two bodies collide, the momentum and the mechanical energy of the system are conserved.

(g) The formula for potential energy is mgh.

(h) The total work done by the external forces on a system of particles is equal to the gain in kinetic energy.

2. Criticise the following statement: 'The deductions from Newton's laws of motion fit the observed facts in Physics, Mechanical Engineering and Astronomy so well that it is inconceivable that the laws could be erroneous.' (See §§ 1·42, 1·63.)

3. A steamer takes time t_1 to travel a distance a up a river flowing with constant speed, and a time t_2 to return. Find the speed (assumed to be always the same) of the steamer in still water. Find also by how much this speed exceeds the speed necessary to cover a distance $2a$ in time $t_1 + t_2$.

[Answers: $\frac{1}{2}a(t_1^{-1} + t_2^{-1})$; $\frac{1}{2}a(t_1 - t_2)^2/\{t_1 t_2(t_1 + t_2)\}$.]

4. A train of mass M travelling at full speed slips a carriage of mass m when distant l from a station. The pull of the engine remains unchanged, and the resistance on any portion of the train is proportional to its weight. Prove that the ratio of the retardation of the slipped carriage to the acceleration of the rest of the train is $(M - m)/m$.

Using a velocity-time diagram, show that if the carriage just reaches the station, the rest of the train will then be at a distance $Ml/(M - m)$ beyond the station.

5. Three trains, A, B, C, move along straight parallel equidistant lines of rails, B being on the middle rails. Initially the rear ends of A, B, C are in a straight line, and the velocities are u, v, w; and the accelerations are a, b, c, all constant. Find the conditions that the rear ends can again be collinear.

[Answer: The expressions $2v - u - w$, $2b - a - c$ must both be zero, or else opposite in sign.]

If v exceeds the mean of u and w, and b exceeds the mean of a and c, and if the trains are of equal length l (their widths being negligible in comparison), prove that B will cease to obscure the visibility of A and C to each other after a time

$$\sqrt{\left\{\left(\frac{2v-u-w}{2b-a-c}\right)^2 + \frac{4l}{2b-a-c}\right\}} - \frac{2v-u-w}{2b-a-c}.$$

6. A load of weight W is raised from rest by a constant vertical force T which acts for a time and then ceases, the load continuing to ascend until it comes to rest again under gravity after rising a total distance s. Prove that the maximum speed is $\sqrt{\{2gs(T-W)/T\}}$, and that the time during which T acts is

$$W\sqrt{(2s)}/\sqrt{\{gT(T-W)\}}.$$

7. One kilowatt-hour is equal to the work which would be done in one hour by a force working at the constant rate of 1,000 watts. An electric motor, 80% efficient, works steadily at 1 h.p. for 40 hours. Find in kilowatt-hours its consumption of electrical energy. [Answer: 37·3.]

8. P and Q are two particles of masses m and m', at a distance r apart. According to Newton's law of gravitation, each particle attracts the other with a force of magnitude Gmm'/r^2, where G is the 'constant of gravitation'. Prove that the dimensions of G are $[M^{-1}L^3T^{-2}]$. Given that the measure of G in c.g.s. units is $6\cdot 67 \times 10^{-8}$, find the measure in m.k.s. units. [Answer: $6\cdot 67 \times 10^{-11}$.]

9. †Given that 1 cm = $3\cdot 28084 \times 10^{-2}$ ft and 1 g = $2\cdot 20462 \times 10^{-3}$ lb, prove that at a place (a little above the Earth's surface) at which g is $31\cdot 998$ ft/sec^2, the measure of 1 joule in ft-lb.wt. is equal to the measure of 1 h.p. in kilowatts correct to five significant figures. [The measure is $0\cdot 74162$.]

10. A train, whose total mass is 200 tons, has a maximum speed of 67·5 m.p.h. on a level track. If the track resistance is 15 lb.wt. per ton, find the horse-power that the engine can supply. [Answer: 540 h.p.]

Prove that at the instant t_1 when the speed is 30 m.p.h. the acceleration is 0·268 ft/sec^2, if the full power is in operation.

Find in lb.wt. the tension in the coupling between the engine and the rest of the train at the instant t_1, given that the mass of the engine is 50 tons.

[Answer: 5062 lb.wt.]

11. A belt conveyer is moving with a uniform horizontal speed of 120 ft/min. A parcel of mass 20 lb is gently placed on the belt. The coefficient of friction is 0·3. Find: (a) the time that elapses before the parcel is at rest relative to the belt; (b) the distance the parcel slides relative to the belt; (c) the energy dissipated during this sliding. [Answers: 5/24 sec; 2½ in; 40 ft-pdl.]

† Due to the late Professor D. M. Y. Sommerville.

12. A mass m falls from rest freely through a distance h, after which it starts to raise a mass M (where $M > m$) initially at rest and connected to m by a light inelastic string passing over a smooth fixed pulley. Find the time that elapses before M returns to its initial position.

$$\left[\text{Answer: } \frac{2m}{M-m}\sqrt{\left(\frac{2h}{g}\right)}.\right]$$

13. A bullet of mass 1 oz moving with a speed of 1600 ft/sec penetrates a block of wood of mass 50 lb. If the block is free to move in the direction of the bullet's motion, find the velocity of the block when the bullet has become embedded, and the loss of kinetic energy.

If the bullet could penetrate 1 ft into a fixed target of the same material, how far short of 1 ft will it penetrate in the present circumstances, the resistance of the target being assumed constant?

[Answers: 2·00 ft/sec; 7·99 × 10⁴ ft-pdl; 0·015 in.]

14. A small sphere A is projected vertically upward with speed v from a point K. Simultaneously, a second equal sphere B is dropped from the point which would otherwise have been the highest point reached by A. Find when and where the spheres collide. [Answer: After time $v/2g$, at height $3v^2/8g$ above K.]

If the coefficient of restitution is e, prove that when A has just reached K once again, B is at height $ev^2(\sqrt{(3 + e^2)} - e)/2g$ above K.

15. A ball is projected vertically upwards with speed V from the ground and rebounds each time it meets the ground. Prove that the ball will be at rest after time $2V/\{g(1 - e)\}$, where e is the coefficient of restitution.

16. A block of mass m falls vertically with speed v on a pile of mass M. There is a coefficient of restitution e, and the resistance of the earth to the movement of the pile is a constant force R. If the pile comes to rest before the block makes a second impact with it, prove that the distance it is driven by the first impact is

$$\frac{m^2(1 + e)^2 v^2}{2(m + M)^2(R/M - g)}.$$

17. A railway carriage of mass M moving with speed V impinges on a second carriage of mass M' at rest. The force necessary to compress the buffers of either carriage through the full extent l is equal to the weight of a mass m. Assuming the buffer-spring to be perfectly elastic, prove that the buffers will be just completely compressed if $V^2 = 2mgl(M^{-1} + M'^{-1})$.

18. A jet of water, of density 62·5 lb/ft³, issues vertically at a speed of 30 ft/sec from a nozzle of section 0·1 sq. in. A ball of mass 1 lb is supported in equilibrium in a smooth vertical tube above the jet by the impact of the water on its under side. Find the height of the ball above the nozzle assuming there is no rebound of the water particles. [Answer: 4 ft 7 in.]

19. A lift of weight W is balanced by a counterpoise of weight W. A man of weight w steps gently into the lift which moves in consequence. Neglecting all friction, prove that, if no further forces are applied to the lift, the man will, inside the lift, experience an apparent gravitational field of intensity $2Wg/(2W + w)$ per unit mass.

*20. One end A of a light inelastic string AB is fixed, and the string passes from A under a smooth movable pulley of weight W, then over a smooth fixed pulley, and supports at B a scale-pan of weight W_1. The portions of the string not in contact with the pulleys are vertical. Inside the scale-pan is a body of weight W_2. Find: (a) the acceleration of the movable pulley; (b) the magnitude of the force between the scale-pan and the body.

[Answers: $(W - 2W_1 - 2W_2)g/(W + 4W_1 + 4W_2)$; $3WW_2/(W + 4W_1 + 4W_2)$.]

*21. A string passing over a smooth fixed pulley supports at its ends two smooth movable pulleys A, B of masses 1, 2 kg, respectively. Over each of A and B passes a string having masses 1, 2 kg at its ends. The strings are light and inelastic and the masses move in vertical lines. Prove that the acceleration of A is $1 \cdot 18$ m/s²; and find the tension of the string joining A and B. [Answer: 4·11 Kg.]

*22. A uniform chain of length 10 ft, of line-density 6 oz/ft, is hanging in equilibrium across a smooth fixed cylinder of horizontal axis and radius 1 ft. The chain begins to slip. At the instant when one end A reaches the cylinder, find the acceleration of the chain and the tension at B, where AB is a horizontal diameter of the relevant section of the cylinder. [Answers: 22 ft/sec²; 13 oz.wt., approx.]

*23. A uniform chain, of line-density λ and length l, is initially in a vertical line and at rest with its lowest point at height h above a fixed horizontal table. The chain is released and let fall on to the table. Find the force on the table at the instant when a length x of the chain (where $0 < x < l$) has reached the table, the impacts being inelastic. [Answer: $\lambda g(3x + 2h)$.]

*24. The resistance of the water to the motion of a certain ship is proportional to the cube of the speed. Prove that after the ship's engines are stopped, the inverse of the speed increases uniformly with respect to the distance described.

If the ship, of mass M, is steaming at a steady speed V, and the engines are stopped, it is found that the speed is reduced to $\tfrac{1}{2}V$ while the ship travels a distance s. Find the power in operation just before the engines were stopped. [Answer: $3MV^3/s$.]

*25. A horse pulls a wagon of 5 tons from rest against a constant resistance of 50 lb.wt. The pull exerted is initially 200 lb.wt., and decreases uniformly with the displacement, until when the displacement is 167 ft the pull is 50 lb.wt. Prove that the speed of the wagon is then about 8·5 ft/sec, and that the elapsed time is about 31 sec.

(Hint.—In the second integration needed, put $x = 334 \sin^2 \theta$, where x is the displacement at time t.)

*26. A particle of mass m moves from rest in a straight line under a constant force P and a resistance of magnitude kv, where k is constant and v is the speed. Prove that the speed will approach, but never attain, the value V, where $V = P/k$.

If x is the distance covered when the speed is v, prove that $(a - bv)^{-1}v\, dv/dx = 1$, where a, b are certain constants. [$a = P/m$; $b = k/m$.]

Hence prove that the distance covered while acquiring from rest a speed of $\tfrac{1}{2}V$ is $(2\log_e 2 - 1)T/P$, where T is the kinetic energy corresponding to the speed V.

*27. A uniform layer of snow whose surface is a rectangle with two sides horizontal, rests on a mountain slope of constant inclination α, the adhesion being just sufficient to hold the snow while at rest. At a certain instant, the uppermost line of snow starts to move downwards and collects with it the snow it meets on the way. Kinetic friction between the snow and slope is negligible. Prove that if v is the speed when a distance x of the slope has been uncovered, then

$$d(x^2v^2)/dx = 2gx^2 \sin \alpha,$$

and hence show that the moving snow has a constant acceleration of $(g/3) \sin \alpha$.

CHAPTER VI

PARTICLE DYNAMICS IN MORE THAN ONE DIMENSION

"A line said to the plane
'I am in doubt if you are sane.
It goes beyond my comprehension
how you dare claim a second dimension.'"
—Reproduced by permission of the Proprietors of PUNCH.

6·1. Introduction to the use of vectors. We now present in outline certain elementary vector properties. The average student need content himself with gaining just a general idea of these properties at first. But he should refer back to this section as each occasion requires when reading the dynamical theory to follow; this should help him to acquire the degree of skill needed in *using* vectors, which is what is primarily needed in understanding later parts of this book.

6·11. Scalars and vectors. The various dynamical entities that have been introduced in the earlier chapters may be classified into: (*a*) those, e.g. mass and kinetic energy, whose specification does not depend on direction; (*b*) those, e.g. displacement, velocity and force, with which a direction has to be specified.

Members of the class (*a*) are *scalars*; and of (*b*), *vectors*.

(It might perhaps be thought that kinetic energy, because of its connection with velocity, belongs to (*b*). But the value of $\frac{1}{2}mv^2$ is positive whatever the direction of a particle's motion, and this magnitude is sufficient to determine the work that can be done by virtue of the motion, without reference to any direction. The vector theory to follow makes the classification into scalars and vectors much more easy to see.)

The full definition of a vector involves: (i) specification of direction as well as magnitude; (ii) addition* according to the 'addition rule of vectors' to be given in § 6·13.

(The necessity of including an addition rule is seen on comparing with §§ 3·16, 4·1 (fourth paragraph). An entity cannot be called a vector unless both the properties (i) and (ii) have been shown, or stipulated, to hold.)

Vectors are denoted in print in bold type, e.g. **F** for force. In manuscript work, the '**F**' may be denoted as 'F'. (A wavy line below a letter indicates to a printer that bold type is required.)

The magnitude of **F** is often denoted as F. (F is, but **F** is not, an ordinary algebraic symbol.)

---
* In Mechanics, vectors are added only when they represent entities of the same dimensions; e.g. two forces or two velocities may be 'added'.

6·12. Free and localised vectors. In addition to (i) and (ii) of § 6·11, it is further necessary, in some contexts, (iii) to specify the *line of action* of a vector; in this case, the vector is said to be *localised* in the line in question. When (iii) is not required the vector is called a *free vector*. In the theory of the dynamics of a single particle, it is sufficient to consider just free vectors. The term 'vector' will be taken to mean 'free vector', unless the contrary is indicated.

6·13. Addition and subtraction of vectors. If **u** and **v** are two vectors, the *sum* of **u** and **v** is *defined* (see § 6·11) to be the vector **z** whose magnitude and direction are determined by the triangle rule exhibited in Fig. 38.

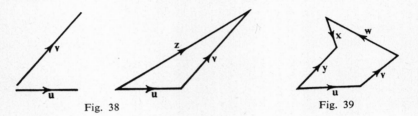

Fig. 38   Fig. 39

Corresponding to Fig. 38, we write *by convention*

$$\mathbf{z} = \mathbf{u} + \mathbf{v}. \tag{i}$$

This equation, and many vector equations to follow, formally resemble algebraic equations in which the symbols stand for ordinary numbers; but the *interpretation* is in accordance with a diagram such as Fig. 38.

Sometimes, corresponding to Fig. 38, **z** is called the *resultant* of **u** and **v**.

It is easy to extend the result to the sum of three or more vectors. Thus e.g. in Fig. 39 (which is not necessarily coplanar) **y** is the resultant of **u**, **v**, **w**, **x**.

By drawing appropriate diagrams, the student can easily see that

$$\mathbf{u} + \mathbf{v} = \mathbf{v} + \mathbf{u}, \tag{ii}$$

$$\mathbf{u} + (\mathbf{v} + \mathbf{w}) = (\mathbf{u} + \mathbf{v}) + \mathbf{w}, \tag{iii}$$

where **u**, **v** and **w** are three vectors.

If **q** is a vector, we denote by '−**q**' the vector whose magnitude is the same as and direction opposite to that of **q**.

If **p** and **q** are two vectors, we define '**p** minus **q**' as '**p** plus (−**q**)'. It then follows from a consideration of Fig. 38 that

$$\mathbf{z} = \mathbf{u} + \mathbf{v} \quad \text{implies} \quad \mathbf{z} - \mathbf{v} = \mathbf{u}. \tag{iv}$$

6·14. Components of a vector.

It can be shown that, in three-dimensional space, a vector can be fully specified by a set of three measures called *components* of the vector.

The components depend on the particular frame of reference used. Throughout this book, the components of a vector will be its projections on some set of rectangular cartesian (or *orthogonal*) axes.

If v_1, v_2, v_3 are the components of the vector **v**, referred to the axes OX, OY, OZ, we may write $\mathbf{v} = (v_1, v_2, v_3)$. Thus

$$\mathbf{v} = (v \cos \theta_1, v \cos \theta_2, v \cos \theta_3),$$

where v is the magnitude of **v**, and θ_1, θ_2, θ_3 are the angles between **v** and OX, OY, OZ, respectively.

If v is in the plane of OXY, v_3 is of course zero. If only the plane OXY is involved (so that the relevant space is just two-dimensional), we may write

$$\mathbf{v} = (v_1, v_2) = (v \cos \theta_1, v \cos \theta_2).$$

6·15. Product of a scalar and vector.

If m is a scalar and **v** a vector, the product $m\mathbf{v}$ (or $\mathbf{v}m$) is defined to be a vector of magnitude mv and direction that of **v**.

If **u**, **v** are two vectors,

$$m(\mathbf{u} + \mathbf{v}) = m\mathbf{u} + m\mathbf{v}; \tag{v}$$

this may be seen from a construction involving two similar triangles whose sides are in the ratio $m : 1$.

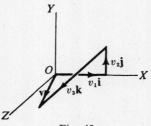

Fig. 40

6·151. The vectors i, j, k.

Corresponding to a given reference frame $OXYZ$, we now introduce **i**, **j**, **k** as *unit vectors* (i.e. vectors of unit magnitude) parallel to OX, OY, OZ, respectively.

If $\mathbf{v} = (v_1, v_2, v_3)$, it is seen from Fig. 40 that

$$\mathbf{v} = v_1\mathbf{i} + v_2\mathbf{j} + v_3\mathbf{k}. \tag{vi}$$

If $\mathbf{v} = 0$ (strictly the '0' should be in bold type, but we frequently do not trouble to do this), then each of v_1, v_2, v_3 is zero; and conversely.

6·16. Scalar product of two vectors. If $\mathbf{u}$ and $\mathbf{v}$ are two vectors, their scalar product, denoted by $\mathbf{u}\cdot\mathbf{v}$ (in some treatments by $\mathbf{uv}$, or by $(\mathbf{uv})$), is defined to be a scalar of magnitude $uv \cos \theta$, where θ is the angle between the directions of $\mathbf{u}$ and $\mathbf{v}$.

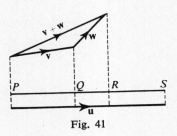

Fig. 41

If $\mathbf{u}$ and $\mathbf{v}$ are parallel, then $\mathbf{u}\cdot\mathbf{v} = uv$. If $\mathbf{u}$ and $\mathbf{v}$ are perpendicular, then $\mathbf{u}\cdot\mathbf{v} = 0$.

By convention, we introduce the notation $\mathbf{v}^2$ to denote $\mathbf{v}\cdot\mathbf{v}$. As a corollary, we have the (not quite trivial) result

$$\mathbf{v}^2 = v^2. \qquad \text{(vii)}$$

It can be shown that if m is a scalar, and $\mathbf{u}$, $\mathbf{v}$ and $\mathbf{w}$ are vectors, then

$$\mathbf{u}\cdot m\mathbf{v} = m\mathbf{u}\cdot\mathbf{v}; \qquad \text{(viii)}$$

$$\mathbf{u}\cdot\mathbf{v} = \mathbf{v}\cdot\mathbf{u}; \qquad \text{(ix)}$$

$$\mathbf{u}\cdot(\mathbf{v} + \mathbf{w}) = \mathbf{u}\cdot\mathbf{v} + \mathbf{u}\cdot\mathbf{w}. \qquad \text{(x)}$$

It is easy to prove (viii) and (ix). To prove (x), we have, using Fig. 41,

$$\mathbf{u}\cdot(\mathbf{v} + \mathbf{w}) = PS.PR$$
$$= PS(PQ + QR)$$
$$= PS.PQ + PS.QR$$
$$= \mathbf{u}\cdot\mathbf{v} + \mathbf{u}\cdot\mathbf{w}.$$

6·161. We have been at pains to produce the formulae (ii), (iii), (iv), (v), (viii), (ix), (x) in order to show that the definitions about vectors lead to formulae analogous to certain of the fundamental formulae of ordinary elementary algebra. We may, therefore, to the extent permitted

6·162. As an example, we shall express **u·v** in terms of the components (u_1, u_2, u_3), (v_1, v_2, v_3).

We note first that
$$\mathbf{i}^2 = \mathbf{j}^2 = \mathbf{k}^2 = 1; \qquad \mathbf{j}\cdot\mathbf{k} = \mathbf{k}\cdot\mathbf{i} = \mathbf{i}\cdot\mathbf{j} = 0. \qquad \text{(xi)}$$
Hence
$$\begin{aligned}\mathbf{u}\cdot\mathbf{v} &= (u_1\mathbf{i} + u_2\mathbf{j} + u_3\mathbf{k})\cdot(v_1\mathbf{i} + v_2\mathbf{j} + v_3\mathbf{k}), \text{ by (vi)},\\ &= u_1v_1\mathbf{i}^2 + u_2v_2\mathbf{j}^2 + u_3v_3\mathbf{k}^2 + u_1v_2\mathbf{i}\cdot\mathbf{j} + \text{ etc., by (viii), (x)},\\ &= u_1v_1 + u_2v_2 + u_3v_3, \text{ by (xi)}.\end{aligned}$$

6·163.

Note that if $m\mathbf{u} = 0$, then either $m = 0$ or $\mathbf{u} = 0$, analogously to elementary algebra. But if $\mathbf{u}\cdot\mathbf{v} = 0$, then $\mathbf{u} = 0$, or $\mathbf{v} = 0$, or $\mathbf{u}$ and $\mathbf{v}$ are perpendicular.

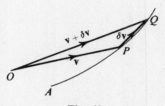

Fig. 42

6·17. Differentiation and integration of vectors.

6·171. Let $\mathbf{v}, \mathbf{v} + \delta\mathbf{v}$ denote the values of a continuously varying vector at times $t, t + \delta t$. Then at time t, the derivative of **v** with respect to t is defined to be $\lim\limits_{\delta t \to 0} \left(\dfrac{1}{\delta t}\delta\mathbf{v}\right)$, and denoted as $\dfrac{d\mathbf{v}}{dt}$ or $\dot{\mathbf{v}}$.

By § 6·15, $(\delta t)^{-1}\delta\mathbf{v}$ is a vector, and it follows that $d\mathbf{v}/dt$ is a vector; (taking the limit does not alter the vector character).

With a view to interpreting $d\mathbf{v}/dt$ geometrically, let the given variable vector be represented at any time t by a straight line with one extremity at a fixed point O (see Fig. 42). Then as t changes, the other extremity traces a path APQ say. If **OP**, **OQ** represent $\mathbf{v}, \mathbf{v} + \delta\mathbf{v}$, then by § 6·13, **PQ** represents $\delta\mathbf{v}$. It follows, taking the limit $\delta t \to 0$, that $d\mathbf{v}/dt$ has the direction of the tangent at P to the curve AP, and magnitude equal to the rate of change of the arc-length AP.

From the definition of differentiation and certain of the formulae (i) to (x), it can be shown that if $\mathbf{u}$, $\mathbf{v}$ are vectors and ρ a scalar, all dependent on t, then

$$\frac{d}{dt}(\mathbf{u} \pm \mathbf{v}) = \frac{d\mathbf{u}}{dt} \pm \frac{d\mathbf{v}}{dt}, \qquad \text{(xii)}$$

$$\frac{d}{dt}(\rho \mathbf{v}) = \frac{d\rho}{dt}\mathbf{v} + \rho\frac{d\mathbf{v}}{dt}, \qquad \text{(xiii)}$$

$$\frac{d}{dt}(\mathbf{u}\cdot\mathbf{v}) = \frac{d\mathbf{u}}{dt}\cdot\mathbf{v} + \mathbf{u}\cdot\frac{d\mathbf{v}}{dt}. \qquad \text{(xiv)}$$

The proofs of these formulae are obtained by changing appropriate symbols from ordinary to bold type in corresponding proofs in elementary calculus and inserting the scalar product dot where needed. This is a consequence of : (a) the formal analogy between the definition of $d\mathbf{v}/dt$ and the definition of differentiation in ordinary elementary calculus, (b) the manipulation analogies pointed out in § 6·161.

A further important property is that

$$\mathbf{v} = (v_1, v_2, v_3) \quad \text{implies} \quad \frac{d\mathbf{v}}{dt} = \left(\frac{dv_1}{dt}, \frac{dv_2}{dt}, \frac{dv_3}{dt}\right), \qquad \text{(xv)}$$

where v_1, v_2, v_3 are the components of $\mathbf{v}$ referred to a fixed reference frame. The result (xv) is obtained using (vi) and (xiii), noting that $\mathbf{i}$, $\mathbf{j}$, $\mathbf{k}$ are constant vectors.

It should be noted that $d\mathbf{v}/dt$ is zero only if $\mathbf{v}$ is constant in both direction and magnitude.

6·172. The integral of $\mathbf{v}$ with respect to t is defined, again analogously to the procedure in elementary calculus, as the limit-sum of terms of the form $\mathbf{v}\delta t$. We shall meet definite integrals of vectors, such as $\int_{t_1}^{t_2} \mathbf{v}\,dt$, and indefinite integrals such as $\int \mathbf{v}\,dt$. These integrals are vectors. (Sometimes, when the context makes the circumstances clear, we do not trouble to show the limits of integration in definite integrals.)

To interpret $\int_{t_1}^{t_2} \mathbf{v}\,dt$ geometrically, the rule exhibited in Figs. 38, 39 is applied. For example, let $\mathbf{v}$ denote in particular the velocity (we shall see in § 6·22 that velocity is a vector) at time t of a point which moves along the curve P_1PP_2 (Fig. 43) between the times $t = t_1$, $t = t_2$. Then $\int_{t_1}^{t_2} \mathbf{v} dt$, being the limit of the vector sum of elements like $\mathbf{v}\delta t$ shown in Fig. 43, is equal to the displacement $\mathbf{P_1P_2}$ during the interval of time.

In §§ 6·171, 6·172, the independent variable has been taken to be the time t; this is very often the case in dynamics. But other continuously varying scalars can also be taken as independent variables.

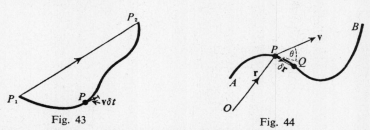

Fig. 43   Fig. 44

6·173. Line-integral of a vector. Let AB be any given curve (Fig. 44), and let P be any point of the curve whose position referred to a fixed origin O is given by $\mathbf{r}$ (see § 6·21). Let $\delta \mathbf{r}$ denote $\mathbf{PQ}$, where Q is a point of the curve near P. Consider a vector (representing e.g. a force, cf. § 6·71) whose magnitude and direction are continuous functions of position, and let $\mathbf{v}$ express the vector at P.

The line-integral of $\mathbf{v}$ along the curve AB is then defined to be the limit-sum of terms of the form $\mathbf{v} \cdot \delta \mathbf{r}$ taken along AB, and is denoted by $\int \mathbf{v} \cdot d\mathbf{r}$. It is also equal to $\int v \cos \theta \, ds$, where θ is the angle between $\mathbf{v}$ and the tangent at P, and δs denotes an element of arc-length.

The line-integral of a vector is a *scalar*.

Theorems on integrals of vectors may be set up analogously to certain theorems in ordinary integration. For example,

$$\int (\mathbf{u} + \mathbf{v}) \cdot d\mathbf{r} = \int \mathbf{u} \cdot d\mathbf{r} + \int \mathbf{v} \cdot d\mathbf{r}. \tag{xvi}$$

6·18. Vector product of two vectors. A pair of vectors, in addition to having a scalar product, has also a vector product which is quite differently defined. The definition of vector product is given in the Appendix (§17.1) and the more important properties listed.

There are several places in Chapter XII (especially where moments of localised vectors and angular momenta are involved) where the use of vector products would both shorten proofs and enable results to be generalised from two to three dimensions. For the sake of interested students, these places are pointed out in the text. But §§ 6·1-6·173 contain all the vector results needed in setting up the theory in Chapters VI to XVI.

6.19. Examples. (1) *If K is any point, and C is a point on the straight line AB such that $\lambda.AC = \mu.CB$, where λ, μ are any scalars and sign significance is attached to the order of the letters in AC, CB, prove that* **KA, KB** *and* **KC** *are connected by*

$$\lambda.\mathbf{KA} + \mu.\mathbf{KB} = (\lambda + \mu)\mathbf{KC}.$$

From Fig. 45, it is seen that

$$\lambda.\mathbf{KA} + \mu.\mathbf{KB} = \lambda(\mathbf{KC} + \mathbf{CA}) + \mu(\mathbf{KC} + \mathbf{CB}). \qquad (i)$$

Since $\lambda.AC = \mu.CB$, and ACB is a straight line, we may write

$$\lambda.\mathbf{CA} + \mu.\mathbf{CB} = 0. \qquad (ii)$$

From (i) and (ii) the result follows.

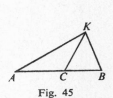

Fig. 45

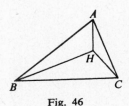

Fig. 46

(2) *Prove by a vector method that the altitudes of any triangle are concurrent.*

Let ABC be any triangle (Fig. 46), and let the altitudes through B, C meet in H.

Since **BH** is perpendicular to **CA**, we have

$$\mathbf{BH \cdot CA} = 0,$$

i.e., $$\mathbf{BH \cdot (CH - AH)} = 0;$$

similarly, $$\mathbf{CH \cdot (AH - BH)} = 0.$$

Adding the last two equations and using the rules for manipulating vectors, we derive

$$\mathbf{AH \cdot (CH - BH)} = 0;$$

thus $$\mathbf{AH \cdot CB} = 0.$$

Since $\mathbf{CB} \neq 0$, it follows that **AH** is either (i) perpendicular to **CB**, or (ii) zero.

In case (i), the altitudes are thus proved concurrent. In case (ii), H is at A, $\angle ABC$ is a right angle and the result is otherwise obvious.

S.P.I.C.E. (α) Ex. (1) will be used later on in this book.

(β) Ex. (2) is a simple illustration of the efficacy of vector methods in solving certain classes of problems.

(γ) The necessity for considering two possible alternatives at the end of Ex. (2) should be noted.

6·2. Motion of a point.

6·21. Position vector. Let P be the position of a moving point at time t. Then the *position vector* **r** of P relative to an origin O is defined by **OP**.

Let O' be another origin of reference (Fig. 47), and let **r**, **r**′ be the position vectors of P relative to O, O'; let **OO**′ = **q**. Then **r** = **q** + **r**′. Hence, by § 6·11, we are justified in calling **r** a vector.

The coordinates (x, y, z) of P referred to orthogonal axes OX, OY, OZ through O are also the components of **r** referred to these axes.

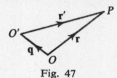

Fig. 47

6·22. Velocity. Let $OXYZ$ be a reference frame taken to be at rest. Referred to this frame, let **r** be the position vector of a moving point P at time t. The (absolute) *velocity* **v** of P at the instant t is then defined as the rate of change of **r** at this instant. Thus

$$\mathbf{v} = \frac{d\mathbf{r}}{dt} (=\dot{\mathbf{r}}). \qquad (24)$$

By § 6·171 (taking **r** in place of **v** in § 6·171; **v** in § 6·171 denotes any vector), the velocity **v** has the direction of the tangent at P to the path of the moving point and magnitude equal to the rate of increase of the arc-length of this path.

Also by § 6·171, since **r** is a vector, **v** is a vector.

By § 6·171 (xv), the components of **v** referred to $OXYZ$ are $(\dot{x}, \dot{y}, \dot{z})$. In the particular case of motion confined to the plane OXY, the components are $(\dot{x}, \dot{y})$; and for motion in the straight line OX, we have the case of § 2·3 with just $\dot{x}$ relevant.

6·23. Acceleration. The (absolute) *acceleration* **f** of the point at the instant t is (as in § 2·4) defined as the rate of change of the (absolute) velocity. Thus

$$\mathbf{f} = \frac{d\mathbf{v}}{dt} (=\dot{\mathbf{v}}) = \frac{d^2\mathbf{r}}{dt^2} (=\ddot{\mathbf{r}}). \qquad (25)$$

Since **v** is a vector, so is **f** (again by § 6·171).

The components of **f** referred to $OXYZ$ are $(\ddot{x}, \ddot{y}, \ddot{z})$, with particular cases as in § 6·22. These components are also equal to $(\dot{v}_1, \dot{v}_2, \dot{v}_3)$, where v_1, v_2, v_3 are the components of **v**.

6·24. Relative motion. Suppose now that the reference frame $OXYZ$ is not necessarily at rest (though OX, OY, OZ remain rigidly mutually perpendicular). Then **r**, **v** and **f** as given in the above formulae are interpreted as the position vector, velocity and acceleration of P relative to the frame $OXYZ$.

In purely kinematical problems, this interpretation suffices for all purposes. Problems on 'relative motion' (so-called) are really problems in which a change from one reference frame to another is involved.

When one reference frame is rotating relatively to a second, the theory needed is too difficult to be included at this stage, and will be deferred to the Appendix, §§ 17·3–17·45. It is to be understood without further mention being made, that in theory and in problems to the end of Chapter XVI, any relative motion of reference frames is purely 'translatory'; i.e., the origins may be in relative motion but there is no change in the inclination of an axis of one frame to an axis of the second frame. In these circumstances, it suffices to speak of velocities, etc., 'relative to the origin'.

We shall in § 6·9 devote a full section to the working of problems in which there are relative translatory motions.

In dynamical theory, however, we shall understand as in Chapter III that, unless the contrary is stated, displacements, velocities and accelerations are referred to an absolutely fixed reference frame.

6·3. Fundamental equation of particle dynamics. We now set out to generalise to the case of three dimensions the theory given in Chapter III on the dynamics of a single particle.

The only addition needed to §§ 3·11, 3·12, 3·13, beyond noting that *mass* is a scalar quantity, is the postulate, based on experimental evidence, that the acceleration of a particle at any instant has the direction of the resultant force.

We generalise § 3·14 by defining the *momentum* and *mass-acceleration* of a particle as the vectors (see § 6·15) $m\mathbf{v}$, $m\mathbf{f}$, respectively, where m, **v** and **f** are the mass, velocity and acceleration.

In place of (8a), (8b) of § 3·15, we now write the *fundamental equation* for a particle in the more general form

$$\mathbf{R} = m\mathbf{f}, \qquad (26a)$$

or
$$\mathbf{R} = \frac{d}{dt}(m\mathbf{v}), \qquad (26b)$$

where **R** is the resultant force on the particle. (As in § 3·13, m is taken to be constant.)

We emphasise that **f** and **v** here denote the *absolute* acceleration and velocity.

6·31. Addition of forces. The vector form in which (26a) and (26b) are written is inadmissible unless force is a vector quantity. This is secured by adding the *postulate*, based on experimental evidence, that forces combine according to the addition rule of vectors. Thus if **R** is equivalent to the combined effects of the forces $\mathbf{F}_1, \mathbf{F}_2, \ldots$, we have

$$\mathbf{R} = \mathbf{F}_1 + \mathbf{F}_2 + \ldots, \qquad (27)$$

which is the generalisation of (9), (§ 3·16).

6·32.

A particle is in *equilibrium* when the resultant force **R** is zero.

6·33. Principle of momentum for a particle. The impulse of a force **F** acting from time $t = 0$ to any general time t is defined (cf. § 3·51) as $\int_0^t \mathbf{F}\, dt$. Thus impulse is a vector quantity. The concepts of instantaneous impulse and impulsive force are similar generalisations of definitions in § 3·54.

The principle of momentum for a particle takes the form (26b), or (cf. (8c))

$$\int_0^t \mathbf{R}\, dt = m\mathbf{v} - m\mathbf{u}; \qquad (26c)$$

i.e., the impulse of the resultant force is equal to the (vector) change in the particle's momentum. The form (26c) is deduced from (26b) by integration as described in § 6·172.

6·34.

No further comment is needed on units and dimensions since units are concerned only with magnitudes.

Note that weight, being a force, is a vector quantity, whereas mass is a scalar. Equation (10) gives only the magnitude of the weight.

All the theory in §§ 3·1 to 3·5 has now been adequately generalised.

In problems involving vector quantities, the student should be at pains to specify directions as well as magnitudes, unless the context makes the direction clear. A phrase such as 'an acceleration of 3 cm/sec²' is to be interpreted as 'an acceleration of magnitude 3 cm/sec²'; in the sequel the words 'of magnitude' will often be taken to be understood.

6·4. The method of resolved parts. The equations in § 6·3, and in the further theory to follow, are vector equations. We now show how to derive equivalent sets of ordinary algebraic equations. This will be a key step in applying the theory to problems. We demonstrate the method in a typical two-dimensional case.

Let OX, OY be a pair of fixed axes at right angles. At any instant let a particle of mass m be subject to the forces $\mathbf{F}_1$, $\mathbf{F}_2$, ..., and let its acceleration be $\mathbf{f}$, all these vectors being in the plane of OXY.

By the fundamental equation (26a), and by (27),
$$\mathbf{F}_1 + \mathbf{F}_2 + \ldots = m\mathbf{f}. \qquad \text{(i)}$$

Let the components (or *resolved parts*), parallel to OX, OY, of $\mathbf{F}_1$, $\mathbf{F}_2$, $\mathbf{f}$ be (X_1, Y_1), (X_2, Y_2), $(\ddot{x}, \ddot{y})$, respectively. Then, by § 6·151 (vi),
$$(X_1\mathbf{i} + Y_1\mathbf{j}) + (X_2\mathbf{i} + Y_2\mathbf{j}) + \ldots = m(\ddot{x}\mathbf{i} + \ddot{y}\mathbf{j}).$$

By the rules for manipulating vectors, we deduce that
$$(X_1 + X_2 + \ldots - m\ddot{x})\mathbf{i} + (Y_1 + Y_2 + \ldots - m\ddot{y})\mathbf{j} = 0.$$

By the last paragraph of § 6·151, it then follows that
$$\left. \begin{array}{l} X_1 + X_2 + \ldots = m\ddot{x} \\ Y_1 + Y_2 + \ldots = m\ddot{y} \end{array} \right\} \qquad \text{(ii)}$$

From (ii) we can, moreover, easily deduce (i). Hence the single vector equation (i) is equivalent to the pair of algebraic equations (ii).

The step from (i) to either of the equations of (ii) is called 'resolving' or 'taking resolved parts' in the particular direction; the step is equivalent to taking projections of the vectors in (i) in that direction.

Resolved parts may be taken in any direction, since OX may be taken in any direction.

If resolved parts are taken in two different directions not at right angles, the resulting pair of equations is again equivalent to (i). The student should convince himself of this.

In a three-dimensional case, there would of course be a set of three equations in place of the pair in (ii).

It is evident that the method of resolved parts may be similarly applied to any vector equation.

6·41.

In some earlier problems (see e.g. Note (ii) and Example (1) of § 3·4), certain forces at right angles to the line of motion were ignored. The method of resolved parts justifies what was done.

6·5. Applications.

6·51. Particle on smooth fixed inclined plane. Let α be the inclination of a plane to the horizontal. If $\tan \alpha = 1/n$, we sometimes say that the inclination is '1 in n'. When α is small it is sufficiently accurate (and often very convenient) in many practical problems to take $\sin \alpha = 1/n$.

Now consider the motion of a particle of mass m down a line AB of greatest slope of the inclined plane. (Motion on inclined planes will be understood in the sequel to be along lines of greatest slope, unless the contrary is stated.)

The forces on the particle are: (a) the force, of magnitude N say, exerted by the plane on the particle and acting at right angles to AB (the plane being *smooth*—see § 3·4 (iii)); (b) the weight of the particle, of magnitude mg, acting vertically downwards. These forces are shown in Fig. 48.

The acceleration of the particle, of magnitude f say, is in the line AB.

By (26a), the resultant of the forces shown in Fig. 48 is equal to the mass-acceleration of the particle. Taking resolved parts along and at right angles to AB, we have

$$mg \sin \alpha = mf, \qquad (i)$$
$$N - mg \cos \alpha = 0. \qquad (ii)$$

By (i), the acceleration is $g \sin \alpha$ in the direction AB. By (ii), the reaction exerted by the plane is $mg \cos \alpha$, or $W \cos \alpha$, where W is the weight of the particle.

Fig. 48   Fig. 49

6·52. Example. *A train of mass* 400 *tons running down an incline of* 1 *in* 100 *at a speed of* 30 *m.p.h. is brought to rest in half a mile by its brakes. Neglecting frictional resistances (apart from the effect of the brakes), find to the nearest ton weight the force, assumed constant, exerted by the brakes.*

During the braking, the forces acting on the train at any instant are W, N, F (pdl) as shown in Fig. 49, where: W is the weight; N is the normal component of the force exerted by the rails; and F is the force due to the brakes.

Let the acceleration be f ft/sec² in the direction BA (taken as standard direction).

The inclination α of the plane is given approximately by $\sin \alpha = 1/100$. Resolving parallel to BA, we then have

$$F - W \sin \alpha = mf, \qquad (i)$$

where m lb is the mass of the train.

Since F, W, α, m are constant, f is constant.

Using '$v^2 = u^2 + 2fs$', with $u = -44$, $v = 0$, $s = -2640$, we obtain
$$f = 44^2/5280.$$

Since $W = mg$, we then have by (i)
$$F - 400 \times 2240 \, g/100 = 400 \times 2240 \times 44^2/5280,$$
yielding
$$F \approx 8 \cdot 6 \times 2240 g \text{ (pdl)}.$$

Hence the required force ≈ 9 tons wt.

6·6. Friction. We now interrupt the main course of the theory by introducing a simplified formal account of *sliding friction*. This will enable us to solve more complicated problems than those attempted so far.

When two bodies A, B are in contact at a point C say (for the present we simplify by assuming the contact to be concentrated at a point, in keeping with the use of *particle* theory), the action, reaction forces between them are not in general along the common normal to the surfaces at C. Let N denote the normal component of the force on A due to B, and F the tangential (or 'frictional') component in the direction shown in Fig. 50. (Acting on B there will be the reaction force at C, of components equal and opposite to N, F; this follows by § 4·2 and the first sentence of § 12·31.)

Sometimes F is called simply the 'friction' or 'friction force'.

Fig. 50   Fig. 51

The following postulates (based on experiment, but more drastically simplified than in many other parts of Mechanics) will be made:

(i) If at the point C of contact, A is moving relatively to B in the direction indicated by the broken arrow in Fig. 50, then: (*a*) F is positive; (*b*) the ratio F/N, $=\mu$ say, depends only on the properties of the materials in contact at C.

(ii) If relative motion as in (i) is on the point of taking place, then $F = \mu N$. (This is called the case of *limiting equilibrium*.)

(iii) If relative motion is neither taking place nor on the point of taking place, then $|F| < \mu N$.

The coefficient μ is called the *coefficient of friction* for the contact. We see that $0 \leqslant \mu \leqslant \infty$. If $\mu = 0$, the contact is *smooth*; if $\mu > 0$, the contact is *rough*. The case $\mu = \infty$ corresponds to the bodies being rigidly attached together at C.

In practice it frequently happens that for the same pair of bodies the value of F/N when relative motion is occurring is significantly less than that in the case of limiting equilibrium. To meet this, two coefficients of friction are sometimes introduced: the 'kinetic coefficient' corresponding to the 'μ' in (i), and the 'static coefficient' corresponding to the 'μ' in (ii) and (iii). We shall here take these coefficients to be equal except where the contrary is stated.

In practice if a body is called 'smooth', all contacts with it are assumed to be smooth. If a body is called 'rough', this means that friction must be considered at each contact with it.

6·61. Angle of friction. Let **CK** represent the force at C on A in either of the cases (i), (ii) above; we have $F = \mu N$ in these cases. Then the angle ϵ between CK and CN is called the *angle of friction*. We see from Fig. 51 that

$$\tan \epsilon = \mu; \tag{28}$$

thus $0 \leqslant \epsilon \leqslant \pi/2$.

Let CK' be drawn on the other side of CN so that $\angle NCK' = \epsilon$. Then in the case (iii) above, the direction of the force at C on A must lie inside the angle KCK'.

(This applies if the conditions are coplanar; more generally, the direction of the force must lie inside the *cone of friction*, i.e. a right circular cone of axis CN and semi-vertical angle ϵ.)

6·62. Application to body on rough fixed inclined plane. Consider next the equilibrium or motion of a body (along a line of greatest slope) on a fixed rough inclined plane of inclination α. As usual, the body will be treated as a particle, the 'common normal' being now the normal to the plane at the point of contact.

As exercises, four cases will be examined, in which the only external forces acting on the body are its weight, mg say, and the force due to the contact with the plane. Let μ and ϵ be the coefficient and angle of friction at the contact. (It is customary, unless the contrary is stated, to take μ as constant throughout any motion that there may be.)

(1) Suppose the body is at rest and not on the point of moving. The forces are as indicated in Fig. 52 (*a*), and
$$F < \mu N.$$
Taking resolved parts along and at right angles to BA, we have
$$F - mg \sin \alpha = 0,$$
$$N - mg \cos \alpha = 0.$$
Hence $\tan \alpha = F/N < \mu = \tan \epsilon.$

Thus the inclination α must be less than the angle of friction.

(2) If the body is on the point of moving down AB, '$<$' has to be replaced by '$=$' in (1); hence α is equal to the angle of friction.

(3) If the body is moving down AB with acceleration f in the direction AB, we have (Fig. 52 (a))

$$F = \mu N,$$
$$F - mg \sin \alpha = -mf,$$
$$N - mg \cos \alpha = 0,$$

yielding
$$f = g(\sin \alpha - \mu \cos \alpha).$$

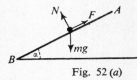 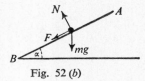

Fig. 52 (a)   Fig. 52 (b)

(4) If the body is moving up BA (having previously been projected up in the direction BA), the positive direction of F is as in Fig. 52 (b). Let f be the acceleration in the direction AB. Then

$$F = \mu N,$$
$$F + mg \sin \alpha = mf,$$
$$N - mg \cos \alpha = 0,$$

yielding
$$f = g(\sin \alpha + \mu \cos \alpha).$$

As corollaries to (3) and (4), we see that $f = g \sin \alpha$ when the plane is smooth, agreeing with the result in § 6·51.

6·63. Example. *A body P of mass m is at the bottom of a fixed rough incline of inclination α and elevation h. A light inelastic string PQ is attached to P, passes over a small smooth fixed pulley at the top of the plane and is attached to a body Q of mass M which is close to the pulley and can move vertically downwards. Initially the system is at rest, and the coefficient of friction between P and the plane is μ. (a) Prove that a necessary and sufficient condition for P to move up the plane is $M > m(\sin \alpha + \mu \cos \alpha)$; (b) assuming this condition to be satisfied, find the speed when Q is level with the initial position of P.*

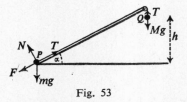

Fig. 53

The forces on P and Q are as indicated in Fig. 53, where T is the tension of the string, and N, F are the normal and frictional components of the force exerted by the plane on P.

Let f be the acceleration of P up the plane.

Then for P, we have
$$N - mg \cos \alpha = 0, \qquad \text{(i)}$$
$$T - F - mg \sin \alpha = mf ; \qquad \text{(ii)}$$
and for Q,
$$Mg - T = Mf. \qquad \text{(iii)}$$
From (ii) and (iii),
$$Mg = F + mg \sin \alpha + (M + m)f. \qquad \text{(iv)}$$
For P to move up the plane, it is necessary and sufficient that $f > 0$.

A sufficient condition for this is that $Mg - mg \sin \alpha$ should exceed the greatest possible value of F, viz. μN; i.e., by (i), that
$$Mg - mg \sin \alpha > \mu mg \cos \alpha. \qquad \text{(v)}$$

The condition is also necessary, since for P to be in motion up the plane we must have $F = \mu N$.

This gives the answer (a).

If (v) is satisfied, we have by (i), (iv) and $F = \mu N$,
$$(M + m)f = Mg - mg(\sin \alpha + \mu \cos \alpha). \qquad \text{(vi)}$$
By '$v^2 = u^2 + 2fs$' (f in (vi) is constant), the required speed, v say, is given by
$$v^2 = 2gh\{M - m(\sin \alpha + \mu \cos \alpha)\}/(M + m).$$

S.P.I.C.E. (α) To emphasise the importance of putting in on a diagram *all* the forces on each of the two particles as a preliminary to using (26a) and the method of resolved parts. This procedure is standard in solving a large class of problems in particle dynamics.

(β) To illustrate the use of the words 'necessary' and 'sufficient'.

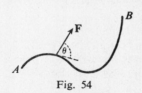

Fig. 54

6·7. Work and power. In §§ 6·7, 6·8, the generalisation of the theory of Chapter III will be completed.

6·71. Work. Suppose a particle moves along any given curve, and let $\mathbf{F}$ denote a force (which may vary in magnitude and direction) acting on the particle during this motion. Then the *work* done by $\mathbf{F}$ is defined as the line integral
$$\int \mathbf{F} \cdot d\mathbf{r} \qquad \text{or} \qquad \int F \cos \theta \, ds \qquad (29)$$
taken along the curve, where θ is the angle between F and the tangent to the curve at the appropriate point (see § 6·173).

This concept may be illustrated by considering a truck moving on rails along the curve AB (Fig. 54) with a horse harnessed at the side exerting the force $\mathbf{F}$ on the truck.

It follows from the definition that work is a *scalar* quantity (by § 6·173).

The formula (29) is the generalisation of (11), to which it reduces when the curve AB is a straight line and $\mathbf{F}$ acts in this line.

Remarks analogous to those in § 3·61 apply to the case when **F** is impulsive.

The work done on a particle by the resultant **R** of forces $\mathbf{F}_1, \mathbf{F}_2, \ldots$, is equal to the sum of the works done by $\mathbf{F}_1, \mathbf{F}_2, \ldots$ For, the work done by $\mathbf{R} = \int \mathbf{R} \cdot d\mathbf{r}$
$$= \int (\mathbf{F}_1 + \mathbf{F}_2 + \ldots) \cdot d\mathbf{r}$$
$$= \int \mathbf{F}_1 \cdot d\mathbf{r} + \int \mathbf{F}_2 \cdot d\mathbf{r} + \ldots$$
This is the generalisation of § 3·62.

6·72. Power. We preserve the definition in § 3·7 of the *power* of a force **F** on a particle at any instant as the rate at which the force is doing work at that instant.

As the generalisation of (12) we have:
$$\text{Power of } \mathbf{F} = \mathbf{F} \cdot \mathbf{v}, \tag{30}$$
where **v** is the velocity of the particle on which **F** acts. This follows from § 6·71, the rate of work of **F** being equal to $\lim_{\delta t \to 0} (\mathbf{F} \cdot \delta \mathbf{r}/\delta t)$.

The power is also equal to $Fv \cos \theta$, where θ is the angle between **F** and **v** at the instant under consideration.

By § 6·16, power is a *scalar* quantity.

6·73. Examples. *A horse is harnessed to the side of a truck that runs on rails, and moves the truck 300 yards in 5 minutes at a uniform speed. The pull of the horse on the truck is 1 cwt and is always inclined at 30° to the rails. Find the rate of work of the horse on the truck.*

Let the rate of work of the horse on the truck $= H$ h.p.
The speed of the truck $= 3$ ft/sec.
Then, by (30),
$$550gH = 112g \times 3 \times \cos 30°.$$
Hence the rate of work in question $= 0·53$ h.p.

(2) *An engine works at the constant rate of H horse-power in drawing a train of total mass M tons up an incline of 1 in n, frictional resistances being q lb.wt. per ton, where q is constant. (a) Prove that the maximum speed that can be generated is approximately $550nH/M(2240 + nq)$ ft/sec. (b) Find the acceleration when the speed is half this maximum. (c) Does this acceleration remain unchanged?*

(*a*) Let V ft/sec be the required maximum speed, P_0 pdl the force then exerted by the engine, and N pdl the normal component of the force exerted by the rails on the train.

The remaining forces are qMg pdl and $2240Mg$ pdl, as indicated in Fig. 55.

By (30),
$$550gH = P_0 V. \tag{i}$$

By (26a), and resolving,
$$P_0 - qMg - 2240Mg \sin \alpha = 0, \qquad \text{(ii)}$$
the right-hand side of (ii) being zero since the acceleration is zero at the maximum speed.

Also
$$\sin \alpha \approx 1/n. \qquad \text{(iii)}$$

From (i), (ii), (iii), we obtain
$$V = 550nH/M(2240 + nq), \qquad \text{(iv)}$$
and hence the answer (a).

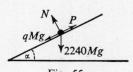

Fig. 55

(b) Let f ft/sec^2 be the acceleration and P pdl the pull of the engine when the speed is v ft/sec. Then in place of (i), (ii), we derive
$$550gH = Pv, \qquad \text{(v)}$$
$$P - qMg - 2240Mg \sin \alpha = 2240Mf. \qquad \text{(vi)}$$

When $v = \tfrac{1}{2}V$, let $f = f_1$ and $P = P_1$. Inserting these values in (v), (vi), and using (iii), (iv), we obtain
$$f_1 = \left(\frac{1}{n} + \frac{q}{2240}\right) g,$$
giving the required acceleration.

(c) Since $f_1 \neq 0$, v is changing at the instant under consideration in (b), and hence also, by (v), P is changing. Hence by (vi) the acceleration is changing at this instant.

S.P.I.C.E. (α) Ex. (1) is a case in which the force doing work is not acting in the direction of the motion of the point of application.

(β) It should be appreciated that the work done by the horse during the motion in Ex. (1) would not be the sole contribution to the net work done on the truck. Indeed, the speed being uniform, it would actually follow from § 6·82 that the work of the *resultant* force on the truck during the motion must be zero. Contributing to this result would be negative work done by other forces such as friction, or the weight of the truck should the truck be ascending an incline.

(γ) Ex. (2) is a standard example giving experience in the use of the fundamental equation (26a), the formula (30) for power, and the method of resolved parts. The method of setting out should be closely studied.

(δ) Ex. (2) is a case in which the acceleration is not constant. The use of formulae such as '$v^2 = u^2 + 2fs$' would be invalid. The argument given in the solution of part (c) should be carefully noted. It is in problems of the type of Ex. (2) that students are often specially tempted to treat the acceleration as constant.

6·8. Energy.

6·81. Kinetic energy. The definition of kinetic energy given in § 3·81 needs no change, and the formula $\tfrac{1}{2}mv^2$ for the kinetic energy of a particle of mass m and speed v is again deducible, by a generalisation of the proof in § 3·811, as follows.

Let $\mathbf{r}_0$ and $\mathbf{v}_0$ be the values at time $t = t_0$ of the position vector $\mathbf{r}$ and velocity $\mathbf{v}$ of the particle. Let $\mathbf{R}$ be the value at time t of a resultant force under which the particle comes to rest, say at time $t = t_1$ and position $\mathbf{r} = \mathbf{r}_1$. Then the K.E. of particle in the position $\mathbf{r}_0$

$$= -\int_{\mathbf{r}_0}^{\mathbf{r}_1} \mathbf{R} \cdot d\mathbf{r}$$

(the line-integral being along the path of the particle)

$$= -\int_{t_0}^{t_1} m \frac{d\mathbf{v}}{dt} \cdot \mathbf{v}\, dt \quad \text{(by (26a) and (24))}$$

$$= \int_{t_1}^{t_0} \frac{d}{dt}(\tfrac{1}{2}m\mathbf{v}\cdot\mathbf{v})\, dt$$

$$= \left[\tfrac{1}{2}mv^2\right]_0^{v_0}$$

$$= \tfrac{1}{2}mv_0^2.$$

This gives the required formula.

Kinetic energy is a *scalar* quantity.

6·82. Principle of energy. The argument in § 3·82 gives again, in the present more general context, the principle of energy, viz.,

$$\left.\begin{array}{l}\textbf{\textit{Work done on particle by}}\\ \textbf{\textit{resultant force}}\end{array}\right\} = \left\{\begin{array}{l}\textbf{\textit{Increase in K.E.}}\\ \textbf{\textit{of particle.}}\end{array}\right. \tag{14a}$$

The corresponding mathematical form is

$$\int_{\mathbf{r}_0}^{\mathbf{r}} \mathbf{R}\cdot d\mathbf{r} = \tfrac{1}{2}mv^2 - \tfrac{1}{2}mu^2.$$

6·83. Conservative forces and potential energy. The definitions of conservative, dissipative and extraneous forces given in § 3·83 are still relevant.

Let $\mathbf{r}_0$ be the position vector, referred to a fixed origin, of a standard position of a particle. Then the potential energy of the particle when in any position $\mathbf{r}$, associated with a conservative field of force $\mathbf{F}$, is given by

$$\int_{\mathbf{r}}^{\mathbf{r}_0} \mathbf{F}\cdot d\mathbf{r} \quad \text{or} \quad \int_{\mathbf{r}_0}^{\mathbf{r}} (-\mathbf{F})\cdot d\mathbf{r}. \tag{31}$$

Potential energy is a *scalar* function of position.

6·831. Potential energy associated with weight. The field of constant gravity, with which the weight of a particle is associated, is conservative (as in the simpler case of § 3·832).

This is proved as follows. The work done by the weight, of magnitude W say, when the particle suffers a small displacement $\delta \mathbf{r}$ is, by the definition of work (29), seen to be equal to $-W\delta x$, where δx is the component of $\delta \mathbf{r}$ in the vertically upward direction. Since W is constant, it follows that if the particle is displaced round any *closed* path in the field the total work done by the weight is zero. Hence the field is conservative.

The formula (16) for the associated potential energy still holds, where $x - x_0$ is the height of the particle above a standard level.

6·832. Potential energies are added as described in § 3·833. And the important alternative form of the principle of energy, viz.:

Increase in $(K.E. + P.E.) = $ *Work done by extraneous forces*
(14b)

may be deduced as in § 3·834.

6·833. Conservation of energy. Every statement in § 3·835 is still relevant.

The context of the second paragraph of § 3·835 (concerning a particle subject only to its weight) may, however, be widened to the case in which the particle is not necessarily moving in a straight line; x_1, x_2 are now interpreted as the coordinates, in the vertically upward direction, of two different positions of the particle.

The equation (32) below is relevant to this case.

6·834. Particle sliding on smooth surface under gravity. Suppose that an object of mass m (treated as a particle) is sliding over a smooth fixed curved surface and is subject only to its weight and the reaction force exerted by the surface.

The reaction force is extraneous, but, by (29) or (30), does zero work during the motion, since it is always at right angles to the velocity.

Hence, by (14b), the mechanical energy is conserved, and so
$$\tfrac{1}{2}mv_1^2 - \tfrac{1}{2}mv_2^2 - mgh = 0;$$
i.e.,
$$v_1^2 - v_2^2 = 2gh, \qquad (32)$$
where v_1, v_2 are the speeds in any two positions P_1, P_2 (not necessarily in a vertical line), and h is the height of P_2 above the horizontal plane through P_1.

6·84. Energy in problems involving strings. Energy methods are often applied at this stage to the motions of systems consisting of two or three particles joined by light strings.

For a perfectly elastic string, it may be shown, by a generalisation of the proof in § 4·61 (i), that the net work of the tension forces in any motion may be taken into account in terms of the potential energy formula $\frac{1}{2}kx^2$, where the extension x of the string is measured along the line of the string, not now necessarily straight.

For an inelastic string, the net work of the tension forces during any motion is zero, provided there are no changes of motion involving impulsive tensions.

The reader who cannot now set down the proofs of these results should be able to do so easily by the time he has read § 13·611.

6·85. Examples. (1) *Solve the part (b) of the Ex. in* § 6·63 *by an energy method.*
Apply (14b), viz.,
 Increase in (K.E. + P.E.) = Work by extraneous forces,
to P, Q and the string in turn, and add.
 Call the resulting equation .... (i)
Fig. 53 (p. 108) shows all the forces on P, Q. The forces on the string are the tension forces and the reaction on the string exerted by the smooth pulley at the top of the inclined plane.

By § 6·84, we may, in forming (i), ignore the tension forces (this includes the T on P and the T on Q shown in Fig. 53).

The work done by N, and the work by the reaction due to the pulley are both zero, by (29).

The works of mg and Mg are taken into account as changes in P.E.

The extraneous force $F = \mu N = \mu mg \cos \alpha$, which is constant. Hence the work done by $F = -\mu mgh \cos \alpha$.

Hence if v is the required speed, the equation (i) is
$$\tfrac{1}{2}mv^2 + \tfrac{1}{2}Mv^2 + mgh \sin \alpha - Mgh = -\mu mgh \cos \alpha.$$
This yields
$$v^2 = 2gh\{M - m(\sin \alpha + \mu \cos \alpha)\}/(M + m),$$
as in § 6·63.

(2) *A ring of mass m can slide on a smooth fixed vertical rod, and is attached to a light inelastic string which passes over a smooth fixed horizontal peg, distant c from the rod, and which has a mass M ($> m$) attached to its other end. Initially the system is at rest with M vertically below the peg, the ring being held at the same level as the peg and the string being taut. The ring is now released. Find how far the ring descends before the system is again at rest.*

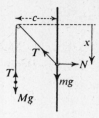

Fig. 56

Fig. 56 shows, in standard notation, the forces on M and m in any general position.

The string is subject to tension forces and the reaction exerted by the peg.

We apply (14b) to the whole moving system:

The net work of the tension forces is zero.

The work done by N and by the reaction due to the peg is zero by (29).

Hence the work done by all the extraneous forces is zero.

Hence the mechanical energy (P.E. + K.E.) is conserved.

Initially and finally, the K.E. is zero.

Hence the final P.E. of the system must be equal to the initial P.E.

Since the string being light has no P.E., we then have:

Loss in P.E. of m = Gain in P.E. of M.

Let x_1 be the required distance. Then
$$mgx_1 = Mg(\sqrt{(c^2 + x_1^2)} - c),$$
yielding $x_1 = 0$ or $x_1 = 2mMc/(M^2 - m^2)$.

The latter value of x_1 is the required answer.

S.P.I.C.E. (α) In both Exs. (1) and (2), the reader should note the degree of care called for in considering the work of *all* the forces present. In both solutions, the algebra used occupies a relatively small space.

(β) The energy method is alternative to the method demonstrated in § 6·63, and is often (but by no means always) the faster. Ex. (2) would be difficult to solve by the method of § 6·63 since the acceleration is not constant; (the student should satisfy himself on this).

On the other hand, there are problems in which the energy method cannot be conveniently used. For example, the work done by friction cannot always be found as readily as in Ex. (1).

(γ) In Ex. (2), the reader should interpret the result $x_1 = 0$.

(δ) The brighter reader should investigate the absurdity of the formal answer in Ex. (2) in the case $M < m$.

(ϵ) It should be noted that the system in Ex. (2) comes to rest only for an instant. After this instant, m would start to ascend.

6·9. Change of reference origin.

6·91. Relative velocity. We now consider reference frames with origins O, O' in relative translatory motion (see § 6·24); and extend the one-dimensional account of relative motion given in § 2·8.

In place of Fig. 6, we construct Fig. 57, with position vectors $\mathbf{q}$, $\mathbf{r}$, $\mathbf{r'}$, replacing the coordinates q, x, x'. In place of $x' = x - q$, we now have $\mathbf{r'} = \mathbf{r} - \mathbf{q}$ (cf. § 6·21).

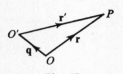

Fig. 57

Differentiating with respect to the time t, we have in place of (7)
$$\mathbf{v}_{P \text{ rel } O'} = \mathbf{v}_{P \text{ rel } O} - \mathbf{v}_{O' \text{ rel } O}, \tag{33a}$$
As in § 2·81, one of the origins, O say, is often treated as being at rest; (33a) is then written in the form
$$\mathbf{v}_{P \text{ rel } O'} = \mathbf{v}_P - \mathbf{v}_{O'}, \tag{33b}$$
where O' is a moving origin. This is the practice in most elementary problems where O is a point fixed on the Earth's surface.

The equation (33) (*a* or *b*) enables us to solve problems on relative velocity. It may be used in three ways:

(*a*) Adding (vectorially) to the velocity (relative to an original origin O) of each body present, a velocity equal and opposite to that (relative to O) of the new reference point O'. Relative to the new reference system, O' is of course now at rest.

(*b*) Piecing together triangles on the basis of (33), using clues contained in the data, and making deductions from the ensuing geometrical constructions.

(*c*) Taking resolved parts of (33). The explicit use of $\mathbf{i}$, $\mathbf{j}$, $\mathbf{k}$ (§ 6·151) is often a help.

6·911. Examples. (1) *Two cars are moving along two straight roads which intersect at right angles, with constant speeds of* 30, 40 *m.p.h., respectively. At a certain instant, the cars are both* 10 *miles from and approaching the intersection. What will be their shortest distance apart, and when will this happen?*

Let P be the car moving at 30 m.p.h, (relative to the ground), Q at 40 m.p.h.; let A, B denote their initial positions, and O the given intersection. Thus in Fig. 58, $AO = BO = 10$ miles.

We shall use the method (*a*), and take a new origin attached to *P*. We thus add to the velocities of *P* and *Q* a velocity of 30 m.p.h. in the direction *OA*.

This results in *P* being now regarded as at rest at *A*, and, by the triangle *BCD* of Fig. 58, in *Q* as having a velocity (relative to *P*) represented by **BD**, i.e. 50 m.p.h. in the line *BD*. (*A*, *B*, *D* are no longer being considered in relation to the ground.)

The required shortest distance is the perpendicular *AN* from *A* to *BD*. From Fig. 58, we have
$$OK = \tfrac{3}{4}OB = 7\tfrac{1}{2} \text{ miles}; \quad \therefore KA = 2\tfrac{1}{2} \text{ miles}.$$
$$\therefore AN = KA \sin \alpha = \tfrac{5}{2} \times \tfrac{4}{5} \text{ miles} = 2 \text{ miles}.$$
Thus the required shortest distance is 2 miles.

Let *t* hours be the time that elapses before this shortest distance is realised. The distance traversed by *Q* relative to $P = BN = \sqrt{(AB^2 - AN^2)} = \sqrt{(10^2 + 10^2 - 2^2)}$ miles $= 14$ miles.
The speed of *Q* relative to $P = 50$ m.p.h.
Hence, by § 2·81, $14 = 50t$.
Thus the required time $= 0·28$ h $= 16·8$ min.

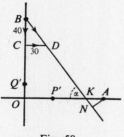

Fig. 58   Fig. 59

(2) *A ship steams due east at the rate of 15 m.p.h. relative to a current which is flowing due north at 4 m.p.h. Find the velocity relative to the ship of a train moving due south at 32 m.p.h.*

We shall use method (*b*). Let *S*, *C*, *T* refer to the ship, current and train, respectively.
Then by the data, $\mathbf{v}_{S \text{ rel } C}$ is represented by 15 m.p.h., $\rightarrow$; $\mathbf{v}_C$ by 4 m.p.h., $\uparrow$; $\mathbf{v}_T$ by 32 m.p.h., $\downarrow$.
By (33), we have
$$\begin{aligned}\mathbf{v}_{T \text{ rel } S} &= \mathbf{v}_T - \mathbf{v}_S \\ &= \mathbf{v}_T - (\mathbf{v}_{S \text{ rel } C} + \mathbf{v}_C) \\ &= \mathbf{v}_T - \mathbf{v}_C - \mathbf{v}_{S \text{ rel } C}.\end{aligned} \quad (\text{i})$$
The vectors $\mathbf{v}_T$, $-\mathbf{v}_C$, $-\mathbf{v}_{S \text{ rel } C}$ are represented in Fig. 59 by **AB**, **BC**, **CD**, respectively.
Hence, by (i), $\mathbf{v}_{T \text{ rel } S}$ is represented by **AD**.
Hence, by Fig. 59, the velocity of the train relative to the ship is 39 m.p.h. in a direction α west of south, where $\alpha = \tan^{-1}(5/12)$.

(3) *When a man runs at 6 m.p.h. due south, the wind appears to him to blow from the east. When the man doubles his speed, the wind appears to blow from the south-east. Find the velocity of the wind.*

By method (c).

Take axes OX, OY towards the east, north, respectively (Fig. 60). Let M and W refer to the man and the wind, respectively.

Then, from the first set of data,
$$\mathbf{v}_M = -6\mathbf{j},$$
$$\mathbf{v}_{W \text{ rel } M} = -a\mathbf{i}; \tag{i}$$

and from the second set
$$\mathbf{v}_M = -12\mathbf{j},$$
$$\mathbf{v}_{W \text{ rel } M} = -b\mathbf{i} + b\mathbf{j}; \tag{ii}$$

where a, b are positive but of (as yet) unknown magnitudes.

In both (i), and (ii), $\mathbf{v}_W = c\mathbf{i} + d\mathbf{j}$ say, is the same.

By (33), $\qquad \mathbf{v}_{W \text{ rel } M} = \mathbf{v}_W - \mathbf{v}_M. \tag{iii}$

Representing the vectors in (iii) in terms of $\mathbf{i}$, $\mathbf{j}$ for the cases (i), (ii), respectively, we have
$$-a\mathbf{i} = c\mathbf{i} + d\mathbf{j} + 6\mathbf{j},$$
$$-b\mathbf{i} + b\mathbf{j} = c\mathbf{i} + d\mathbf{j} + 12\mathbf{j}.$$

Taking resolved parts of these equations (i.e. equating coefficients of $\mathbf{i}$, $\mathbf{j}$) we have
$$-a = c, \quad 0 = d + 6,$$
$$-b = c, \quad b = d + 12;$$

yielding $a = 6$, $b = 6$, $c = -6$, $d = -6$.

Hence $\qquad\qquad \mathbf{v}_W = -6\mathbf{i} - 6\mathbf{j}.$

Hence the velocity of the wind (relative to the ground) is $6\sqrt{2}$ m.p.h. from the north-east.

By method (b).

In the first case, $\mathbf{v}_M$ is represented (Fig. 61) by **AB**, where $AB = 6$ units, and $\mathbf{v}_{W \text{ rel } M}$ has the direction BH, where $\angle ABH = 90°$.

In the second case, $\mathbf{v}_M$ is represented by **AC**, where $AC = 2AB$, and $\mathbf{v}_{W \text{ rel } M}$ has the direction CK, where $\angle ACK = 45°$.

By (iii) above, $\mathbf{v}_W$ is the resultant of $\mathbf{v}_M$ and $\mathbf{v}_{W \text{ rel } M}$, and is the same in both cases. On considering the triangles involved, we see that the data can be met only if $\mathbf{v}_W$ is represented by **AN**, where N is the intersection of BH, CK.

Hence, the velocity of the wind (relative to the ground) is, by Fig. 61, $6\sqrt{2}$ m.p.h. from the north-east.

S.P.I.C.E. (α) The above examples have been selected to illustrate the three methods (*a*), (*b*), (*c*). Some examples are better done by method (*a*), others by (*b*) or (*c*); the student has to learn by experience when to use (*a*) and when (*b*) or (*c*).

(β) The difference between the methods (*b*) and (*c*) is that (*b*) relies on skill in fitting triangles together, while (*c*) reduces the problem to a piece of algebra. In practice (*b*) often involves less computation but more hard thinking than (*c*); (*c*) gives a reliable routine process which the average student will find safer to use than (*b*).

6·921] DYNAMICS OF A SINGLE PARTICLE 119

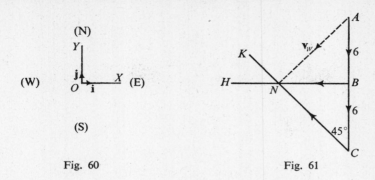

Fig. 60   Fig. 61

(γ) It may be noted that Ex. (1) could have been solved quite independently of equation (33). Thus if P', Q' (Fig. 58) are the positions of P, Q after any time t hours, we have
$$P'Q'^2 = OP'^2 + OQ'^2 = (10 - 30t)^2 + (10 - 40t)^2.$$
By differentiating with respect to t, or otherwise, it is found that $P'Q'^2$ is a minimum when $t = 0.28$, and that then $P'Q' = 2$, leading to the answers previously found.

6·92. Relative acceleration. On differentiating (33a) with respect to t, we have
$$\mathbf{f}_{P\ \text{rel}\ O'} = \mathbf{f}_{P\ \text{rel}\ O} - \mathbf{f}_{O'\ \text{rel}\ O}, \qquad (34a)$$
where $\mathbf{f}$ denotes acceleration.

It was pointed out in § 3·15 that it is important in dynamical problems to refer to an origin absolutely fixed. The form most usually needed then is
$$\mathbf{f}_P = \mathbf{f}_{P\ \text{rel}\ O'} + \mathbf{f}_{O'}, \qquad (34b)$$
where O' is a moving origin, and $\mathbf{f}_P$, $\mathbf{f}_{O'}$ are absolute accelerations.

6·921. Example. *A particle of mass m moves down the inclined face (inclination α) of a smooth wedge of mass M which is free to move on a smooth fixed horizontal table. Find the acceleration of the wedge and the reaction force between the wedge and the particle.*

(It is necessary to assume here that the theory of the dynamics of a single particle can be applied to the wedge in the circumstances of this problem. The assumption is justified by the theory of Chapter XIII.)

Fig. 62(a) shows (in usual notation) the forces on the particle; Fig. 62(b) shows the forces on the wedge (including the force N exerted by the particle on the wedge and the reaction N' due to the table).

In Fig. 62(c), f denotes the acceleration of the wedge, and f' the acceleration of the particle relative to the wedge. By (34), the (absolute) acceleration of the particle is the resultant of the two vectors f, f' shown in Fig. 62(c).

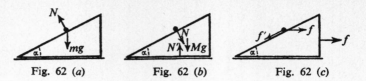

Fig. 62 (a)   Fig. 62 (b)   Fig. 62 (c)

For the particle, using (26a) and resolving along and at right angles to the inclined plane, we have

$$mg \sin \alpha = m(f' - f \cos \alpha), \qquad \text{(i)}$$
$$mg \cos \alpha - N = mf \sin \alpha. \qquad \text{(ii)}$$

For the wedge, resolving horizontally and vertically, we have

$$N \sin \alpha = Mf, \qquad \text{(iii)}$$
$$N \cos \alpha + Mg - N' = 0. \qquad \text{(iv)}$$

The equations (i), (ii), (iii), (iv) are sufficient to determine the four unknowns f, f', N, N'. Here we need only f and N, which are given from (ii), (iii) by

$$f = \frac{mg \cos \alpha \sin \alpha}{M + m \sin^2 \alpha}; \qquad N = \frac{Mmg \cos \alpha}{M + m \sin^2 \alpha}.$$

S.P.I.C.E. (α) In this problem it is desirable to show in separate diagrams *all* the forces on the particle, *all* the forces on the wedge, and the accelerations of the wedge and the particle. The absolute acceleration of the particle is represented as the resultant of two vectors.

(β) In spite of the complexity of the problem, the simple direct use of the fundamental equation, resolved in two directions for each 'particle' present, delivers all the possible dynamical equations.

(γ) In this problem, the energy of the whole system is conserved. An energy equation could therefore be readily formed and used in place of one of the equations (i) to (iv). But in this instance it happens that the course of the solution would then be a little less rapid.

EXAMPLES VI

1. A, B, C, D are four points, not necessarily coplanar, and E, F, G, H are the midpoints of AB, BC, CD, DA. Prove by a vector method that EG, FH meet and bisect each other.
(Hint: First show that **EF** = **HG**.)

*2. The direction-cosines of two lines are (l_1, l_2, l_3), $(\lambda_1, \lambda_2, \lambda_3)$, respectively; (i.e. l_1, l_2, l_3 are the cosines of the angles between the first line and OX, OY, OZ, respectively; etc.). Prove, by considering the scalar product of two unit vectors in the lines, that the angle θ between the lines satisfies the relation $\cos \theta = l_1 \lambda_1 + l_2 \lambda_2 + l_3 \lambda_3$.

*3. Find a unit vector which is at right angles to both of the vectors $\mathbf{i} - 3\mathbf{j} + \mathbf{k}$ and $2\mathbf{i} + \mathbf{j} - \mathbf{k}$.
(Hint: Use the result that the scalar product of two vectors at right angles is zero.)
[Answer: $(2\mathbf{i} + 3\mathbf{j} + 7\mathbf{k})/\sqrt{62}$.]

4. Prove that the resolved part of $\mathbf{u}$ in the direction of $\mathbf{a}$ is given by the vector
$\left(\dfrac{\mathbf{u} \cdot \mathbf{a}}{a^2}\right)\mathbf{a}$.

5. If $\mathbf{u} \cdot \mathbf{v} = \mathbf{u} \cdot \mathbf{w}$ and $\mathbf{u} \neq 0$, prove that $\mathbf{v} = \mathbf{w} + \mathbf{x}$, where $\mathbf{x}$ is a vector of undetermined magnitude, perpendicular to $\mathbf{u}$. (Note:—It does not follow from the data that $\mathbf{v} = \mathbf{w}$ necessarily.)

6. If $\mathbf{u}$, $\mathbf{v}$ are two vector functions of a scalar variable t, and if $\mathbf{u}$, $\mathbf{v}$ are always parallel, prove that $\mathbf{u}\dfrac{d\mathbf{v}}{dt} = u\dfrac{dv}{dt}$.
(Hint: Put $\mathbf{u} = \phi \mathbf{v}$, where ϕ is a scalar function of t.)

7. Forces of magnitudes $F_1, F_2, \ldots$, act in directions inclined to the x-axis at $\alpha_1, \alpha_2, \ldots$. Prove that their resultant has magnitude
$$\sqrt{\{(\Sigma F \cos \alpha)^2 + (\Sigma F \sin \alpha)^2\}},$$
and acts in a direction inclined to the x-axis at $\tan^{-1}\{(\Sigma F \sin \alpha)/(\Sigma F \cos \alpha)\}$.

8. Two smooth fixed inclined planes A, B meet at their highest points in a horizontal line, and have their lowest points at the same level. A particle is projected with speed V from the base of A up a line of greatest slope, and at the top is smoothly constrained to descend down a line of greatest slope of B without losing speed at the turn. What is the speed at the base of B? [Answer: V.]

9. A mass m is drawn up a smooth fixed plane of inclination† $30°$ by means of a mass M which descends vertically, the masses being joined by a light inelastic string which passes over a small smooth pulley fixed at the top of the plane. If the acceleration of either mass is $\tfrac{1}{4}g$, find the ratio of the masses. [Answer: $1:1$.]

10. C is a fixed vertical circle, whose highest and lowest points are A, B, respectively. Prove that the times taken by a particle to slide down smooth inclined planes AP, QB, are equal, where P, Q are any points of the circumference of C.

11. Find the straight line of quickest descent (down a smooth inclined plane) from a given point A to a given straight line l, where A, l are in a vertical plane.
[Answer: Draw AB horizontally to cut l in B; take P in l below the level of B and distant AB from B; then AP is the required line.]

12. C is a fixed vertical circle whose lowest point is B. A is a fixed point outside C in the plane of C above the level of B. Find a point P of C so that the time taken for a particle to slide down a smooth inclined plane AP is the least possible.
[Answer: P is at the intersection of C and AB.]

13. In sliding down a fixed rough plane of vertical height h and inclination α, a body acquires a speed v. Find the coefficient of friction.
[Answer: $(1 - v^2/2gh) \tan \alpha$.]

† The words 'to the horizontal' are often understood in contexts like this.

14. A mass P is initially at rest on a rough horizontal plane. The plane is slowly tilted until P starts to move, and is then kept fixed. The static and kinetic coefficients of friction are 0·4, 0·3 respectively. Find how far P moves during the first $2\frac{1}{2}$ seconds of its motion. [Answer: 9·3 ft.]

15. A truck, starting from rest, runs down an incline of 1 in 100 for a distance of $\frac{1}{2}$ mile and then runs up an equal incline, coming to rest after 250 yards. Assuming that there is no energy dissipated at the instant when the velocity changes its direction, find, in lb.wt. per ton mass of the truck, the frictional resistance, assumed constant, that has been opposing the motion. [Answer: 12·5 lb.wt. per ton.]

16. A motor car of mass 1 ton runs at a steady speed of 40 m.p.h. down an incline of 1 in 20 with the clutch in neutral. It is subject to resistances proportional to the square of the speed. Find the horse-power in operation when it runs steadily at 30 m.p.h. up the same incline. [Answer: 14 h.p.]

17. The total mass of a cyclist and his machine is M lb and the resistance to motion is proportional to the speed. The cyclist can ride without pedalling down an incline of 1 in m at a steady speed of U ft/sec. If he ascends an incline of 1 in n at a steady speed of V ft/sec, find the horse-power he is putting into operation.

Evaluate when $M = 180$, $m = 25$, $n = 200$, and the speeds are 20, 10 m.p.h., respectively.

$$\left[\text{Answers: } \frac{MV}{550U}\left(\frac{V}{m}+\frac{U}{n}\right) \text{ h.p.; } 0\cdot12 \text{ h.p.}\right]$$

18. A motor car runs up an incline at a steady speed of 20 m.p.h., the power in operation being 12 h.p. It runs down the same incline at the same steady speed with 5 h.p. in operation. The resistance is always proportional to the square of the speed. Prove that these data are sufficient to determine the h.p. in operation when the car runs at any given steady speed on the level.

In particular, find the h.p. in operation when the speed is (a) 20 m.p.h., (b) 30 m.p.h., on the level. [Answers: 8·5 h.p.; 28·7 h.p.]

19. A locomotive, working at the constant rate of 500 h.p., pulls a train of total mass 200 tons along a level track, the resistances being 16 lb.wt. per ton. Find the acceleration at the instant when the speed is 30 m.p.h. Does this acceleration remain unchanged ?

At what steady speed can the locomotive pull the train up an incline of 1 in 100, the power and the resistance being as before ? Prove that this speed could not be attained in a finite time, starting from rest.

[Answers: 0·218 ft/sec²; no; 24·4 m.p.h.]

20. Three equal masses are attached to points P, Q, R of a light inextensible string PQR hanging over two small smooth fixed horizontal pegs A, B, which are at the same level and 60 cm apart. Initially the system is at rest with the middle mass held at the midpoint of AB and the string taut. If the middle mass is now released, find how far it descends before coming instantaneously to rest. [Answer: 40 cm.]

21. In a mine haulage track inclined at 30° to the horizontal, two trucks are connected by a wire cable which passes round a drum at the top of the track so that the loaded truck when descending hauls up the empty truck. The length of the track is 90 yards, and the mass of either truck is ½ ton when empty and 2 tons when loaded. Neglecting friction and the mass of the cable, find the speed the loaded truck would acquire if there were no braking.

If there were no braking for the first half of the distance, what constant force acting up the track on the loaded truck would then bring the system to rest by the time the loaded truck reached the bottom? [Answers: 49·1 m.p.h.; 1½ tons wt.]

22. An aeroplane has a speed of 360 m.p.h. in still air. When the wind blows from the east, the velocity of the aeroplane as observed from the ground is 270 m.p.h. towards the north-east. Find the speed of the wind in the vicinity of the aeroplane. [Answer: 114 m.p.h.]

23. A boat is to be rowed in a straight line, with speed 8 ft/sec relative to the current, from a point A of one bank to the immediately opposite point B of the other bank of a river flowing (at all points) at 3 m.p.h., the banks being taken to be straight and parallel and 200 yards apart. How far upstream from A is the point of the opposite bank towards which the bow of the boat should initially be directed? [Answer: 132 yards.]

24. If the bow of the boat in Ex. 23 is initially directed towards the immediately opposite point B, other features being as in Ex. 23, how far downstream from A will be the point where the boat reaches the opposite bank? [Answer: 110 yards.]

Prove that in this case the least possible effort is expended in getting the boat across the river (the speed of 8 ft/sec relative to the current being assigned).

25. A steamer is sailing at 14 knots S.E., and a vane on the mast-head points (into the wind) towards E.N.E. The steamer then turns and sails N.E. with the same speed, and the vane points towards N.N.E. What is the velocity of the wind? [Answer: 14 knots from the north.]

26. A man travelling west at 4 m.p.h. finds that the wind appears to blow from the south. On doubling his speed, he finds that the wind appears to blow from the south-west. Find the velocity of the wind. [Answer: $4\sqrt{2}$ m.p.h. from the south-east.]

27. An aeroplane flies daily a straight line course from A to B and back. In calm weather the speed is u and the time for the double journey is T. On a certain day the velocity of the wind is v in a direction inclined at θ to AB. Prove that on the outward and homeward journeys, the aeroplane has to be headed in directions equally inclined to AB, and find the inclination. Find also the time for the double journey.

$$\left[\text{Answers: } \sin^{-1}\left(\frac{v}{u}\sin\theta\right); \ \frac{Tu\sqrt{(u^2 - v^2\sin^2\theta)}}{u^2 - v^2}. \right]$$

28. An aeroplane travels in still air at 240 m.p.h. Neglecting time lost in turns, find how long will it take to fly round a square course of side-length 30 miles if there is a wind blowing at 60 m.p.h. parallel to a diagonal. [Answer: $31\frac{1}{2}$ minutes.]

29. A man seeks to cross a road, walking with constant speed u, as far as possible in front of a tram moving along the road with constant speed v, where $v > u$. What should be the angle between the direction the man takes and the direction of the road?

[Answer: $\cos^{-1}(u/v)$.]

30. A battleship which can steam at 32 knots sights an enemy vessel at a distance of 20 nautical miles due west. If the enemy vessel steams due south at 40 knots, find: (*a*) the course the battleship should steer in order to get as close as possible to the vessel; (*b*) the shortest distance between the two vessels if the battleship steers this course.   [Answers: (*a*) 53°·1 south of west; (*b*) 12 nautical miles.]

CHAPTER VII

DYNAMICS OF A PARTICLE MOVING IN A CIRCLE

"Instead of debating whether it (centrifugal force) is a force 'seeking the centre' or 'fleeing from the centre', we should expedite its 'flight' from the subject of (elementary) mechanics altogether."

—E. G. PHILLIPS, *Mathematical Gazette*.

7·1. Angular velocity and acceleration of a point.

7·11. Angular velocity of a point about a point. Let O be a given point, and OX any straight line through O of fixed direction. Let P be any position of a point which moves in a plane containing OX. At time t, let $\angle XOP = \theta$, with sense attached corresponding to the direction of the arrow in Fig. 63. Then the *angular velocity* ω of the moving point about O is defined as the rate of change of θ; i.e.

$$\omega = \frac{d\theta}{dt} \quad (=\dot{\theta}). \tag{35}$$

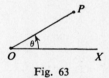

Fig. 63

The theoretical unit of angle is the radian (rad), and angle is dimensionless.

The theoretical unit of angular velocity is the rad/sec, and the dimensions are $[T^{-1}]$.

When the direction of increase of θ is not being considered, we may speak of *angular speed*.

If the direction of OX is not fixed, $d\theta/dt$ gives the angular velocity of P about O, *relative to the direction* OX. Following remarks made in § 6·24, we shall, however, until the Appendix, understand that reference lines such as OX keep fixed directions, unless we indicate otherwise.

7·12. Angular acceleration of a point about a point. This is defined, in the circumstances of § 7·11, to be $d\omega/dt$, i.e. $\dot{\omega}$. Thus

$$\dot{\omega} = d^2\theta/dt^2 = \ddot{\theta}.$$

A formula of high importance is

$$\ddot{\theta} = \frac{d}{d\theta}(\tfrac{1}{2}\dot{\theta}^2), \tag{36}$$

which is analogous to (3), and is proved by replacing x by θ in the proof of (3).

The theoretical unit of angular acceleration is the rad/sec², and the dimensions are [T^{-2}].

7·13. The details in §§ 7·11, 7·12 are relevant to motion in any (plane) curve, and are not restricted to circular motion which will be the exclusive concern of §§ 7·2-7·6.

It is to be noted that when we speak of 'velocity' and 'acceleration' without the adjective 'angular', we mean the velocity and acceleration as defined in Chapter VI.

7·14. Example. *A point is moving in a circle. Prove that its angular velocity about the centre of the circle is double its angular velocity about any fixed point on the circumference.*

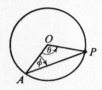

Fig. 64

Let O be the centre of the circle, A the fixed point on the circumference, and P any position of the moving point.

Let θ, ϕ be as in Fig. 64.

By Fig. 64, $\qquad \theta + 2\phi = \pi.$

Differentiating with respect to the time t, we have $\dot{\theta} + 2\dot{\phi} = 0;$

i.e. $\qquad \dot{\theta} = -2\dot{\phi}.$

This gives the required result, since $\dot{\theta}$ and $-\dot{\phi}$ are the angular velocities of P about O and A, in the senses of θ increasing, ϕ decreasing, respectively, i.e. in the same (anticlockwise) sense in both cases.

S.P.I.C.E. (α) To emphasise that the angular velocity of a moving point can be specified only in relation to a given reference point.

(β) To demonstrate the use of arrows as in Fig. 64 showing the sense of increase of angles. Each arrowhead in Fig. 64 is on a moving line, while the tail-end is on a fixed line. This way of drawing arrows corresponds to that demonstrated in Chapter II for rectilinear motion, and facilitates the fixing of signs in problems.

7·2. Circular motion: preliminary relations. Let P be the position at time t of a point describing a circle of centre O and radius a. Let OX be a fixed line cutting the circumference in A, and let the arc $AP = s$, $\angle AOP = \theta$, as indicated in Fig. 65. Then $s = a\theta$. Differentiating successively with respect to t, we have

$$v = a\omega, \tag{37}$$
$$\dot{v} = a\dot{\omega}, \tag{38}$$

where v is the speed of P, with positive or negative sign attached according as the arc AP in Fig. 65 is increasing or decreasing; $\dot{v}$ is the rate of increase of the speed (*note*: speed, not velocity); and ω, $\dot{\omega}$ are the angular velocity, angular acceleration of P about O.

It will be seen in § 7·3 that $\dot{v}$ is equal to the tangential component of the acceleration of P; $\dot{v}$ is *not* the magnitude of the acceleration of P.

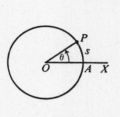

Fig. 65

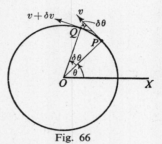

Fig. 66

7·3. Circular motion: acceleration components. We now proceed to find resolved parts of the acceleration of the point moving as in § 7·2: (i) along the tangent at P in the sense of θ increasing; (ii) along the normal in the sense towards O.

Let Q be the position, and $v + \delta v$ the speed (with attached sign), of the moving point at time $t + \delta t$. (See Fig. 66.)

The resolved parts of the increase in velocity during the time δt are:

(i) $(v + \delta v) \cos \delta \theta - v$;   (ii) $(v + \delta v) \sin \delta \theta$.

Now $\sin \delta \theta \approx \delta \theta$, when $\delta \theta$ is small. Also $\cos \delta \theta \approx 1 - \frac{1}{2}(\delta \theta)^2$; for $\cos \delta \theta = 1 - 2 \sin^2(\frac{1}{2}\delta \theta) \approx 1 - 2(\frac{1}{2}\delta \theta)^2$.

Hence, neglecting small quantities of the second and higher orders, we have that the resolved parts of the velocity change during the interval δt are: (i) δv; (ii) $v \delta \theta$.

Divide by δt, and let $\delta t \to 0$. It then follows from the definition of acceleration and the last sentence of § 6·23 that the resolved parts of the acceleration of the moving point are: (i) dv/dt; (ii) $v\omega$.

Using (37) and (38), we then have:

$$\left.\begin{array}{l}\text{\textbf{Tangential component of acceleration}}\\ \text{\textbf{(\textit{in sense of} θ \textit{increasing})}}\end{array} = \dot{v} = a\dot{\omega} = a\ddot{\theta};\\ \begin{array}{l}\text{\textbf{Normal component of acceleration}}\\ \text{\textbf{(\textit{in sense towards} O)}}\end{array} = v\omega = \dfrac{v^2}{a} = a\omega^2 = a\dot{\theta}^2.\right\} \tag{39}$$

7·4. Uniform circular motion. This is the particular case in which v is constant. By (37), this entails that ω is constant, and vice versa.

In uniform circular motion, the tangential component of the acceleration is zero. The acceleration is of constant magnitude v^2/a (or $a\omega^2$, etc.), and is always directed towards O.

The *period* of the motion is defined to be the time of a single revolution. The period is equal to $2\pi/\omega$, since the angle described in a unit of time is ω.

In problems on uniform circular motion, practical units of angular velocity such as the 'revolution per minute' are often used; 1 rev/min = $2\pi/60$ rad/sec.

At this stage, the term 'centrifugal force' is sometimes introduced. The student is referred to the quotation at the head of this chapter, and to § 17·4 (which is not 'elementary').

7·5. Circular motion: dynamical equations. Let m be the mass of a particle moving in a circle. At time t, let F, N be the tangential and normal components (in the same senses as in § 7·3) of the resultant force on the particle. Then, by the fundamental equation (26a), taking resolved parts tangentially and normally, we have

$$\left. \begin{array}{l} F = ma\ddot{\theta} \\ N = ma\dot{\theta}^2 \end{array} \right\} \qquad (40)$$

The equations (40) cover the dynamical theory of circular motion of a particle.

7·51.

The elementary problems on circular motion to be treated in Chapter VII are of two main types: (*a*) *uniform* circular motion; (*b*) motion in a vertical circle in which the only force doing work is a particle's weight.

In case (*a*), $\ddot{\theta} = 0$ and hence F is zero. In case (*b*), the mechanical energy is conserved and we may use the energy equation (see §§ 6·833, 6·834)

$$v_1^2 - v_2^2 = 2gh \qquad (32)$$

in place of $F = ma\ddot{\theta}$. Thus in the present chapter, the first of equations (40) will not often be needed in the form given.

The second equation of (40) will in Chapter VII usually be taken in the form $N = mv^2/a$.

In Chapter XIII, important use will be made of both the equations (40) in the form given above.

7·6. Examples. (1) *A light string OP is fixed at the end O, and is attached at the other end P to a particle which is moving uniformly in a* horizontal circle whose centre A is vertically below and distant h from O. Prove that the period of this motion is $2\pi\sqrt{(h/g)}$.

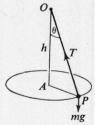

Fig. 67

Let ω be the angular speed of P about A, and θ the angle AOP.

The forces on the particle are the tension T of the string and its weight, mg say.

The acceleration of the particle is $AP.\omega^2$, i.e. $h \tan \theta.\omega^2$, in the direction PA.

By the fundamental equation (26a), we have, on resolving vertically and along PA,
$$T \cos \theta - mg = 0,$$
$$T \sin \theta = mh \tan \theta.\omega^2.$$

Eliminating T, we obtain $\omega^2 = g/h$.

The required period $= 2\pi/\omega = 2\pi\sqrt{(h/g)}$.

(2) *A train is travelling round a horizontal curve with uniform speed v. The radius of the curve (treated as a circular arc) is a. The rails are distant b apart. Find the elevation of the outer above the inner rail if there is no lateral force on the rails.*

Evaluate when the speed, radius of the curvature and distance between rails are 40 m.p.h., $\frac{1}{4}$ mile, 4' $8\frac{1}{2}$", respectively.

 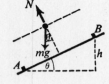

Fig. 68

The forces on the train are its weight, mg say, and the force N exerted by the rails A, B. By the data, N is at right angles to AB.

The acceleration of the train is v^2/a, horizontally to the left in Fig. 68.

Let θ, h be as in Fig. 68.

Taking resolved parts parallel to AB, we have
$$mg \sin \theta = m(v^2/a) \cos \theta,$$
whence
$$\tan \theta = v^2/ag.$$
Hence $\quad h = b \sin \theta \approx b \tan \theta = bv^2/ag.$

The numerical data are $v = \tfrac{2}{3} \times 88$, $a = 1320$, $b = 4{\cdot}71$, in f.p.s. units, and give $h = 0{\cdot}384$ (ft).

Thus the required elevation is about 4·6 inches.

(3) *A light inelastic string of length l initially vertical and at rest has a particle P attached at its lower end and is attached to a fixed point at its upper end. P is now given a horizontal velocity of magnitude u. Prove that if P describes a complete circle without the string slackening, u must exceed the speed which a free particle would acquire in falling vertically from rest through a distance 5l/2.*

Fig. 69

Let O be the centre of the circle, and A the initial position of P. Let v be the speed of P when the angle AOP is θ.

The forces on P are its weight mg and the tension T of the string, as shown in Fig. 69.

The component of the acceleration of P in the direction PO is v^2/l.

Resolving in the direction PO, we then have
$$T - mg \cos \theta = mv^2/l. \tag{i}$$

Since only the weight of P does work, (32) gives
$$u^2 - v^2 = 2gl(1 - \cos \theta). \tag{ii}$$

Eliminating v^2 between (i) and (ii), we obtain
$$T = mg \cos \theta + mu^2/l - 2mg(1 - \cos \theta)$$
$$= mg(u^2/lg + 3 \cos \theta - 2).$$

If the string does not become slack, $T > 0$ for all θ.

This requires u^2/lg to be greater than the maximum value of $2 - 3 \cos \theta$. This maximum occurs when $\theta = \pi$; then $2 - 3 \cos \theta = 5$.

Hence we require $u^2/lg > 5$, i.e. $u^2 > 2g(5l/2)$, from which the answer follows.

(4) *A smooth uniform hoop of mass M, threaded by two small heavy smooth rings each of mass m, is standing in a vertical plane on a horizontal table. The system is smoothly restricted to motion in this plane. Initially the system is at rest with the rings held at the top of the hoop. The rings*

are now released so that they fall down opposite sides of the hoop. Find the condition that the hoop will leave the table, and (assuming the condition satisfied) the position of the rings at the instant when this occurs.

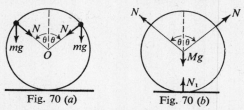

Fig. 70 (a)   Fig. 70 (b)

Symmetry about the vertical through the centre O of the hoop will be preserved (see *S.P.I.C.E.* (ζ) below).

Let θ be the angle described by either ring, and v its speed, when in any general position, up to the instant when the hoop starts to rise.

Consider first either one of the rings.

The forces on it are N, mg, as shown in Fig. 70 (a). (N will be negative during at least part of the motion.)

The normal component of the acceleration of the ring is v^2/a towards O, where a is the radius of the hoop.

Resolving normally, and using the energy equation, we have
$$N + mg\cos\theta = mv^2/a,$$
$$v^2 = 2ga(1 - \cos\theta),$$
whence $\qquad N = mg(2 - 3\cos\theta).$ \hfill (i)

Now consider the hoop.

The forces on it are: the forces of magnitude N exerted by the two rings; N_1 vertically upwards exerted by the table; and the weight Mg—as shown in Fig. 70 (b).

The resultant upward force on the hoop
$$= N_1 + 2N\cos\theta - Mg$$
(by (i)) $\quad = N_1 + mg(4\cos\theta - 6\cos^2\theta) - Mg.$ \hfill (ii)

This resultant force must, by (26a), be zero till the hoop starts to rise. At this instant, $N_1 = 0$.

Hence the hoop will be on the point of rising if and when there exists a real value of θ such that (ii), with $N_1 = 0$, is zero;

i.e. such that $\quad 6m\cos^2\theta - 4m\cos\theta + M = 0;$

i.e. such that $\quad \cos\theta = \tfrac{1}{3}\{1 \pm \sqrt{(1 - 3M/2m)}\}.$ \hfill (iii)

This will be the case if $m \geqslant 3M/2$.

The required condition is therefore $m > 3M/2$.

When the condition is satisfied, the position of the rings is given by the lesser value of θ in (iii), viz.
$$\theta = \cos^{-1}[\{1 + \sqrt{(1 - 3M/2m)}\}/3].$$

(The greater value of θ given by (iii) has no relevance to the problem, because of changed dynamical circumstances after the ring leaves the table.)

(5) *Treating the Earth as a spherically symmetrical body of radius 6370 km rotating about its axis (assumed to be fixed) with a uniform angular velocity of 1 revolution per day, investigate the approximate effect of this rotation on the acceleration due to gravity at a point of the Earth's surface in latitude ϕ.*

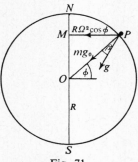

Fig. 71

Let O denote the centre of the Earth, N and S the north and south poles, R m the Earth's radius, and Ω rad/sec the angular speed of the Earth.

Thus $R = 6\cdot 37 \times 10^6$, and $\Omega = 2\pi/(24 \times 60^2)$.

Consider a particle of mass m falling freely near a point P fixed on the Earth's surface in latitude ϕ.

Let the 'acceleration due to gravity' be g in a direction inclined at β to OP as in Fig. 71. The acceleration g is *relative* to the Earth's surface (see § 2·61).

The absolute acceleration of the particle is, by (34b), the resultant of the 'acceleration due to gravity' and the absolute acceleration of P; the latter is, by (39), $R\cos\phi.\Omega^2$ in the direction PM.

Let mg_0 be the magnitude of the force on the particle due to the Earth's attraction.

On account of the assumed spherical symmetry of the Earth, this force acts along PO, and mg_0 is independent of ϕ.

This force is the only force acting on the particle.

Using (26a), and resolving at right angles to and along PO, we have
$$0 = m(g\sin\beta - R\cos\phi.\Omega^2.\sin\phi), \qquad \text{(i)}$$
$$mg_0 = m(g\cos\beta + R\cos\phi.\Omega^2.\cos\phi). \qquad \text{(ii)}$$

The relevant numerical values give $R\Omega^2/g \approx 0\cdot 0034$, which is small.

By (i) $\qquad \sin\beta = (R\Omega^2/g)\sin\phi\cos\phi;$

hence $\qquad \beta \approx 0\cdot 0034 \sin\phi\cos\phi.$ $\qquad \text{(iii)}$

By (ii) $\qquad g_0 = g\cos\beta + R\Omega^2\cos^2\phi.$

Hence g_0 is the value of g at the poles; (by (iii), β is zero at N and S).

Since β is small, $\cos\beta \approx 1$ (see § 7·3). Hence
$$g \approx g_0\{1 - (R\Omega^2/g_0)\cos^2\phi\}$$
$$\approx g_0(1 - 0·0034\cos^2\phi). \qquad \text{(iv)}$$
(iii) and (iv) give the required effect in direction and magnitude, respectively.

S.P.I.C.E. (α) It will be noticed that simple direct use of the fundamental equation (26a) continues to be a key step in solving problems on the dynamics of a particle.

(β) The mechanism in Ex. (1) is sometimes called a *conical pendulum*. It will be noticed from the solution that if the angular speed ω is increased, then h is decreased. This is the principle underlying the use of *governors*.

(γ) In Ex. (2), we resolved in one direction only, namely at right angles to the direction of N. A second resolution would have involved N, the value of which is not given and not sought in the problem. The student should check that the same answer would be given, though with slightly more effort, by resolving say horizontally and vertically. It is good technique to avoid, so far as possible, the introduction into the equations of unknown quantities whose values are not being sought. (In Ex. (1), incidentally, we could have avoided introducing T by resolving at right angles to OP.)

(δ) In Exs. (3) and (4), it is well to appreciate that there are non-zero tangential as well as normal components of acceleration. In these Exs., the tangential resolution was replaced by use of the energy equation.

(ϵ) In Ex. (4), the student should appreciate the need for using two diagrams in order to show fully the forces on various members of the system.

(ζ) We invoked a principle of symmetry in Ex. (4) in a way involving a further element of intuition. The principle of symmetry is commonly appealed to in mechanics and will be used in this book. Solutions obtained in this way can always be derived by other means, though usually less rapidly; to solve Ex. (4) without the use of symmetry would involve theory not yet given.

Another use of symmetry occurs in Ex. (5).

(η) In Ex. (5), we have gone one step deeper than our usual working hypothesis (§ 3·3) that the Earth's surface is fixed. It is necessary in Ex. (5) to appreciate that g is the magnitude of a *relative* acceleration (see also §§ 3·31 (vi) and 17·9).

The use of 'absolute acceleration' in Ex. (5) rests on the stated assumption in the data that NS is fixed; this of course is still an over-simplification.

(θ) The result (iv) of Ex. (5) differs from the more accurate formula given in § 2·62, (ii). This is because the rotation of the Earth is only one among a number of influences affecting local values of g.

EXAMPLES VII

1. A point P is moving in a circle of centre O with uniform angular speed ω about O. N is the foot of the perpendicular from any position of P to a fixed diameter of the circle. Find the acceleration of N. [Answer: $\omega^2 ON$ towards O.]

2. A point P moves in a straight line with constant velocity. Prove that its angular velocity about a fixed point O (outside the given line) is inversely proportional to the square of the distance OP.

3. A point P is describing a plane curve, the speed at any instant being v, and the angular speed about a given fixed point O being ω. Prove that $r^2\omega = pv$, where $r = OP$ and p is the perpendicular from O to the tangent of the curve at P.

(Hint:—Show that each of $\frac{1}{2}r^2\omega$, $\frac{1}{2}pv$ is equal to the 'areal velocity' about O; the areal velocity is the rate at which the area bounded by OA, OP and the arc AP is increasing, where A is a fixed point on the given plane curve.)

4. Find the angle at which a highway should be banked at a place where the radius of curvature is 220 yards, assuming the most frequent speed of cars on the highway to be 45 m.p.h. [Answer: $11°·6$.]

5. A particle is placed on a rough horizontal plate (coefficient of friction $= 0·6$) at a distance of 15 cm from a vertical axis about which the plate can rotate. Find in revolutions per minute the greatest steady angular speed the plate can have without the particle slipping. [Answer: $6\sqrt{g/\pi}$.]

6. A motor-car of mass 8 cwt. crosses a bridge whose vertical section is convex upwards, the radius of curvature at the highest point A being 242 ft. Find the speed of the car at A if the wheels just cease to press on the ground.

Construct a graph from which the force F lb.wt. between the car and the bridge at A can be determined for any given value of the speed V m.p.h. at A.

[Answers: 60 m.p.h.; the graph is part of the parabola $F = 112(8 - V^2/450)$.]

7. A number of particles, whose action on one another is negligible, move with constant speeds in circles which have a common centre O. On each particle there acts a force towards O directly proportional to its mass and inversely proportional to the square of the distance from O. Prove that the squares of the periods of revolution are proportional to the cubes of the radii of the circles.

8. According to Newton's law of gravitation, the force of gravitational attraction between two particles of masses m, m' is Gmm'/r^2, where G is the constant of gravitation and r is the distance between the particles. Given that $G = 6·67 \times 10^{-11}$ m.k.s. units, $g = 9·80$ m/s², and the radius of the Earth is 6370 km, prove that the mean density of the Earth is about $5·5$ g/cm³. (It may be assumed that the attraction of the Earth at an external point is the same as that due to a particle of mass equal to that of the Earth, situated at the Earth's centre.)

Treating the Earth as a particle revolving in a circle of radius $1·5 \times 10^{11}$ m about the Sun as centre, the period of revolution being 1 year, find the mass of the Sun.

[Answer: 2×10^{30} kg.]

9. A particle, attached by a light inelastic string of length l to a fixed point O, describes with uniform speed a horizontal circle. The string is just capable of supporting n times the weight of the particle without breaking. Prove that the period T of the circular motion cannot be less than $2\pi\sqrt{(l/ng)}$.

If $l = 6$ ft and $T = 2$ sec, prove that the inclination of the string to the vertical is about $57°$, and that the tension is about $1·85$ times the weight of the particle.

10. If the number of (steady) revolutions per minute of a conical pendulum is increased from 75 to 80, find the rise in the level of the bob.
[Answer: Approx. $\frac{3}{4}$ in.]

11. A particle of mass $\frac{1}{4}$ oz rests on a smooth horizontal disc, and is attached by two strings each of length 4 ft to the extremities of a diameter of the disc. The disc is made to revolve steadily about a vertical axis through its centre at 100 rev/min. Prove that the tensions in the strings are equal and find them.
[Answer: 0·107 lb.wt.]

12. Two particles P, Q of masses 1, 2 lb are attached to a light inelastic string OPQ, the end O of which is fixed; $OP = 1$ ft and $PQ = 3$ ft. The system is revolving (the string remaining straight) on a smooth fixed horizontal plane, the period being 2 sec. Find the tensions in the portions OP, PQ of the string. [Answer: $9\pi^2$, $8\pi^2$ pdl.]

13. A particle P, attached by a light inelastic string to a fixed point A, is describing uniformly a horizontal circle whose centre O is vertically below A and distant h from A. Prove that the angular speed of P about O is $\sqrt{(g/h)}$.

P is now joined by a second light inelastic string, of the same length as the first string, to a fixed point B vertically below A and O and distant $2h$ below A; and the angular speed of P about O is now made equal to $3\sqrt{(g/h)}$ and kept steady at this new value. Find the ratio of the tensions in the two strings in the new motion.
[Answer: 5 : 4.]

14. On a horizontal curve of radius a, a railway track is banked so that there is no lateral force on the rails when the speed of a passing train is V. If the speed is v, and if V^2 is small compared with ag, prove that the outward lateral force on the rails exerted by a carriage of mass M is approximately $M(v^2 - V^2)/a$.
[Note:—First show that the force is equal to $Mg(v^2 - V^2)(V^4 + a^2g^2)^{-\frac{1}{2}}$.]

15. A Watt's Governor consists of a rhombus $ABCD$ of light smoothly-jointed rods each of length l, attached at A to a fixed point of a vertical revolving shaft of fixed axis, and smoothly attached at C (below A) to a collar of mass m which can slide smoothly up or down the shaft. The attachment at A is such that $ABCD$ revolves with the shaft, but AB and AD can rotate smoothly about A in the plane of $ABCD$. At each of B and D, the rhombus carries a heavy mass M (treated as a particle). Find the angle BAD when the shaft has a steady angular speed Ω about its axis.
[Answer: $2\cos^{-1}\{(M+m)g/Ml\Omega^2\}$.]

16. A particle P is just displaced from rest on the top of a smooth fixed sphere of centre O. Find the angle between OP and the vertical at the instant when the particle leaves the sphere. [Answer: $\cos^{-1}(2/3)$.]

17. A bead sliding on a fixed smooth vertical circular hoop of radius a has speed V at the lowest point. Prove that the bead makes complete revolutions if $V^2 > 4ag$, and that the force exerted by the bead on the hoop is always radially outwards if $V^2 > 5ag$.

18. A bead sliding on a fixed smooth vertical circular wire is making complete revolutions. If the ratio of the greatest to the least speed is 2 : 1, find the ratio of the greatest to the least force between the bead and the wire. [Answer: 19 : 1.]

19. The idea has been put forward of having an observing station revolving round the Earth like a satellite. If the altitude of such a satellite were 26,000 miles, find approximately the speed it would need to have. (Take the Earth's radius as 4000 miles and the value of g at the Earth's surface as 32 ft/sec^2; assume that g varies inversely as the square of the distance from the Earth's centre; assume that the orbit of the satellite is circular and that there is negligible resistance.) [Answer: 6500 m.p.h.]

20. A particle P is suspended by two light inelastic strings AP, BP attached to fixed points A, B. Initially the system is in equilibrium and at rest, the inclinations of AP, BP to the vertical being α, β. If the string BP is suddenly cut, prove that the tension in AP is suddenly changed in the ratio

$$\sin(\alpha + \beta)\cos\alpha/(\sin\beta).$$

*21. A particle P of mass m is on the inside surface of a rough hollow circular cylinder of internal radius a, and is at rest relative to the cylinder while the cylinder rotates about its axis, which is fixed horizontally, with constant angular speed Ω. Prove that the coefficient of friction μ cannot be less than β, where $\beta = g/\sqrt{(\Omega^4 a^2 - g^2)}$.

If $\mu = \beta$, prove that slipping is on the point of occurring when the angular displacement of P from the upward vertical is $\cos^{-1}(g/a\Omega^2)$.

*22. A rough hollow cone (coefficient of friction μ), whose internal semi-vertical angle is α and whose vertex is lowermost, is spinning with uniform angular speed Ω about its axis which is fixed vertically. A particle remains at relative rest on the inside surface of the cone. Find the range of possible heights of the particle above the level of the vertex, given that μ is less than both $\tan\alpha$ and $\cot\alpha$.

$$\left[\text{Answer:—Between } \beta g/\Omega^2 \text{ and } \gamma g/\Omega^2, \text{ where } \beta, \gamma = \frac{\cot\alpha \mp \mu}{\tan\alpha \pm \mu}.\right]$$

Discuss the interpretations when $\mu > \tan\alpha$ and when $\mu > \cot\alpha$.

CHAPTER VIII

SIMPLE HARMONIC MOTION

"Amongst the most important classes of motions which we have to consider in Natural Philosophy, there is one, namely Harmonic Motion, which is of such immense use . . . that we make no apology for entering here into considerable detail."

—THOMSON AND TAIT.

8·1. Definitions.

8·11. Simple harmonic motion. Let O be a fixed point on the straight line $X'OX$, and let P be the position at time t of a point moving along $X'OX$. Let $OP = x$. Then the point is moving in *simple harmonic motion* about O as *centre* if

$$x = a \cos (nt + \epsilon), \tag{41a}$$

where a, n, ϵ are constants, a, n being taken positive.

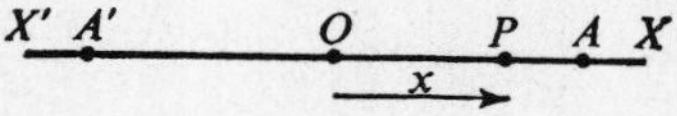

```
X'  A'           O        P   A   X
●───●────────────●────────●───●───
              x →
```

Fig. 72

8·12. Remarks.

(i) If the arc-length from a fixed point to a moving point on a curve satisfies a relation of the form (41a), we speak of 'curvilinear simple harmonic motion'.

If an angular coordinate satisfies a relation of the form (41a) (see e.g. § 8·9), we may speak of 'angular simple harmonic motion'. (The word 'angular' is usually omitted.)

(ii) In place of (41a), we could equally well have written

$$x = a \sin(nt + \epsilon'), \tag{41b}$$

or

$$x = A \cos nt + B \sin nt, \tag{41c}$$

where ϵ', A, B are constants. It is seen that the various constants are connected by the relations

$$\epsilon' = \epsilon + \pi/2,$$
$$A = a \cos \epsilon, \quad B = -a \sin \epsilon.$$

8·13. Amplitude; phase; epoch. The maximum value of x, viz. a, is called the *amplitude* of the motion. If $OA = a$ and $OA' = -a$ (Fig. 72), then the motion is clearly confined to the range of positions from A' to A.

The *phase* of the motion is $nt + \epsilon$. The initial phase, ϵ, is sometimes called the *epoch*; (sometimes the epoch is taken as ϵ').

8·2. Velocity and acceleration at any instant.

On differentiating (41a) successively with respect to t, we obtain

$$\dot{x} = v = -an \sin(nt + \epsilon),$$
$$\ddot{x} = f = -an^2 \cos(nt + \epsilon),$$

where v and f are the velocity and acceleration at time t.

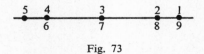

Fig. 73

8·3. Period and frequency.

Consider the motion over a time-interval during which the phase $nt + \epsilon$ changes from $2\pi k$ to $2\pi(k + 1)$, where k is any integer. The table below contains some corresponding representative values of $nt + \epsilon$, x, v, f over this interval.

	(1)	(2)	(3)	(4)	(5)	(6)	(7)	(8)	(9)
$nt+\epsilon$	$2\pi k$	$2\pi k+\pi/4$	$2\pi k+\pi/2$	$2\pi k+3\pi/4$	$2\pi k+\pi$	$2\pi k+5\pi/4$	$2\pi k+3\pi/2$	$2\pi k+7\pi/4$	$2\pi(k+1)$
x	a	$a/\sqrt{2}$	0	$-a/\sqrt{2}$	$-a$	$-a/\sqrt{2}$	0	$a/\sqrt{2}$	a
v	0	$-an/\sqrt{2}$	$-an$	$-an/\sqrt{2}$	0	$an/\sqrt{2}$	an	$an/\sqrt{2}$	0
f	$-an^2$	$-an^2/\sqrt{2}$	0	$an^2/\sqrt{2}$	an^2	$an^2/\sqrt{2}$	0	$-an^2/\sqrt{2}$	$-an^2$

In Fig. 73, the positions corresponding to the numbers 1 to 9 in the table are indicated. By following out the values of v and f in these positions in Fig. 73, the student should get a general idea of the nature of simple harmonic motion.

It is evident that the motion will repeat itself after the time-interval involved above. This time-interval is called the *period T* of the motion, and is given by

$$T = 2\pi/n. \qquad (42)$$

The result (42) is seen immediately by noting that in (41a) (or b or c) and the formulae of § 8·2, if t is increased by $2\pi/n$, the values of x, v, f are all unaltered.

The motion during one period is called one *oscillation*.

The *frequency* (of dimensions $[T^{-1}]$) is the number of oscillations per unit time, and is equal to $n/2\pi$.

8·4. Speed and acceleration in any position.

Eliminating t between (41a) and each of the formulae in § 8·2 in turn, we derive

$$v^2 = n^2(a^2 - x^2),$$
$$\ddot{x} = -n^2 x. \tag{43}$$

Note that the sign of v is not determined if only the position is given ; (hence the term *speed* in the heading).

8·5. Alternative investigation of simple harmonic motion.

The equation (43), with n^2 an assigned positive constant, is commonly taken in place of (41a) in defining simple harmonic motion.

(41a) (or b or c) can be deduced from (43) by integration, the constants a, ϵ entering as integration constants.

(This has already been demonstrated in the example of § 2·71. The step from (43) to (41a) may be alternatively made in one line, using a known method in Differential Equations.)

All the earlier formulae may now be deduced as before from (41a).

8·51. Application to problems.

In many problems, the form (43), with n^2 a positive constant determined by the data, emerges as a result of applying the dynamical theory of Chapter VI in the usual way. When this happens, we recognise that the motion is simple harmonic, and we can then write down any one of the forms such as (41a).

The constants a, ϵ (or one of the alternative pairs (a, ϵ'), (A, B)) are determined if the values of x and v are known at some definite instant.

For example, suppose that initially $x = 5$, $v = 0$. Then, using (41a), we have $5 = a \cos \epsilon$, $0 = -an \sin \epsilon$. The latter gives $\sin \epsilon = 0$ ($a = 0$ is incompatible with $5 = a \cos \epsilon$); hence $\epsilon = 0$; ($\epsilon = \pi$ would make $\cos \epsilon$ negative). Hence $x = 5 \cos nt$.

Secondly, suppose that when $t = 0$ the point is passing through O with positive velocity. Then, using (41a), we have that $0 = a \cos \epsilon$ and $-an \sin \epsilon$ is positive. We derive $\epsilon = -\pi/2$, and hence $x = a \sin nt$.

The student should obtain these results using (41c) instead of (41a). The use of (41c) is actually much the simpler in these cases; we used (41a) in order to give a little experience in the care needed in handling the trigonometry involved.

8·6. Simple harmonic motion as projection of uniform circular motion.

Let Q be the position at time t of a point describing a circle of radius a with uniform angular velocity ω about the centre O. Let P be the projection of Q on any diameter $A'OA$, let $\angle AOQ = \theta$, and let $OP = x$.

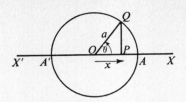

Fig. 74

Thus $\dot{\theta} = \omega$, where ω is constant.

Hence $\theta = \omega t + \epsilon$, where ω, ϵ are constants.

Hence $x = a \cos \theta = a \cos (\omega t + \epsilon)$.

Thus the projection of Q moves along $A'OA$ in simple harmonic motion of amplitude equal to the radius of the circle, the 'n' of earlier formulae being equal to the angular velocity of Q about O.

It is useful to note that the speed of P at O, viz. an, and the acceleration of P at A, viz. $-an^2$, are rapidly determined by considering the motion of the auxiliary point Q.

8.7. Examples.
(1) *A point moving with simple harmonic motion starts from a point 5 cm from the centre of the motion with a speed of 1 cm/sec. The period is 11 sec. Find the maximum speed and acceleration.*

In standard notation, we have $x = a \cos (nt + \epsilon)$. (i)

By (42), $n = 2\pi/11 = 4/7$ (approximately).

Differentating (i) twice, and substituting $n = 4/7$, we have

$$\dot{x} = -\frac{4a}{7} \sin \left(\frac{4t}{7} + \epsilon\right),$$ (ii)

$$\ddot{x} = -\left(\frac{4}{7}\right)^2 a \cos \left(\frac{4t}{7} + \epsilon\right).$$

Hence the maximum speed and acceleration are $4a/7$ cm/sec and $16a/49$ cm/sec², respectively.

To determine a, we have, using (i) and (ii) and the initial conditions, viz. $|x| = 5$ and $|\dot{x}| = 1$ when $t = 0$:

$$5 = a|\cos \epsilon|,$$

$$1 = \frac{4a}{7} |\sin \epsilon|.$$

Hence $a^2 = 25 + 49/16 = 449/16.$

Hence the maximum speed $= \sqrt{449}/7$ cm/sec
$= 3 \cdot 03$ cm/sec;

and the maximum acceleration $= (4/49)\sqrt{449}$ cm/sec²
$= 1 \cdot 73$ cm/sec².

(2) *The speed v cm/sec of a point moving along the x-axis is given by $v^2 = 36 - 6x - 2x^2$, where x is in cm. Prove that the motion is simple harmonic, and find the period and the amplitude.*

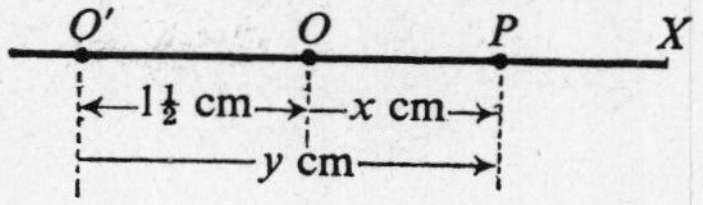

Fig. 75

Differentiating the given equation with respect to x (using (3)), we have
$$2\ddot{x} = -6 - 4x;$$
i.e. $\quad\quad\ddot{x} = -2(x + 3/2);$
i.e. $\quad\quad\ddot{y} = -2y,$ (i)
where $y = x + 3/2$.

Thus, by (43), the moving point oscillates with simple harmonic motion about O', where O' is distant $1\frac{1}{2}$ cm from O on the negative side.

By (i), the period $= 2\pi/\sqrt{2}$ sec $= \pi\sqrt{2}$ sec.

Since $v^2 = -2(x - 3)(x + 6)$, v is zero when $x = 3$ and when $x = -6$. Hence the amplitude $= 4\frac{1}{2}$ cm.

S.P.I.C.E. (α) It will be noticed that in solving Ex. (1) we did not quote formulae such as those in § 8·2, but deduced them from (41a). The average student is advised to follow this procedure rather than overburden his memory with formulae.

(β) In Ex. (2), the existence of simple harmonic motion was detected through the emergence of an equation of the form (43).

(γ) In Ex. (2), the centre of the motion is not at the original origin. The process of changing the origin should be noted.

(δ) In the enunciation of Ex. (2), the units of v, x need to be specified because the numerical coefficients in the given equation must (by dimension considerations) represent measures of *dimensioned* quantities.

8·8. Application to oscillations of particle attached to light perfectly elastic string. We consider the following problem: Let a light vertical perfectly elastic string of unstretched length a and modulus λ be fixed at one end O and carry a particle of mass m at its other end. We investigate the motion of the system in a vertical line through O, the string being assumed to be always taut.

Let $OA = a$, and let x be the distance of the particle below A at time t (Fig. 76).

The forces on the particle are its weight mg, and the tension, F say, of the string; by (17b), $F = \lambda x/a$.

By the fundamental equation (26a),
$$mg - \lambda x/a = m\ddot{x};$$
i.e. $\quad\quad\quad\quad\ddot{x} = -(\lambda/ma)(x - mga/\lambda);$

i.e. $\quad\quad\quad\quad\ddot{y} = -n^2 y,$

where $n^2 = \lambda/ma$, and $y = x - mga/\lambda$. It is seen that y is the distance below C, where C is distant mga/λ below A.

Hence the particle moves in simple harmonic motion, with period $2\pi/n = 2\pi\sqrt{(ma/\lambda)}$, about C as centre.

It is immediately seen that C is the position in which the particle would rest in equilibrium.

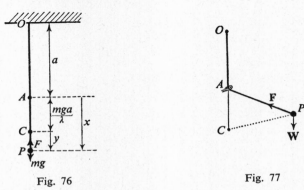

Fig. 76   Fig. 77

8·81. Remarks.

(i) If the string became slack during part of the motion, the particle would move with constant acceleration g during this part, and not in simple harmonic motion.

(ii) The student should check that the dimensions of the formula for the period, $2\pi\sqrt{(ma/\lambda)}$, are those of time.

(iii) In the matter of dimensions, it is interesting to note further that the formula for the period (apart from the value of the dimensionless factor 2π) is indicated by dimension considerations alone.

For, suppose the period $= cm^x a^y \lambda^z$, where c is a dimensionless constant. Equating dimensions we have
$$[T] = [M]^x [L]^y [ML/T^2]^z.$$

Equating powers of [T], [M], [L], we have
$$1 = -2z, \quad\quad 0 = x + z, \quad\quad 0 = y + z,$$
yielding $\quad\quad x = \tfrac{1}{2}, \quad\quad y = \tfrac{1}{2}, \quad\quad z = -\tfrac{1}{2},$

and hence $\quad\quad T = cm^{\frac{1}{2}} a^{\frac{1}{2}} \lambda^{-\frac{1}{2}} = c\sqrt{(ma/\lambda)}.$

(iv) An alternative direct method of solving the above problem would be to take the origin at the equilibrium position C at the outset.

8·82. Examples.

(1) *A particle of mass m on a smooth horizontal table is joined by two light perfectly elastic strings, of unstretched lengths a, a', and moduli λ, λ', to the fixed points O, O' of the table. The particle moves in the line OO', the strings being always taut. Find the period of the motion.*

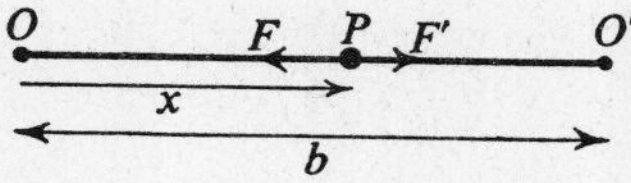

Fig. 78

Let $OO' = b$, and let $OP = x$, where P is the position of the particle at time t.

The horizontal forces on the particle are the tensions F, F' of the strings. By (17b),
$$F = \lambda(x-a)/a; \qquad F' = \lambda'(\overline{b-x}-a')/a'.$$
By (26a), we have
$$m\ddot{x} = -F + F'$$
$$= -\lambda x/a + \lambda + \lambda'(b-a')/a' - \lambda'x/a';$$
i.e. $\qquad \ddot{x} = -m^{-1}(\lambda/a + \lambda'/a')(x-c),$

where c is a constant which we need not evaluate.

Hence $\qquad \ddot{y} = -n^2 y,$

where $\qquad y = x - c,\qquad$ and $\qquad n^2 = (\lambda/a + \lambda'/a')/m.$

Hence the motion is simple harmonic, of period $2\pi \sqrt{\left\{ m\left(\dfrac{\lambda}{a} + \dfrac{\lambda'}{a'}\right)^{-1} \right\}}.$

(2) *A light perfectly elastic string of unstretched length a and modulus λ has one end fixed at O. The other end is passed through a small smooth fixed ring at A distant a vertically below O, and is joined to a particle of mass m. The particle is drawn aside from the vertical OA and released from rest. Prove that the particle moves in a straight line with simple harmonic motion, and find the period.*

Let P be the position of the particle at time t.

The forces on the particle are its weight **W** of magnitude mg, and the tension, **F** say, of magnitude $\lambda.PA/a$, (PA being the extension of the string); these forces act in the directions shown in Fig. 77.

Now $\mathbf{PA} = (a/\lambda)\mathbf{F}$, and if we construct **AC** vertically downwards of magnitude $(a/\lambda)mg$, then **PC** must be (a/λ) times the resultant force on the particle.

Hence, by (26a), the acceleration of the particle $= (\lambda/ma)\mathbf{PC}$.

By construction, C is a fixed point. Thus the acceleration of the particle is always towards the fixed point C and of magnitude equal to n^2 times the distance from C, where $n^2 = \lambda/ma$.

Since P starts from rest, it must therefore move in the straight line through C, in simple harmonic motion, with period $2\pi\sqrt{(ma/\lambda)}.$

S.P.I.C.E. (α) In Ex. (1), labour was saved by not evaluating the constant c. This constant would give the centre of the motion. The student should solve Ex. (1) independently, taking as origin the equilibrium position of P; this position is also the centre of the motion.

(β) A special point of Ex. (2) is that it has first to be proved that the particle moves in a straight line. The solution is facilitated by a neat use of vector addition.

8·9. Simple pendulum. A simple pendulum consists of a light inelastic string (of length l say), fixed at one end O and attached at the other end to a particle (of mass m say) which moves under gravity in a vertical circle.

(A pendulum used in practice is of course more complicated than this artificially simplified conception.)

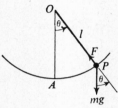

Fig. 79

Let the position P of the particle be given by θ, as in Fig. 79, where A is the equilibrium position.

By (26a), resolving at right angles to the tension F of the string, and using (39), we have

$$mg \sin \theta = -ml\ddot{\theta};$$

i.e. $$\ddot{\theta} = -(g/l) \sin \theta.$$

This equation is relevant to the whole circular motion of the particle. But in the case of the simple pendulum there is some special interest in problems where θ always remains small enough for its square to be disregarded. Since in this case $\sin \theta \approx \theta$, we have approximately

$$\ddot{\theta} = -n^2\theta,$$

where $n^2 = g/l$.

This equation gives (angular) simple harmonic motion of period $2\pi\sqrt{(l/g)}$.

8·91. Remarks.

(i) In any general case of simple harmonic motion of period $2\pi/n$, the expression g/n^2 is, on the basis of § 8·9, often called the length of the *equivalent simple pendulum* (E.S.P.).

Exs. VIII] SIMPLE HARMONIC MOTION 145

(ii) If the period is 2 seconds, the pendulum is said to beat seconds.

(iii) It can be shown (from an analysis of the accurate equation $\ddot{\theta} = -(g/l)\sin\theta$) that the formula $2\pi\sqrt{(l/g)}$ for the period, T say, is too small by approximately $(\alpha^2/16)T$, where α radians is the (angular) amplitude, assumed to be not too great.

(iv) When the pendulum is in equilibrium, the tension F is equal and opposite to the weight mg; hence the use of the *plumb-bob* to determine the vertical direction at any place.

(v) In problems involving the pendulums of clocks, it is a matter of interest to find the effect of small changes in l and (or) g on the period or the frequency of oscillations.

Let T be the period, and ν the number of oscillations per unit time. Then
$$1/\nu = T = 2\pi\sqrt{(l/g)}.$$
If l, g vary, we have, taking logarithms and then forming differentials,
$$-d\nu/\nu = dT/T = \tfrac{1}{2}dl/l - \tfrac{1}{2}dg/g.$$
Hence if δ denotes a small increase, we have approximately
$$-\frac{\delta\nu}{\nu} = \frac{\delta T}{T} = \tfrac{1}{2}\frac{\delta l}{l} - \tfrac{1}{2}\frac{\delta g}{g}.$$

8·92. Example. *A pendulum clock gains* 30 *seconds per day. Find the alteration necessary in the length of the pendulum which should beat seconds.*

We require
$$\delta\nu/\nu = -30/(24 \times 60^2),$$
where $\delta\nu$ is the increase (negative in this case) needed to the frequency. In this problem, $\delta g = 0$. So we have
$$\frac{\delta l}{l} = -\frac{2\delta\nu}{\nu} = \frac{60}{24 \times 60^2}.$$
Hence the length must be increased by approximately $l/1440$, where l is given by $2 = 2\pi\sqrt{(l/g)}$, i.e. $l = g/\pi^2$.

The alteration needed is thus found to be an increase of about 0·027 in.

EXAMPLES VIII

1. A point moving in simple harmonic motion has a speed of 5 ft/sec when passing through the centre O of its path, and a period of π seconds. Find its speed and acceleration when 1 ft 6 in from O. [Answers: 4 ft/sec; 6 ft/sec².]

2. A point moving with simple harmonic motion of period 3·14 sec is observed to have a velocity of 16 cm/sec away from the centre O of the motion when the distance from O is 6 cm. Find the amplitude of the motion, and the time that elapses before the point first comes instantaneously to rest. [Answers: 10 cm; 0·46 sec.]

3. The displacement x at time t of a particle moving in a straight line is given by $x = a\cos(nt + \epsilon)$. Find the forms to which this relation reduces, given that, initially: (a) $\dot{x} = 0$, $x = -5$; (b) $x = 0$, and the velocity is negative.
[Answers: $x = 5\cos(nt + \pi)$; $x = a\sin(nt + \pi)$.]

4. The speed v ft/sec of a point which is moving along the x-axis is given by $v^2 = -9x^2 + 18x + 27$, where x is measured in feet. Prove that the motion is simple harmonic, and find the centre of the motion, the amplitude and the frequency.
[Answers: At $x = 1$; 2 ft; $3/2\pi$ sec^{-1}.]

5. A particle of mass m makes N simple harmonic oscillations per unit time. Prove that its kinetic energy when distant x from the centre O is less than that at O by $2\pi^2 N^2 m x^2$.

6. A shaking table moves in a horizontal plane with simple harmonic motion of frequency N min^{-1}. If the value of the coefficient of friction is μ, find the greatest allowable amplitude of the motion if an object placed on the table is not to slip.
[Answer: $893\mu/N^2$ m.]

7. The rise and fall of the tide at a certain harbour may be taken to be simple harmonic, the interval between successive high tides being 12 h 20 min. The harbour entrance has a depth of 34 ft at high tide, and 14 ft at low tide. If low tide occurs at noon on a certain day, find the earliest time thereafter that a ship drawing 29 ft can pass through the entrance. [Answer: 4 h $6\frac{8}{9}$ min, p.m.]

8. A particle P is initially distant b from a fixed point O and moving with velocity V towards O. When $OP = x$, the acceleration of P is μx towards O. Find the elapsed time when P first reaches O.

$$\left[\text{Answer: } \frac{1}{\sqrt{\mu}} \sin^{-1}\left(\frac{b\sqrt{\mu}}{\sqrt{(V^2 + \mu b^2)}}\right).\right]$$

9. A mass of 10 lb is suspended from a light vertical spring fixed at the upper end, producing an extension of 10 in. If the mass is pulled down through a further distance 1 in and then released, find (a) the speed of the mass when $\frac{1}{2}$ in above the lowest point of its path; (b) the tension of the spring when the mass is highest.
[Answers: $1/\sqrt{5}$ ft/sec; 9 lb.wt.]

10. A mass of 10 lb is attached at B to a light vertical spring AB, where A is fixed; and is set vibrating vertically, making 100 oscillations in 33·3 sec. The mass is removed. Find the force required to stretch the spring through 1 inch from its natural length.
[Answer: $30\pi^2/g$ lb.wt.]

11. A light vertical perfectly elastic string fixed at its upper end supports a mass 8 lb at its lower end. The system is initially in equilibrium, the extension of the string being $1\frac{1}{2}$ in. What is the greatest vertical instantaneous impulse that can be given to the mass if the string is never to become slack? [Answer: 16 lb-ft/sec.]

12. A light perfectly elastic string of natural length a and modulus λ is fastened at one end O to a fixed point of a smooth horizontal table, and a particle of mass m is attached to the other end. Initially the particle is held at rest on the table with the string stretched to a total length of $2a$, and then released. Find the elapsed time when the particle reaches O. [Answer: $(\frac{1}{2}\pi + 1)\sqrt{(ma/\lambda)}$.]

13. A rigid ball is dropped from a height h on to a fixed horizontal plate from which the ball rebounds without loss of energy. During the contact with the plate, the surface of the plate is depressed through a distance y and then fully recovers its previous form. Assuming that the ball moves in simple harmonic motion during the contact and that y is small compared with h, find approximately: (a) the maximum acceleration of the ball during this motion; (b) the time during which the contact lasts.

[Answers: (a) $2gh/y$; (b) $\pi y/\sqrt{(2gh)}$.]

14. A particle P of mass m is joined to two fixed points A, B by two light, equal and completely similar perfectly elastic strings AP, PB. The system is on a smooth fixed table, and initially P is at rest in the equilibrium position, i.e. the midpoint of AB, the tension in each string being T. If P now receives a small horizontal impulse at right angles to AB, prove that P will move approximately in simple harmonic motion of period $2\pi\sqrt{(mb/2T)}$, where b is the initial length of either string.

If the initial speed imparted to P is u, find the displacement of P after time t.

[Answer: $u\sqrt{(mb/2T)} \sin \{\sqrt{(2T/mb)}t\}$.]

*15. A particle of mass m moves in a straight line joining two fixed centres, O, O', each of which attracts it with a force of magnitude $m\mu$ times the distance. If x is the distance of the particle from O, prove that $\ddot{x} + 2\mu x = \mu l$, where $l = OO'$. Hence find the period of the motion.   [Answer: $2\pi/\sqrt{(2\mu)}$.]

*16. A particle P moving in a straight line l is attracted towards a centre Q with a force of magnitude $\mu.PQ$ per unit mass. If Q is moved in the line l with constant velocity, prove that, relative to Q, P moves in simple harmonic motion with the same period as if Q were fixed.

(Hint: Let ξ be the position coordinate of Q referred to a fixed origin O in the line l, and let $OP = x$; and then show that $d^2(x - \xi)/dt^2 = -\mu(x - \xi)$.)

*17. If circumstances are as in Ex. 16 except that Q now has a fixed acceleration α, prove that P moves in simple harmonic motion relative to Q, the centre of the motion being at a point distant α/μ behind Q, and the period being $2\pi/\sqrt{\mu}$.

(Hint: With notation as in Ex. 16, show that
$$d^2(x - \xi + \alpha/\mu)/dt^2 = -\mu(x - \xi + \alpha/\mu).)$$

18. A clock with a seconds pendulum gains a minute a day. What alteration should be made to the pendulum length?

[Answer: The length should be increased by about 0·054 in.]

19. A seconds pendulum is found to lose 20 beats per day when taken to the top of a mountain. Find the change in the value of g.

[Answer: A reduction of 0·015 ft/sec².]

20. Assuming that outside the Earth (here treated as spherically stratified) the value of g varies inversely as the square of the distance from the Earth's centre, prove that the number of beats made in a given time by a simple pendulum of given length when it is raised from the Earth's surface to a height H is decreased by the fraction H/R, approximately, where R is the Earth's radius.

CHAPTER IX

FREE MOTION OF A PARTICLE UNDER CONSTANT GRAVITY

"You may watch falling bodies for an eternity, but without Mathematics mere watching will yield no law of gravitation."
—G. H. Lewes, *Life and Works of Goethe*.

9·1. Preliminary remarks. (i) This chapter is concerned with the special problem of the motion of a projectile above the Earth's surface. We artificially simplify by treating the projectile as a particle, taking the acceleration of gravity to be constant in magnitude and direction, neglecting motion of the Earth's surface, and neglecting air-resistance.

(The less simplified case of a rigid-body projectile is considered in § 13·(10)3, Ex. (1).)

(ii) The motion must be in a vertical plane. For, at any instant during the flight, let P be the position of the projectile and $\mathbf{v}$ its velocity. Let S be the vertical plane containing P and the vector $\mathbf{v}$ drawn through P. The only force on the projectile is its weight, and this acts in the plane S. Hence, by (26a), any change of velocity can be only in the plane S. Hence the projectile must always be in this plane.

(iii) It follows from § 6·833 that the formula (32), viz., $v_1^2 - v_2^2 = 2gh$, is relevant to the projectile motion we consider.

(iv) The theory of the present chapter is in effect a generalisation of § 2·5 to more than one dimension. Indeed, we could proceed from the vector equation $\ddot{\mathbf{r}} = \mathbf{g}$, which is the generalised form of $\ddot{x} = f$ (§ 2·5), where $\mathbf{r}$ represents the position vector of the projectile and $\mathbf{g}$ the acceleration due to gravity in magnitude and direction. By integration, we could derive formulae such as

$$\mathbf{v} = \mathbf{u} + \mathbf{g}t, \qquad \mathbf{r} - \mathbf{r}_0 = \mathbf{u}t + \tfrac{1}{2}\mathbf{g}t^2, \qquad \mathbf{v}\cdot\mathbf{v} = \mathbf{u}\cdot\mathbf{u} + 2\mathbf{g}\cdot(\mathbf{r} - \mathbf{r}_0).$$

The student who is adequately versed in vector analysis may be interested to derive and interpret these equations; the last equation is a disguised form of (32).

9·2. Path of a projectile.

9·21. Equations of motion. At some definite point, O say, of the path (or *orbit*) of a projectile, let the velocity be V in a direction inclined at α to the horizontal. Let OX, OY be horizontal and vertical axes in the plane of the motion.

Let t be the time measured from the instant when the projectile was at O, and let P, of coordinates x, y, be the position at any time t.

The initial conditions are that $x = 0$, $y = 0$, $\dot{x} = V\cos\alpha$, $\dot{y} = V\sin\alpha$, when $t = 0$.

The only force on the projectile is its weight mg vertically downwards.

By (26a), resolving horizontally and vertically, we have the cartesian equations of motion, viz.,

$$m\ddot{x} = 0, \qquad m\ddot{y} = -mg. \tag{44}$$

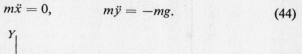

Fig. 80

9·22. Velocity at time t.
Integrating (44) with respect to t, we have
$$\dot{x} = C_1, \qquad \dot{y} = -gt + C_2,$$
where C_1, C_2 are integration constants. The initial conditions give
$$V\cos\alpha = C_1, \qquad V\sin\alpha = 0 + C_2.$$
Hence the horizontal and vertical components of velocity are
$$\dot{x} = V\cos\alpha, \qquad \dot{y} = V\sin\alpha - gt.$$

9·23. Position at time t.
Integrating again with respect to t, we have
$$x = Vt\cos\alpha + C_3, \qquad y = Vt\sin\alpha - \tfrac{1}{2}gt^2 + C_4,$$
where C_3, C_4 are integration constants. The initial conditions give
$$0 = 0 + C_3, \qquad 0 = 0 + 0 + C_4.$$
Hence $\qquad x = Vt\cos\alpha, \qquad y = Vt\sin\alpha - \tfrac{1}{2}gt^2.$

9·24. Cartesian equation of path.
The last pair of equations in § 9·23 constitutes the 'parametric equations' of the orbit in terms of the parameter t. Eliminating t between this pair of equations, we have the required cartesian equation, viz.,

$$y = x\tan\alpha - \tfrac{1}{2}gx^2 V^{-2} \sec^2\alpha. \tag{45}$$

9·25. Position of highest point of path.
At the highest point, we have $\dot{y} = 0$, and hence by § 9·22, $t = (V/g)\sin\alpha$.

Substituting this value of t into the formulae of § 9·23, we find the coordinates (x', y') of the highest point, O' say, to be

$$x' = (V^2/g)\sin\alpha\cos\alpha; \qquad y' = (V^2/2g)\sin^2\alpha.$$

9·26. Properties of the path.

(i) The equation (45) may, on rearranging the algebra, be written in the form
$$4a(y - y') = -(x - x')^2,$$
where x', y' are as in § 9·25, and $4a$ (the *latus rectum*) is found from (45) to be equal to $(2V^2/g)\cos^2 \alpha$. This is the equation of a parabola, concave downwards, with vertex at O' and axis $(O'N)$ vertical.

(ii) It also follows that $2g(y' + a) = V^2 \sin^2 \alpha + V^2 \cos^2 \alpha = V^2$. Hence the speed at any point O of the path is equal to the speed that would be acquired by another particle falling from rest from A to O, where A is vertically above O and is on the *directrix* of the parabola, i.e. the line perpendicular to the axis $O'N$ and distant a above O', as shown in Fig. 81.

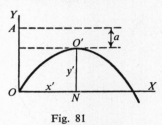

Fig. 81

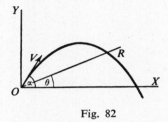

Fig. 82

9·3. Estimation of range.

9·31. Range on inclined plane.

Let OR be the range on an inclined plane through O of inclination θ (θ may be positive or negative), of a particle projected from O with velocity given by V, α (Fig. 82).

OR is determined by substituting $x = OR \cos \theta$, $y = OR \sin \theta$ into (45), yielding
$$OR \sin \theta = OR \cos \theta \tan \alpha - \tfrac{1}{2}gV^{-2}OR^2 \cos^2 \theta \sec^2 \alpha;$$
and hence $OR = 0$ (which is irrelevant), or
$$OR = (2V^2/g)(\cos \theta \tan \alpha - \sin \theta)\sec^2 \theta \cos^2 \alpha$$
$$= \frac{2V^2 \cos \alpha}{g \cos^2 \theta}(\cos \theta \sin \alpha - \sin \theta \cos \alpha)$$
$$= \frac{2V^2 \cos \alpha \sin(\alpha - \theta)}{g \cos^2 \theta}.$$

9·32. Maximum range, given V, θ.

By elementary trigonometry, it follows that
$$OR = \frac{V^2\{\sin(2\alpha - \theta) - \sin \theta\}}{g \cos^2 \theta}.$$

Hence if we keep V, θ fixed and allow the initial direction α to vary, OR is a maximum when $\sin(2\alpha - \theta)$ is a maximum; i.e. when

$2\alpha - \theta = \frac{1}{2}\pi$; i.e. when $\alpha = \theta + \frac{1}{2}(\frac{1}{2}\pi - \theta)$; i.e. when the initial direction bisects $\angle ROY$.

The maximum range, OR_0 say, in these circumstances is then given by

$$OR_0 = \frac{V^2(1 - \sin\theta)}{g\cos^2\theta} = \frac{V^2}{g(1 + \sin\theta)}.$$

9·33. Range on horizontal plane. In the particular case where $\theta = 0$, we have

$$OR = \frac{2V^2 \cos\alpha \sin\alpha}{g}; \quad OR_0 = \frac{V^2}{g};$$

and the initial direction for the maximum range is inclined at $\pi/4$ to the horizontal.

These results are immediate deductions from §§ 9·31, 9·32. Alternatively, they are very simply derived on putting $y = 0$ in (45).

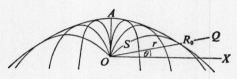

Fig. 83

9·4. The enveloping parabola. Consider particles projected (in an assigned vertical plane) from the fixed point O with an assigned speed V, but with different initial directions, resulting in a *family* of parabolic paths.

For any given θ, there will be a corresponding path S to give the maximum range OR_0 on a plane of inclination θ, as indicated in Fig. 83. By § 9·32, the locus of R_0 as θ varies is given by

$$r = \frac{V^2}{g(1 + \sin\theta)}, \tag{46}$$

where we have put $OR_0 = r$.

(46) is the polar equation of a parabola* whose vertex is at A (Fig. 83), where A (which also corresponds to the 'A' in Fig. 81) is the highest point of that particular path whose initial direction is vertically upwards.

This parabola, called the *enveloping parabola*, divides the given vertical plane into an inner region, all points of which can be reached by a

---

* In standard notation of analytical geometry, (46) would be replaced by $r = V^2/\{g(1 + \cos\theta)\}$, where θ is here the complement of the θ in (46).

152 PARABOLIC MOTION [9.4—

projectile from O with the given initial speed; and an outer region, no points of which can be so reached. This property follows from § 9·32; for if Q is any point in OR_0 produced, Q could not be reached since OR_0 is the maximum range in the direction θ; on the other hand, any point between O and R_0 can clearly be reached.

The cartesian equation of the enveloping parabola referred to axes OX (horizontal) and OY (vertically upwards) is

$$x^2 + 2V^2 y/g - V^4/g^2 = 0.$$

For (46) gives

$$V^2/g - r \sin \theta = r,$$

i.e.

$$V^2/g - y = \sqrt{(x^2 + y^2)};$$

on squaring, we get the result.

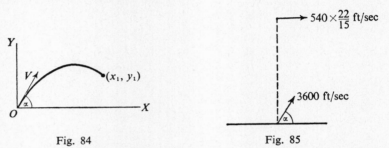

Fig. 84   Fig. 85

9·5. Examples. (1) *Find the direction in which a projectile must be aimed from a given point O with given initial speed V in order to pass through a given point (x_1, y_1), referred to horizontal and vertical axes OX, OY through O.*

Let α, as in Fig. 84, give the required direction.
Starting from (44), we find as in § 9·2 that the equation of the path is

$$y = x \tan \alpha - \tfrac{1}{2} g x^2 V^{-2} \sec^2 \alpha.$$

Since (x_1, y_1) is to be on this path, we require

$$y_1 = x_1 \tan \alpha - \tfrac{1}{2} g x_1^2 V^{-2} \sec^2 \alpha.$$

In this equation, α is the only unknown; rearranging the equation as a quadratic in $\tan \alpha$, we have

$$\tfrac{1}{2} g x_1^2 V^{-2} \tan^2 \alpha - x_1 \tan \alpha + y_1 + \tfrac{1}{2} g x_1^2 V^{-2} = 0,$$

whence

$$\tan \alpha = \frac{V^2}{g x_1} \pm \frac{1}{x_1} \sqrt{\left\{ \frac{V^4}{g^2} - \frac{2V^2}{g} y_1 - x_1^2 \right\}}.$$

Provided the expression under the square root sign is positive, this equation gives two real values of α, both possible answers.

(2) *A shot is fired from the ground at an aircraft, flying at 540 m.p.h., at the instant when the aircraft is vertically overhead. The initial speed of the shot is 3600 ft/sec. Find the direction in which the shot must be aimed. If the height of the aircraft is 2 miles, find the time that elapses before it is struck.*

Let α as in Fig. 85 give the required direction.

It is sufficient for the shot to hit the aircraft that the horizontal component of the velocity of the shot should be equal to the speed of the aircraft.

Thus we require $3600 \cos \alpha = 540 \times 22/15$, yielding $\alpha = 77°\cdot 3$. Let t sec be the time that elapses before the aircraft is struck.

Considering the motion of the shot, resolved vertically, we have
$$2 \times 5280 = 3600 \sin \alpha \cdot t - \tfrac{1}{2}gt^2,$$
where $\alpha = 77°\cdot 3$.

This gives $t = 3\cdot 1$ or 216.

The earlier time, $3\cdot 1$ sec, is relevant; at this instant the shot is still rising. (At the later time the shot would be falling and would strike the aircraft again—if perchance the first impact should have caused no interference in the flight of either aircraft or shot.)

(3) *A certain cricketer is capable of catching a ball with equal ease at any height from the ground between y_1 and y_2, where $y_1 > y_2$. Show that for a hit which gives a ball a (horizontal) range R and greatest height h, the fieldsman should estimate his position on the field in the plane of the ball's flight within a distance $\tfrac{1}{2}R\{\sqrt{(1 - y_2/h)} - \sqrt{(1 - y_1/h)}\}$.*
Evaluate when $R = 100$ yards, $h = 75$ ft, $y_1 = 6$ ft, $y_2 = 1$ ft.

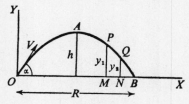

Fig. 86

Let OAB in Fig. 86 be the path of the ball.

The fieldsman should be between M and N when catching the ball, where M, N are as shown.

Starting from (44), we derive the equation of the path OAB in the usual form
$$y = x \tan \alpha - \tfrac{1}{2}gx^2 V^{-2} \sec^2 \alpha. \qquad \text{(i)}$$

The coordinates of the points A, P, Q, B,—i.e. $(\tfrac{1}{2}R, h)$, (x_1, y_1), (x_2, y_2), $(R, 0)$, respectively, where $OM = x_1$, $ON = x_2$,—satisfy the equation (i).

Hence, writing b for $\tan \alpha$ and c for $\tfrac{1}{2}gV^{-2} \sec^2 \alpha$, we have
$$h = \tfrac{1}{2}Rb - \tfrac{1}{4}R^2 c, \qquad \text{(ii)}$$
$$y_1 = x_1 b - x_1^2 c, \qquad \text{(iii)}$$
$$y_2 = x_2 b - x_2^2 c, \qquad \text{(iv)}$$
$$0 = Rb - R^2 c. \qquad \text{(v)}$$

By (v), since $R \neq 0$, we have $b = Rc$, and hence, by (ii), $h = \tfrac{1}{4}R^2c$.
Hence the required distance MN

$$= x_2 - x_1$$
$$= \frac{b + \sqrt{(b^2 - 4cy_2)}}{2c} - \frac{b + \sqrt{(b^2 - 4cy_1)}}{2c}$$

(by (iv), (iii); the positive value of the square root is relevant in each case since in this problem each of x_1, x_2 exceeds $\tfrac{1}{2}R$)

$$= \sqrt{\left(\frac{b^2}{4c^2} - \frac{y_2}{c}\right)} - \sqrt{\left(\frac{b^2}{4c^2} - \frac{y_1}{c}\right)}$$
$$= \sqrt{(\tfrac{1}{4}R^2 - y_2 R^2/4h)} - \sqrt{(\tfrac{1}{4}R^2 - y_1 R^2/4h)},$$

which gives the answer.

In the numerical case, we have $R = 300$, $h = 75$, $y_1 = 6$, $y_2 = 1$; this gives MN approximately equal to 5 ft.

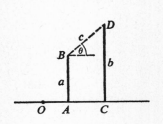

Fig. 87

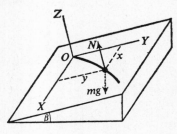

Fig. 88

(4) *Two parallel vertical walls are of heights a, b, and the distance between their top edges is c. A stone projected from the ground in a vertical plane at right angles to the walls is required to clear both walls. Prove that the least speed of projection needed is $\sqrt{\{g(a + b + c)\}}$.*

Let AB, CD, represent the walls, as in Fig. 87.
Let V_O be the required least speed of projection from the ground.
The corresponding speed V_B at B is then, by (32), given by
$$V_O^2 = V_B^2 + 2ag. \qquad (i)$$
Since $2ag$ is constant, V_B is the least possible speed at B.
Since V_B is the least possible speed at B, the particle will just clear the wall CD at D, and BD will be the maximum range in the direction BD.
Hence if θ is the inclination of BD to the horizontal, we require by (46)
$$BD = \frac{V_B^2}{g(1 + \sin \theta)};$$
i.e. $$V_B^2 = g.BD(1 + \sin \theta)$$
$$= gc\left(1 + \frac{b - a}{c}\right).$$

Hence, by (i), $V_O^2 = g(c + b - a + 2a)$, which gives the answer.

(5) *A particle is in motion, not necessarily along a line of greatest slope, on a smooth fixed inclined plane of inclination β. The particle is subject to no forces except its weight and the reaction exerted by the surface. Investigate the motion.*

Take axes OX horizontal in the given inclined plane, OY upward along a line of greatest slope, and OZ at right angles to the plane.

The forces on the particle in any position are mg, N, as indicated in Fig. 88.

Resolving along OX, OY, OZ, we have

$$m\ddot{x} = 0, \qquad m\ddot{y} = -mg\sin\beta, \qquad 0 = N - mg\cos\beta.$$

The first pair of these would be the same as (44) if we replaced g in (44) by $g\sin\beta$.

It follows that the path of the particle is a parabola. If V is the speed at O and α the angle between OX and the velocity at O, the equation of the new parabola is given by replacing g by $g\sin\beta$ in (45).

The third equation gives N if desired.

S.P.I.C.E. (α) It will be noticed that, except in Ex. (4), arguments were based directly on the equations of motion (44). When (45) is needed to solve a problem, it should be derived from (44), rather than quoted from memory.

(β) Ex. (1) gives an alternative means of finding the cartesian equation of the enveloping parabola. From the answer to Ex. (1), it follows that the point (x_1, y_1) can or cannot be reached in the given circumstances according as $x_1^2 + 2V^2 y_1/g - V^4/g^2$ is less than or greater than zero. Thus the equation $x_1^2 + 2V^2 y_1/g - V^4/g^2 = 0$ gives those points (x_1, y_1) which can be just reached.

(γ) Ex. (2) could have been solved using the result in Ex. (1). But it will be noticed how easily it was solved by simply resolving horizontally.

(δ) In practical problems such as Ex. (2), we have no choice at this stage to do other than adopt the simplifications referred to in § 9·1(i). In the subject of Ballistics, the motion of projectiles through the air is studied in much greater detail, and with recourse to a much wider range of experimental results than we have appealed to.

(ε) Ex. (3) is a slightly more difficult problem. It will be noticed nevertheless that it was not found necessary to quote the formula for range.

(ζ) In Ex. (4), good use was made of the polar equation of the enveloping parabola. This example could also have been solved in a number of other ways.

(η) Ex. (5) is a simple three-dimensional problem. The student should discern from Ex. (5) the more general result that if a particle is subject to any force which is constant in magnitude and direction and to no other forces, then it describes a parabola. (The parabola of course degenerates to a straight line if the initial velocity of the particle is in the direction of the force.)

EXAMPLES IX

1. A man can just throw a ball 49 yards vertically upwards. How far can be throw it on a horizontal plane? [Answer: 98 yards.]

2. The path of a projectile has its end points on a horizontal plane. Prove that the number of feet in the greatest altitude reached is given approximately by four times the square of the number of seconds in the time of flight.

3. A shell on striking the (horizontal) ground bursts and scatters its fragments all with speed V. Find the area within which a man runs the risk of being hit by a fragment. [Answer: $\pi V^4/g^2$.]

4. A bombing aircraft flying horizontally 1000 ft above sea-level at 400 knots is overtaking a vessel steaming in a straight line at 30 knots. At what distance astern of the vessel should a bomb be released to hit the vessel?
[Answer: 1650 yards.]

5. Two heavy bodies are projected from the same point, at the same instant, in the same direction, with different velocities. Prove that the line joining them remains fixed in direction.

6. A projectile leaves a point O on a horizontal plane, and after $1\frac{1}{2}$ sec just clears a vertical wall whose height above the plane is 24 ft and which is 12 ft distant from O. The wall is perpendicular to the plane of motion. Find: (a) where the particle strikes the horizontal plane; (b) the direction of motion at O; (c) the direction of motion at the point where the wall is cleared.
[Answers: 8 ft beyond the wall; $\tan^{-1} 5$ above, and $\frac{1}{4}\pi$ below the horizontal.]

7. A gun is fired from a point 3000 ft above sea-level. The maximum height reached by the shell is 4000 ft above sea-level, and the point at which it hits the sea is at an angle of depression $\tan^{-1}\frac{1}{4}$ from the gun. Find the angle of elevation of the gun, assumed fixed. [Answer: $26°\cdot 6$.]

8. A slope on a hill is a plane of inclination $30°$. From a point of the slope two stones are projected, one up and one down the hill, the initial speed being the same and the inclination being $45°$ above the horizontal in both cases. Prove that one of the ranges is nearly $3\frac{3}{4}$ times that of the other.

9. Two marksmen, using similar rifles and cartridges, fire simultaneously, one horizontally from the top A of a tower of height h and one from a point B directly in front of the base. The shots simultaneously hit a target which is in the vertical plane through A and B, is at the same level as B and is distant c from the tower. Prove that the distance of the second marksman from the tower is $c - \sqrt{(c^2 - h^2)}$.

10. An aircraft is flying with a constant velocity of magnitude U at constant height h. A gun is fired point blank at the aircraft after it has passed directly over the gun and when its angle of elevation as viewed from the gun is α. Prove that the shell will hit the aircraft provided $2(V\cos\alpha - U)U\tan^2\alpha = gh$, where V is the initial speed of the shell.

11. A particle is projected at an angle α to the horizontal up an inclined plane of inclination β. Prove that the particle will strike the plane at right angles if $\tan\alpha = \cot\beta + 2\tan\beta$.

12. A number of particles are projected from the same point with the same initial speed V. Prove that at any subsequent time t, they lie on the surface of a sphere of radius Vt.

*13. A projectile, projected from a fixed point O with given speed V, will hit a target B if it leaves O at an angle of elevation α. If a small error is made whereby the angle of elevation is increased by θ degrees, prove that the projectile will pass at a height $\pi b \theta \sec^2 \alpha \{1 - (gb/V^2) \tan \alpha\}/180$, approximately, over the target, where b is the horizontal component of the distance OB.

14. A gun is firing from sea-level out to sea. It is then mounted in a battery at height h above sea-level and fired with the same initial speed V and inclination α. Prove that its range is increased approximately by the fraction $gh/2V^2 \sin^2 \alpha$. (It is assumed that gh is small compared with $V^2 \sin^2 \alpha$.)

15. A particle P, hanging by a light inelastic string of length 21 inches from a fixed point O, is projected horizontally from its lowest equilibrium position with a speed of 14 ft/sec. Find the height of P above the horizontal through O: (a) when the string slackens; (b) when the height of P is greatest.

[Answers: (a) $10\tfrac{1}{2}$ in; (b) 14·4 in.]

*16. A battleship is steaming ahead with speed U. A gun is mounted on the battleship so as to point straight backwards, and is set at an inclination α. If V is the speed of projection of a shell relative to the gun, prove that the range is $(2V/g)(V\cos\alpha - U)\sin\alpha$. Also, show that the angle of elevation for maximum range is $\cos^{-1}\{U + \sqrt{(U^2 + 8V^2)}/4V\}$.

17. Prove that the maximum range on a horizontal plane of projectiles fired from a point O at a height h above the plane is $2\{c(c+h)\}^{\frac{1}{2}}$, where the speed of projection is that due to a fall through a vertical distance c, and the range is measured from the point of the plane vertically below O.

18. Find the highest point that a projectile with initial speed V can reach on a vertical wall distant c from the point of projection. [Answer: $(V^4 - g^2c^2)/2V^2g$.]

*19. A particle is projected from the highest point of a hemispherical mound of radius a. Prove that it cannot clear the mound unless its initial speed exceeds $\sqrt{(\tfrac{1}{2}ag)}$.

*20. A particle is projected from a point A so as to pass through a point B. C is the point vertically below B and at the same level as A. Prove that the least speed of projection is $\sqrt{\{g(AB + BC)\}}$.

CHAPTER X

INTERLUDE

"It may be right to go ahead, I guess
It may be right to stop, I guess;
Also, it may be right to retrogress."
—Lewis Carroll.

10·1. In Chapter VI, the theory given in Chapters II and III was generalised to the case of the three-dimensional motion of a particle, the topics concerned being those listed in § 5·1, (i) to (v).

The additional notions of angular velocity and angular acceleration of a moving point about a given point were introduced in the early sections of Chapter VII. It should be noted that with these notions, only the case of coplanar motion was considered.

(In the three-dimensional treatment, angular velocity has the character of a vector—see Appendix, § 17·2. But it is not necessary to emphasise the vector aspect in the case of motion in one plane, since the angular velocity vector is always perpendicular to this plane.)

The later sections (§§ 7·2-7·6) of Chapter VII (on circular motion), and the whole of Chapters VIII (on simple harmonic motion) and IX (on projectile motion) were concerned with *particular* problems and applications of the theory contained in Chapter VI and in § 7·1. The topics selected are ones of much importance in practice; moreover, the details on circular motion and simple harmonic motion are important in connection with theory in later chapters.

10·2. The formulae (24), (25), (26a, b, c), (27), (29), (30), (31), (33) of Chapter VI are the generalised forms of (1), (2), (8a, b, c), (9), (11), (12), (15), (7), respectively, of Chapters II and III. The formulae (14a), (14b) of Chapter III were shown in Chapter VI to be still valid in three dimensions. The special energy equation (32) for the case in which the only work done is that of constant gravity, is remembered through its resemblance to (6). Equation (34) on relative acceleration is analogous to (33) on relative velocity. This, with the addition of the simple friction formula (28), accounts for all the formulae in Chapter VI which need to be remembered.

In Chapter VII, (35) and (36) are analogous to (1) and (3) of Chapter II. The formulae (37), (38), (39) give the key to problems on the circular motion of a particle; and the equations (40) are immediate deductions from (39).

In problems on simple harmonic motion (Chapter VIII), the crucial formulae are (41a) and (43). The formulae (41b, c) and (42) may be immediately derived from (41a). In regard to other formulae on the theory of simple harmonic motion, it is normally better to remember how to derive them rapidly from (41a) than to try to remember the formulae themselves. The formula for the period of a simple pendulum should be remembered, and possibly that for the period of a vibrating particle attached to a vertical elastic string as in § 8·8.

In problems on projectile motion, it is best to work from the equations of motion (44), and to be able to derive other equations such as (45) from (44). In problems where a maximum range (or a minimum speed of projection) is involved, the equation (46) of the enveloping parabola is often useful.

With the help of the above remarks, the reader is now advised to prepare a summary of the theory of the dynamics of a single particle.

10·3. The three following chapters, XI to XIII, will be concerned with more complicated systems than single particles.

Chapter XI is concerned with two key characteristics of the distribution of mass in such systems, namely 'centre of mass' and 'moment of inertia'. The account of centre of mass given is the generalisation to three dimensions of the account in § 4·11. The theory in Chapter XI is needed as a preliminary to the dynamical theory in Chapters XII and XIII. In Chapter XI, the centres of mass and moments of inertia of a number of bodies of common simple shapes are determined; these results are needed in solving particular problems in later chapters.

In Chapter XII, the theory of Chapter IV on systems of particles is generalised to the case of coplanar motion. The further notion of angular momentum with attending theory also needs to be introduced. In some sections of Chapter XII, the results hold generally in three dimensions, but the main purpose of Chapter XII is to give the basic theory needed for treating two-dimensional dynamical problems.

The theory of Chapter XII is applied in this book to two types of dynamical systems: (a) systems consisting of a small number of particles; (b) rigid bodies. Systems of the type (a) are considered towards the end of Chapter XII. The type (b) are the subject matter of Chapter XIII;

and the student should give special attention to mastering the *theory* of Chapter XII before proceeding to Chapter XIII.

10·4. It is part of the plan of this book to cater for students who may wish to proceed as soon as possible to the chapters on Statics and Hydrostatics without reading at this stage all the dynamical theory of Chapters XII and XIII.

As a preliminary to Chapter XV on Statics, such students should now read: §§ 11·1-11·5 on centres of mass; the opening paragraphs of Chapter XII on the term 'two-dimensional'; §§ 12·1-12·23 (but § 12·21 may be omitted) on localised vectors and sets of forces in two dimensions; the definition of the term 'rigid body' in § 13·1; and § 13·3 on pin-joints. A further remark special to the needs of this class of students will be made in § 15·1A.

As a preliminary to § 16·314 in Chapter XVI on Hydrostatics, §§ 11·6-11·(11) on moments of inertia also need to be read.

10·5. The following are some miscellaneous examples on the topics of Chapter VI to IX. They are mostly a little more difficult than Exs. VI-IX.

EXAMPLES X

1. Write amended statements where needed in place of the following:

(*a*) If ω is the angular velocity and α the angular acceleration of a point when θ is the angular coordinate, and if ω_0 is the initial value of ω, then $\omega^2 = \omega_0^2 + 2\alpha\theta$.

(*b*) The acceleration of a point describing a circle with uniform velocity is constant.

(*c*) If a point is moving with speed v in a circle of radius a, its acceleration is v^2/a towards the centre.

(*d*) The acceleration of a particle describing a parabola under constant gravity is constant.

(*e*) A particle of mass m kg is joined by a light taut string to a fixed centre O, and is moving with uniform speed v m/s in a circle of radius a m and centre O. If the string were suddenly cut, the particle would start to move under the centrifugal force mv^2/a N in a direction directly away from O.

[(*d*) alone requires no amendment.]

2. *ABC* is a triangle and *P* is any point of *BC*. If **PQ** represents the resultant of forces represented by **AP**, **PB**, **PC**, prove that *Q* is fixed independently of the position of *P*.

3. **AB** is a fixed chord of a given circle C, and **AP** a variable chord. The resultant of forces represented by **AB** and **AP** is represented by **AR**. Prove that the locus of R is a circle of radius equal to the radius of C.

Deduce a construction for finding the position of P so that AR is a maximum.

[Answer: The required position is on the line parallel to AB which passes through the point at which the second circle, of centre O say, is cut by AO produced.]

*4. If **r** is the position vector of a point P, t is a scalar variable, and **a** and **b** are constant vectors, show that the equation $\mathbf{r} = \mathbf{a} + \mathbf{b}t$ represents a straight line, and write down the corresponding cartesian equations.

[Answer: $(x - a_1)/l_1 = (y - a_2)/l_2 = (z - a_3)/l_3$,
where $\mathbf{r} = (x, y, z)$, $\mathbf{a} = (a_1, a_2, a_3)$, $\mathbf{b} = (l_1, l_1, l_3)$.]

5. AB, AC are two fixed inclined planes sloping upwards from A, AB being above AC; AB is rough, the angle of friction being equal to the angle BAC, while AC is smooth. Prove that if particles move from rest at B, C, respectively, they will arrive at A (a) in the same time if BC is perpendicular to AC, (b) with the same speed if BC is perpendicular to AB.

6. A cyclist, whose mass together with that of his bicycle is 175 lb, works always at the same rate while riding, and rides at speeds of 30 ft/sec on the level, and 20 ft/sec up an incline of 1 in 100. The resistance to his motion may be assumed constant. Find: (a) his steady speed down an incline of 1 in 200; (b) the resistance.

[Answers: 40 ft/sec; 3·5 lb.wt.]

7. Prove that no force, however large, which has a component vertically downwards, can move a body which is lying on a fixed rough horizontal plane unless its line of action is inclined to the vertical by an angle exceeding the angle of friction.

8. Two trains A, B are moving with equal constant speeds on two straight railway lines, respectively, which meet at right angles. At a certain instant, A is at the junction and B is moving towards the junction. Prove that the trains are nearest to each other when they are equally distant from the junction.

9. Points P and Q describe circles of the same centre O and radii $2a$ and a, with speeds V and $2V$ respectively, in the same sense of rotation. Prove that when the relative velocity of P to Q is parallel to PQ the angle POQ is $\tan^{-1} \frac{3}{4}$, and that the relative speed is $3V/\sqrt{5}$.

10. A cricket ball reaches a batsman with a horizontal velocity of 70 ft/sec (assumed to be parallel to the length of the pitch); after the stroke the ball has a horizontal velocity of 80 ft/sec in a direction inclined at 30° to the pitch, so as to be approximately towards mid-on. The ball's mass is $5\frac{1}{2}$ oz, and contact with the bat lasts 1/40 second. Find the force, assumed constant, of the bat on the ball during the contact.

[Answer: 62 lb.wt., inclined at about 16° to the pitch.]

11. Two particles P and Q are simultaneously released from rest at the same point A. P slides down a fixed smooth inclined plane, and Q falls freely under gravity. Prove that, relative to P, Q moves in a straight line perpendicular to the inclined plane.

*12. On a smooth fixed inclined plane of inclination α, there is placed a smooth wedge of mass M and angle α in such a way that the upper face of the wedge is horizontal. A particle of mass m is placed on this horizontal face, and the system is initially at rest. Prove that, in the subsequent motion, the particle moves in a straight line with acceleration $(M+m)g \sin^2 \alpha/(M+m \sin^2 \alpha)$.

*13. A smooth prism of equilateral cross section and mass M is placed with one rectangular face on a fixed smooth table. It carries a small smooth pulley fixed to its top edge; and two masses m_1, m_2, where $m_1 > m_2$, are attached to the ends of a light inelastic string which passes over the pulley and is always in a vertical plane at right angles to the top edge of the wedge. If f is the acceleration of the wedge when the system is free to move, and f_1 the acceleration of m_1 relative to the wedge, prove that
$$Mf + (m_1 + m_2)(f - \tfrac{1}{2}f_1) = 0,$$
$$(m_1 - m_2)g\sqrt{3}/2 = (m_1 + m_2)(f_1 - \tfrac{1}{2}f).$$
Hence find the horizontal component of the acceleration of the highest point of the string.
$$\left[\text{Answer:} \quad \frac{\sqrt{3}(m_1 - m_2)(2M + m_1 + m_2)g}{(m_1 + m_2)(4M + 3m_1 + 3m_2)}.\right]$$
Also show that the force on the table is less than $(M + m_1 + m_2)g$.

14. A railway gauge, of width b, is taken round a curve of radius r. It is required that a train travelling round the curve at speed v_1 shall exert the same lateral force on the inner rail as a train travelling at speed v_2 (where $v_2 > v_1$) exerts on the outer rail. Prove that the elevation of the outer above the inner rail should be $b(v_1^2 + v_2^2)/2gr$, approximately.
Evaluate for the case: $b = 4$ ft $8\tfrac{1}{2}$ in; v_1, $v_2 = 30$, 60 m.p.h.; $r = \tfrac{1}{4}$ mile. [Answer: 3·2 in.]

15. A train is moving round a circular curve of radius 1000 yards at a steady speed of 60 m.p.h. A light inelastic string is attached to the roof of a carriage and has a mass attached at its other end. At what steady inclination to the vertical will the string hang? [Answer: 4°·6.]

16. A particle of weight W, attached to a fixed point by a light inelastic string, describes a circle in a vertical plane. The tension of the string when the particle is at the highest point of the orbit is kW, and when at the lowest point it is lW. Prove that $l = k + 6$.

17. A light string ABC passes through a small smooth heavy ring B, and has its ends fixed at two points A, C, the point C being vertically below A. If the ring is whirled round so as to describe a horizontal circle with constant angular velocity Ω, prove that
$$2(c + a)gb = \Omega^2\{(c + a)^2 - b^2\}(c - a),$$
where BC, CA, $AB = a$, b, c, respectively.

18. A thin smooth tube, bent to form a circle of radius 9 ft, is fixed in a vertical plane. Two elastic particles, for which $e = 0·5$, of masses $2m$, m, are simultaneously released from rest inside the tube at opposite ends of the horizontal diameter. Prove that after the second impact between the particles, the heavier will move from the lowest point of the tube through an arc subtending an angle of $\cos^{-1} \tfrac{3}{4}$ at the centre of the circle.

19. A particle moving in a straight line with simple harmonic motion has speeds v_1, v_2, when its displacements from the centre are x_1, x_2, respectively. Prove that the period of the motion is $2\pi\sqrt{\{(x_1^2-x_2^2)/(v_2^2-v_1^2)\}}$.

20. A horizontal shelf moves vertically in simple harmonic motion in which the amplitude is a. A particle is placed on the shelf when the shelf is in its highest position. The particle and shelf are seen immediately to separate, but rejoin at the instant when the shelf next reaches its lowest position. Prove that the speed of the shelf when passing through its mean position is $\tfrac{1}{2}\pi\sqrt{(ag)}$.

21. Solve the parts (b), (c), (d) of Ex. (2) of § 3·94, independently of the principle of energy.

22. A gun has a range R on a horizontal plane. If the gun is to send a shot over a ridge of height h above the level of the gun, prove that the horizontal distance of the gun from the ridge must be less than $\sqrt{\{R(R-2h)\}}$.

23. Particles of mud are thrown tangentially from a car travelling at 30 m.p.h from points of the back tyres where the tangent makes an angle of 45° or less with the road. The diameter of the tyres is 2 ft. The lowest point of the wind-screen of a following car also travelling at 30 m.p.h. is 4 ft above the road level. Prove that if the wind-screen is splashed it is within about 56 ft of the leading car.

*24. If $\mathbf{r}$ is the position vector of a point P, t is a scalar variable, and $\mathbf{a}$, $\mathbf{b}$, $\mathbf{c}$ are constant vectors all in one plane, show that the equation $\mathbf{r} = \mathbf{a} + \mathbf{b}t + \mathbf{c}t^2$ represents a parabola.

(Hint:—Let $\mathbf{r} = x\mathbf{i} + y\mathbf{j}$, $\mathbf{a} = a_1\mathbf{i} + a_2\mathbf{j}$, etc.; take resolved parts of the given equation, and then eliminate t.)

Use this result to prove that the path of a particle under constant gravity (or other constant field of force) is a parabola.

*25. Taking the equation of motion of a projectile in the vector form $\ddot{\mathbf{r}} = \mathbf{g}$, derive the equation $v_1^2 = v_2^2 + 2gh$ (as referred to in § 9·1).
(Hint:—First derive the form $2\mathbf{v}\cdot\dot{\mathbf{v}} = 2\mathbf{g}\cdot\mathbf{v}$.)

26. OR is the horizontal range of a projectile projected from O. The line joining R to any position P of the projectile in its flight meets the vertical through O in Q. Prove that Q ascends with a constant velocity of magnitude equal to the initial vertical component of the projectile's velocity.

27. From the rim of a wheel of radius a, which is rotating about a fixed horizontal axis with constant angular velocity Ω, particles fly off tangentially with speeds $a\Omega$. For a particle flying off at a point where the radius is at an angle θ behind the upward vertical radius, find the greatest height above the centre of the wheel that is attained. Find for what value of θ this height is a maximum; and show that when θ has this value, the highest point attained is vertically above the centre of the wheel.

[Answers: $a\cos\theta + (a^2\Omega^2/2g)\sin^2\theta$; $\cos^{-1}(g/a\Omega^2)$.]

*28. A boat is rowed across a river flowing uniformly at speed V. The speed of the boat relative to the water is also V, and the boat is always steered towards a fixed point S on the bank. Prove that the path of the boat is a parabola with S as focus, and that the landing point will be the vertex of the parabola.

*29. A shell of mass m is fired from a gun of mass M placed on a smooth fixed horizontal plane, and the barrel is inclined at a fixed angle β to the horizontal. Prove that if the speed of the shot just after leaving the gun is v, the range will be

$$\frac{2v^2(1 + \mu) \tan \beta}{g\{1 + (1 + \mu)^2 \tan^2 \beta\}},$$

where $\mu = m/M$.

(Hint:—First show that, just after leaving the gun, the inclination α of the velocity of the shot to the horizontal is given by $\tan \alpha = (1 + \mu) \tan \beta$.)

*30. Prove that if the distance of a projectile from the point of projection is always increasing, the initial inclination of the velocity to the upward vertical must be greater than $\sin^{-1} \frac{1}{3}$.

*31. A particle moves in a semicircular path under a force perpendicular to the bounding diameter. Prove that the force is inversely proportional to the cube of the distance from the particle to the diameter.

CHAPTER XI

CENTRES OF MASS AND MOMENTS OF INERTIA

"One should be concerned not merely with the weight of one's body, but with how this weight is distributed".
—*Australian Women's Weekly.*

11·1. Centre of mass of a system of particles.

11·11. Position of centre of mass. We now generalise the definition of centre of mass given in § 4·11. At any time t, let the particles of a system, of masses $m_1, m_2, \ldots$, be at points (Fig. 89) whose position vectors referred to any origin O are $\mathbf{r}_1, \mathbf{r}_2, \ldots$, respectively. Then the position of the centre of mass, G say, is defined to be given by the position vector $\bar{\mathbf{r}}$, where

$$\bar{\mathbf{r}} = \frac{m_1\mathbf{r}_1 + m_2\mathbf{r}_2 + \ldots}{m_1 + m_2 + \ldots} = \frac{\Sigma(m\mathbf{r})}{\Sigma m}, \text{ say.} \qquad (47a)$$

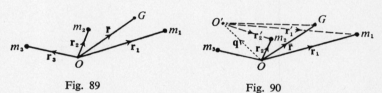

Fig. 89   Fig. 90

It is necessary (cf. § 4·11) to check that this definition gives a point whose position is independent of the particular origin O. Let G' be the position of the centre of mass given by the formula (47a) when a different origin O' is taken. Then

$$\mathbf{O'G'} = (m_1\mathbf{r}_1' + m_2\mathbf{r}_2' + \ldots)/(m_1 + m_2 + \ldots).$$

But (see Fig. 90) $\mathbf{r}_1' = \mathbf{O'O} + \mathbf{r}_1, \quad \mathbf{r}_2' = \mathbf{O'O} + \mathbf{r}_2, \ldots$

Hence $\mathbf{O'G'} = \mathbf{O'O} + (m_1\mathbf{r}_1 + m_2\mathbf{r}_2 + \ldots)/(m_1 + m_2 + \ldots)$
$= \mathbf{O'O} + \mathbf{OG} = \mathbf{O'G}.$

Hence G and G' are one and the same point.

11·12. Velocity and acceleration of centre of mass. Let O be fixed. By (24), (25), § 6·2, the velocity $\bar{\mathbf{v}}$ and acceleration $\bar{\mathbf{f}}$ of G are found, by differentiating (47a), to be (in obvious notation) given by

$$\bar{\mathbf{v}} = \frac{\Sigma(m\mathbf{v})}{\Sigma m}; \qquad \bar{\mathbf{f}} = \frac{\Sigma(m\mathbf{f})}{\Sigma m}. \qquad (48)$$

11·121. Example. *Two objects P, Q, of masses 2, 1 lb, are released from rest at the same instant. P slides from a point A at the ridge of a house down the sloping roof to a point B on the eaves. Q falls vertically from B to C on the ground and reaches C at the instant when P reaches B. Find the locus of the centre of mass of P and Q.*

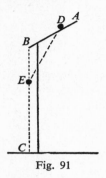

Fig. 91

The accelerations of P, Q are both constant in magnitude and direction (see § 6·51 if necessary).

Hence, by the second of (48), the acceleration of the C.M. is constant in direction; (the magnitude is not required).

Initially the C.M. is at D in AB, where $AD = \frac{1}{3}AB$, and at rest; finally it is at E in BC, where $BE = \frac{1}{3}BC$.

Hence the required locus must be the straight line DE.

S.P.I.C.E. This example exhibits the power of vector methods. By using vectors, we solved it without having to make any algebraical calculations.

11·13. Theorem. In finding the centre of mass of a system of particles we may, in using (47a), replace any group of particles by a single 'equivalent particle' whose mass is equal to the total mass of the group and whose position is at the centre of mass of the group.

To prove this theorem, let m_1', m_2', ..., m_k', be the masses of the particles of the group; m_s the total mass of these particles; $\mathbf{r}_1'$, etc., their position vectors (referred to O); and $\mathbf{r}_s$ the position vector of their centre of mass. Then, by (47a), $\mathbf{r}_s = (m_1'\mathbf{r}_1' + \ldots + m_k'\mathbf{r}_k')/m_s$. Again by (47a), the centre of mass of the whole system is given by

$$\bar{\mathbf{r}} = \frac{m_1\mathbf{r}_1 + m_2\mathbf{r}_2 + \ldots + (m_1'\mathbf{r}_1' + \ldots + m_k'\mathbf{r}_k') + \ldots}{m_1 + m_2 + \ldots + (m_1' + \ldots + m_k') + \ldots}$$

$$= \frac{m_1\mathbf{r}_1 + m_2\mathbf{r}_2 + \ldots + m_s\mathbf{r}_s + \ldots}{m_1 + m_2 + \ldots + m_s + \ldots},$$

which proves the theorem.

11·14. Cartesian formulae. By taking resolved parts of (47a), we see that the coordinates $\bar{x}$, $\bar{y}$, $\bar{z}$ of the centre of mass of a system of particles of masses $m_1, m_2, \ldots$, at the points (x_1, y_1, z_1), (x_2, y_2, z_2), ..., (referred to axes OX, OY, OZ), are given by

$$\bar{x} = \frac{\Sigma(mx)}{\Sigma m}, \quad \bar{y} = \frac{\Sigma(my)}{\Sigma m}, \quad \bar{z} = \frac{\Sigma(mz)}{\Sigma m}. \tag{47b}$$

Analogous formulae give components of the velocity and acceleration of the centre of mass.

11·141. For the particular case of a pair of particles (in the plane OXY say), we have

$$\bar{x} = (m_1 x_1 + m_2 x_2)/(m_1 + m_2),$$
$$\bar{y} = (m_1 y_1 + m_2 y_2)/(m_1 + m_2).$$

Thus the centre of mass divides the join of the positions of the two particles inversely in the ratio of their masses.

11·142.
Because of an analogy with formulae for taking moments in the case of parallel forces (see § 12·15), $\Sigma(mx)$, $\Sigma(my)$ are sometimes in practice called 'first moments'. Cf. 'second moments' (§ 11·6).

11·2. Continuously distributed mass-systems. Let V be the volume of a region containing a system of particles, and let δM be the total mass of the particles inside any element δV of V. Then if the particles are so closely packed that the limit as $\delta V \to 0$ of $\delta M/\delta V$ is finite at every point of V, we say that V is occupied by a continuously distributed mass-system.

There are logical questions connected with extending the theory for systems consisting of finite numbers of particles ('discrete systems') to cases of continuously distributed mass-systems. We shall here assume, without going into detail, that such postulates are laid down as will validate the application to continuously distributed mass-systems of all theorems proved in this book for general discrete systems of particles.

The above limit, i.e. dM/dV, gives the *density* (volume-density), ρ say, at a point. The dimensions of density are $[ML^{-3}]$, and the m.k.s., c.g.s. and f.p.s. units are the kg/m³, the g/cm³, and the lb/ft³.

We shall meet bodies we call *laminas* in which the mass is continuously distributed in a plane; for example, a circular disc will be usually assumed to be 'thin' and treated as a lamina.

(The terms 'thin' and 'thick' in relation to thickness are analogous to the terms 'light' and 'heavy' in relation to mass—see § 3·19 (iii).)

For a lamina, we refer to the *surface-density*, or mass per unit area, σ say, of dimensions $[ML^{-2}]$, given by $\sigma = dM/dS$, where S denotes surface area.

We shall usually treat rods, wires and chains (or 'heavy' strings) as having their mass continuously distributed in a line. We then refer to the *line-density*, or mass per unit length, λ say, of dimensions $[ML^{-1}]$, given by $\lambda = dM/ds$, where s denotes the relevant arc-length.

For a three-dimensional continuous body, we have, in place of (47b),

$$\bar{x} = \lim \frac{\Sigma(x \delta M)}{\Sigma(\delta M)}$$

$$= \lim \frac{\Sigma(\rho x \delta V)}{\Sigma(\rho \delta V)};$$

i.e.,
$$\bar{x} = \frac{\int \rho x \, dV}{\int \rho \, dV}, \qquad (47c)$$

and similar expressions for $\bar{y}$, $\bar{z}$, where ρ is the density at a typical point (x, y, z) inside V.

In the cases of laminas and rods, we have analogous formulae with ρ replaced by σ, λ respectively and dV by dS, ds respectively. In all cases, the denominators in formulae like (47c) are equal to the total mass of the body concerned.

In using (47c), it is permissible, by § 11·13, to take as typical element of volume one which has one or two of its three dimensions finite (i.e. not small)—see e.g. §§ 11·32, 11·352. It is important to note that in such cases the x in the right-hand side of (47c) is a coordinate *of the centre of mass of the element taken*.

If the body is *uniform*, i.e. if the density is constant, *but not otherwise*, ρ may be taken outside the integral sign in both numerator and denominator of (47c) and then cancelled.

11·21. Centroid. The *centroid* of a *volume* is defined so as to coincide with the centre of mass of a material of uniform density occupying the volume; the coordinates of the centroid are given by formulae such as (47c), with ρ replaced by unity. Analogous remarks apply to the centroid of an *area* or *arc*.

(The term *centre of gravity* will be referred to in § 12·23.)

11·22. Example. *The distance of the centre of mass of a rod AB from the end B is one-eighth of the length of the rod. The line-density is known to be proportional to a certain power of the distance from A. Find the power in question.*

Let x be the distance from A of any point P of the rod; $\bar{x}$ the distance from A of the C.M., G; and l the length of the rod.

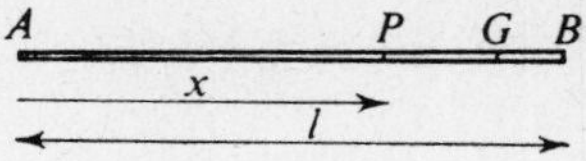

Fig. 92

The line-density at P, λ say, $= \beta x^n$ where β is constant and n is the required power.

By a formula of the type (47c), we then have

$$\tfrac{7}{8}l = \bar{x} = \frac{\int_0^l \lambda x\,dx}{\int_0^l \lambda\,dx}$$

$$= \frac{\int_0^l \beta x^{n+1}\,dx}{\int_0^l \beta x^n\,dx}$$

$$= \frac{\beta l^{n+2}/(n+2)}{\beta l^{n+1}/(n+1)}$$

$$= \frac{n+1}{n+2}\,l.$$

Hence $n = 6$.

S.P.I.C.E. In most of the cases to follow, the density is constant and could be cancelled in formulae such as (47c). The above example is included to emphasise that when the density is not constant it would be invalid to cancel it in this way.

11·3. Centres of mass in particular cases.

11·31. Uniform symmetrical bodies. The centre of mass of a uniform rod is at its centre, O say. For the rod may be divided up into pairs of elements, symmetrically situated about O; the centre of mass of each pair is at O; the result then follows by the grouping theorem of § 11·13.

It follows similarly that the centre of mass of any uniform symmetrical body, e.g. a uniform circular or elliptical ring or disc, uniform rectangular plate or prism, uniform sphere or ellipsoid, is at the centre of symmetry.

11·32. Uniform triangular lamina. Let D be the midpoint of the side BC of the lamina ABC, and take axes AX, AY as shown in Fig. 93 (page 170).

Let $AP = x$, $PQ = \delta x$, and let RPS, TQV be parallel to BDC.

The matter enclosed between RS and TV may, to sufficient accuracy, be treated as rod whose centre of mass is at P, i.e. at the point $(x, 0)$. The corresponding element of mass is proportional to $RS.PQ$, i.e. to $AP.PQ$, i.e. to $x\delta x$.

Hence, the lamina being uniform, the coordinates $(\bar{x}, \bar{y})$ of its centre of mass are given by

$$\bar{x} = \frac{\int_0^{AD} x.dx}{\int_0^{AD} dx} = \frac{\frac{1}{3}AD^3}{\frac{1}{2}AD^2} = \tfrac{2}{3}AD,$$

$$\bar{y} = \frac{\int_0^{AD} 0.dx}{\int_0^{AD} dx} = 0.$$

Thus the centre of mass G is on AD and such that $AG = \tfrac{2}{3}AD$.

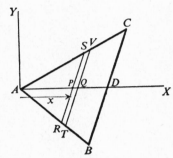

Fig. 93

11·321.
We could, of course, have obtained the result of § 11·32 more rapidly using the property of concurrency of the medians of a triangle. We have given the above proof in order to demonstrate the use of analytical methods. We have incidentally in effect proved that the medians are concurrent.

11·322.
It may be checked that the centres of mass of the following systems coincide: (a) a uniform triangular lamina ABC; (b) a set of three particles of equal mass at A, B, C; (c) a set of three particles of equal mass at the midpoints of BC, CA, AB.

11·33. Uniform pyramid or cone of matter. By taking an origin O at the vertex and the x-axis along OH, where H is the centroid of the base, it follows by the method used in § 11·32 that the centre of mass G of a uniform pyramid or cone is on OH and such that $OG = \tfrac{3}{4}OH$. The only modifications needed to the proof in § 11·32 arise from taking as element a 'lamina' parallel to the base, instead of a 'rod' as in § 11·32. The mass of the element is then proportional to $x^2 \delta x$, in place of $x \delta x$. The result $\bar{x} = \tfrac{3}{4}OH$ is now quickly obtained.

11·331.
For the particular case of a tetrahedron, we have incidentally shown in effect that the four lines of the type OH (§ 11·33) are concurrent.

11·34. Uniform wire in shape of circular arc. Let a be the radius of the arc and 2α the angle it subtends at the centre of curvature O. Take axes as shown in Fig. 94. Let P, Q be two neighbouring points of the arc such that $\angle XOP = \theta$, $\angle POQ = \delta\theta$, arc $PQ = \delta s$. Let λ be the line-density.

By symmetry, $\bar{y} = 0$.

The mass of the element PQ is $\lambda \delta s$. Hence $\bar{x} = \int x \lambda \, ds / \int \lambda \, ds$, where x is the abscissa of P. Putting $x = a \cos \theta$, $ds = a d\theta$, we have, λ being constant,

$$\bar{x} = \frac{\int_{-\alpha}^{\alpha} a^2 \cos \theta \, d\theta}{\int_{-\alpha}^{\alpha} a \, d\theta} ;$$

and, hence, on carrying out the integrations,

$$\bar{x} = \frac{a \sin \alpha}{\alpha} = \frac{\text{radius} \times \text{chord } AB}{\text{arc } AB}.$$

11·341. In particular, if the arc is a semicircle, $\bar{x} = 2a/\pi$.

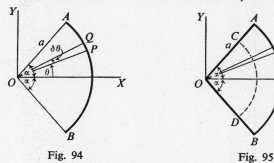

Fig. 94   Fig. 95

11·35. Uniform lamina in shape of sector of circle. Let a be the radius of the sector, and 2α the angle it subtends at the centre of curvature O.

Divide the sector up into equal elementary sectors of which OPQ in Fig. 95 is typical. The centres of mass of all these elements lie (to sufficient accuracy) on the dotted arc CD of radius $\tfrac{2}{3}a$, and their masses are equal. Hence the centre of mass of the sector AOB coincides with that of a uniform wire along the arc CD.

Hence the centre of mass of the sector AOB (referred to OX, OY of Fig. 95) is given by

$$\bar{x} = \frac{2a \sin \alpha}{3\alpha} ; \quad \bar{y} = 0.$$

11·351. In particular, for a uniform semicircular lamina, $\bar{x} = 4a/3\pi$.

11·352. Uniform semicircular lamina. We now obtain the result in § 11·351 independently of § 11·35.

Take axes as in Fig. 96, and take as element the 'rod' $PRSQ$, where $OP = x$, $PQ = \delta x$, and PR, QS are parallel to OY. Let σ be the surface-density.

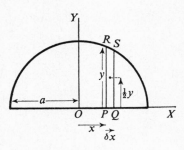

Fig. 96

To sufficient accuracy, the mass of the element is $\sigma y \delta x$, and the coordinates of its centre of mass are $(x, \tfrac{1}{2}y)$, where (x, y) are the coordinates of R and so satisfy the equation $x^2 + y^2 = a^2$.

Hence

$$\bar{y} = \frac{\int_{-a}^{a} \sigma y dx \cdot \tfrac{1}{2}y}{\tfrac{1}{2}\pi a^2 \sigma};$$

(in this expression the significance of the '$\tfrac{1}{2}$' in the numerator should be noted—see the penultimate paragraph of § 11·2; in the denominator we have simply set down an expression for the mass of the lamina).

Hence, σ being constant, we have

$$\bar{y} = \{\int_{-a}^{a} (a^2 - x^2)dx\}/\pi a^2,$$

which gives

$$\bar{y} = 4a/3\pi.$$

By symmetry, $\bar{x}$ is zero.

11·36. Uniform distribution of mass over surface of a sphere. Take axes as in Fig. 97, the origin O being the centre of the sphere. Let P, Q be two neighbouring points on OX such that $OP = x$, $PQ = \delta x$; and consider the element of area on the surface of the sphere cut off between the planes RS, TU which pass through P, Q, respectively, and are at right angles to OX.

To sufficient accuracy, this element of surface area $= 2\pi y.\delta s$, where $y = PS$, $\delta s = SU$ (Fig. 97). But $y.\delta s = a \sin \theta.\delta s \approx a\delta x$, where a is the radius of the sphere. Hence the element of area $\approx 2\pi a \delta x =$ the area cut off by the two planes on the surface of an enveloping cylinder of axis OX and radius a.

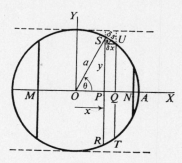

Fig. 97

Consider now a uniform mass-distribution over the surface of this enveloping cylinder, the surface-density being the same as that of the given distribution of the sphere. It follows that the elements of mass cut off on the surfaces of the sphere and cylinder between the planes RS, TU are equal. Also the centres of mass of these elements coincide.

Now take any two planes at right angles to OX, cutting OX in points M, N, which are a finite distance apart. It then follows from the last paragraph and the grouping theorem of § 11·13, that the centre of mass of the zone cut off between these planes on the surface of the sphere is the same as for the corresponding zone of the enveloping cylinder, i.e. is the midpoint of MN.

11·361. Uniform hemispherical shell.

By 'shell' is meant a continuous distribution of mass over a surface, usually curved. In the case of a hemispherical shell we do not (in the absence of special mention) include the diametral plane which would make the shell 'closed'.)

Let O be the centre of curvature, and A the point in which the axis of symmetry cuts the shell. It follows from § 11·36 that the centre of mass of the shell is the midpoint of OA.

11·37. Uniform solid hemisphere.
Let O be the centre of curvature and a the radius, and let OX be taken along the axis of symmetry. Divide the hemisphere up into elements approximating to hemispherical shells, a typical element being of radius x and thickness δx, as sketched in Fig. 98.

To sufficient accuracy, the mass of the typical element is $2\pi x^2 \delta x . \rho$, where ρ is the (volume-) density; and, by § 11·361, its centre of mass is on OX, distant $\frac{1}{2}x$ from O.

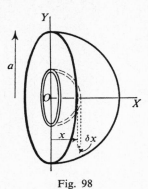

Fig. 98

The centre of mass of the solid hemisphere is therefore given by

$$\bar{x} = \frac{\int_0^a 2\pi x^2 dx . \rho . \tfrac{1}{2}x}{\text{Mass of hemisphere}}$$

$$= \frac{\tfrac{1}{4}\pi a^4 \rho}{\tfrac{2}{3}\pi a^3 \rho}$$

$$= \tfrac{3}{8}a.$$

11·4. Theorems of Pappus. Suppose a curve is revolved through 2π radians about an axis which is in the plane of the curve but does not

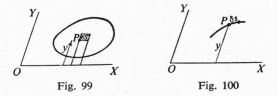

Fig. 99   Fig. 100

cut the curve (though it may meet it—see § 11·41). The theorems of Pappus state:

(i) Volume generated = Area enclosed by curve × Length of path of centroid of area.

(ii) Surface area generated = Length of arc of curve × Length of path of centroid of arc.

In (i) the curve must of course be closed; in (ii), this need not be the case.

To prove (i), let OX be the axis about which the curve revolves, and take OY perpendicular to OX in the initial plane of the curve. Divide the area, S say, enclosed by the curve into elements such as δS, indicated in Fig. 99, y being the ordinate of a point of S. The volume generated
$$= \int 2\pi y dS, \text{ taken over the area } S,$$
$$= 2\pi \int y dS$$
$$= 2\pi \bar{y} S,$$
where $\bar{y}$ is the ordinate of the centroid of the area S. This proves (i).

To prove (ii), we proceed analogously, using Fig. 100, and show that the surface area generated $= 2\pi \bar{y} s$, where $\bar{y}$ is here the ordinate of the centroid of the arc, and s is the arc-length.

11·41. We illustrate the use of the first theorem by finding again the centre of mass of a uniform semicircular lamina.

Let a semicircular area of radius a be revolved through 2π radians about its diameter, and so generate a solid sphere. Then, by § 11·4 (i), $\frac{4}{3}\pi a^3 = \frac{1}{2}\pi a^2 . 2\pi \bar{y}$, where $\bar{y}$ is the distance of the centroid of the area from the centre of curvature. This gives $\bar{y} = 4a/3\pi$, in agreement with §§ 11·351, 11·352.

Similarly, using (ii), the formula $2a/\pi$ for the distance of the centre of mass of a uniform semicircular wire from the centre of curvature may be derived.

11·42.

The theorems of Pappus may conversely be used in finding volumes and surfaces of revolution. The student may be interested to show in this way that the volume of an anchor ring formed by revolving a circle of centre C and radius a about a line in the plane of the circle distant d from C (where $d > a$) is $2\pi^2 a^2 d$; also that the surface area of the anchor ring is $4\pi^2 ad$.

11·5. Examples on centre of mass. In §§ 11·3, 11·4, we have shown how to determine the centres of mass of many of the bodies to be met in later dynamical and statical problems. The centres of mass of bodies of other shapes may be determined by the methods indicated in these sections. We give two further examples.

(1) *A uniform wire is in the shape of a regular hexagon, of side-length a, with one side removed. Find the position of the centre of mass of the wire.* Takes axes as in Fig. 101, AF being the missing side.

Consider the complete hexagon in which the missing side has been restored. This hexagon,
$$ABCDEFA \text{ of mass } 6k, \text{ and C.M. at } O,$$
is made up of
$$ABCDEF \text{ of mass } 5k, \text{ and C.M. at } G \text{ say, and}$$
$$AF \quad\quad \text{ of mass } k, \text{ and C.M. at } H \text{ (Fig. 101)},$$
where k is a proportionality constant. (We have in Fig. 101 put G to the right of O, but expect the ensuing algebraic equation to deliver a negative value of OG.)

Hence $\quad\quad 0 =$ abscissa of C.M. of complete hexagon
$$= \frac{5k.OG + k.OH}{6k},$$
which gives $OG = -OH/5 = -\sqrt{a}/10$.

Hence G is on OX, distant $a\sqrt{3}/10$ to the left of O.

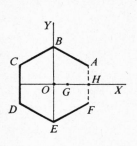
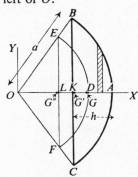

Fig. 101   Fig. 102

(2) *Find the centre of mass of a uniform segment, of thickness h, of a sphere of radius a.*

First method.

Let O be the centre of curvature, and OA the axis of symmetry of the segment.

We shall refer to solids and surfaces of revolution about OA by letters in the plane section shown in Fig. 102. We shall refer e.g. to the given spherical segment as ABC.

Let $OBAC$ be the (spherical) sector corresponding to the given segment. The centroid of $OBAC$ coincides (cf. § 11·35) with the centroid of the spherical surface EDF, where $OD = \frac{3}{4}a$, and is therefore at G', where $G'D = \frac{1}{2}LD = \frac{1}{2}(\frac{3}{4}h)$; thus $OG' = \frac{3}{4}a - \frac{3}{8}h$.

The volume V' of the sector is given by
$$\frac{V'}{\frac{4}{3}\pi a^3} = \frac{S'}{4\pi a^2},$$
where $S' =$ area of the spherical surface $BAC = 2\pi ah$ (by theory given in § 11·36). Hence $V' = \frac{2}{3}\pi a^2 h$.

The volume V'' of the cone OBC
$$= \tfrac{1}{3}\pi KB^2 . OK = \tfrac{1}{3}\pi\{a^2 - (a-h)^2\}(a-h)$$
$$= \tfrac{1}{3}\pi h(a-h)(2a-h).$$

The centroid G'' of the cone is given by $OG'' = \tfrac{3}{4}(a-h)$.

Let G be the centroid of the segment ABC. Then, regarding the sector as made up of the segment and the cone, we have

$$\tfrac{3}{4}a - \tfrac{3}{8}h = \frac{V''\tfrac{3}{4}(a-h) + (V'-V'')OG}{V'}$$

$$= \frac{\tfrac{1}{4}\pi h(a-h)^2(2a-h) + \tfrac{1}{3}\pi h^2(3a-h)OG}{\tfrac{2}{3}\pi a^2 h},$$

which gives
$$OG = \frac{3(2a-h)^2}{4(3a-h)}.$$

Second method.

Taking axes OX, OY as in Fig. 102, and considering the revolution about OX of the shaded element of area, we see that the volume V of the segment of the sphere

$$= \int_{a-h}^{a} \pi(a^2 - x^2)dx.$$

$$= \pi\left[a^2x - \tfrac{1}{3}x^3\right]_{a-h}^{a},$$

which reduces to $\tfrac{1}{3}\pi h^2(3a-h)$.

Further, the expression $\int x dV$ for this volume

$$= \int_{a-h}^{a} \pi(a^2-x^2)x\, dx$$

$$= \pi\left[\tfrac{1}{2}a^2x^2 - \tfrac{1}{4}x^4\right]_{a-h}^{a},$$

which reduces to $\tfrac{1}{4}\pi h^2(2a-h)^2$.

Hence
$$OG = \frac{\int x dV}{V} = \frac{3(2a-h)^2}{4(3a-h)}.$$

S.P.I.C.E. (α) In Ex. (1), the handling of signs should be noted. In harder problems, it is not always obvious whether the centre of mass is on the one or the other side of an origin.

(β) Ex. (1) could be solved thus: The given piece
$ABCDEF$ of mass $5k$, and C.M. at G
may be regarded as made up of
$ABCDEFA$ of mass $6k$, and C.M. at O,
together with AF of mass $-k$, and C.M. at H.

Hence
$$OG = \frac{6k.0 + (-k).OH}{5k} = -\tfrac{1}{5}OH;\text{ etc.}$$

A similar process could have been applied in solving Ex. (2) by the first method.

G

(γ) It is interesting to compare the methods of solving Ex. (2) with and without the direct use of the calculus.

(δ) We actually found the position of the centroid of the volume in Ex. (2), but this is also the position of the C.M. of the mass since the density is stated to be uniform.

11·6. Moment of inertia of a system of particles about a given line. Let r be the (perpendicular) distance of a typical particle, of mass m, of a system of particles from a given line. The moment of inertia I (sometimes called the 'second moment'—cf. § 11·142) of the system about the line (or *axis*) is defined by

$$I = \Sigma(mr^2). \tag{49a}$$

The radius of gyration k of the system about the line is defined by $k^2 = I/M$, where $M = \Sigma m$, the total mass. Thus

$$I = Mk^2. \tag{49b}$$

For continuously distributed mass-systems, we have formulae of the type

$$I = \int r^2 dM, \qquad k^2 = \int r^2 dM / \int dM; \tag{49c}$$

in these formulae we may put dM equal to ρdV, σdS or λds, according as we have a volume-, surface-, or line-distribution of mass, respectively.

The dimensions of moment of inertia are [ML^2].

Corresponding to the use of the term 'centroid', we can have second moments of volume, area or arc about a line, defined by formulae of the form $\int r^2 dV$, $\int r^2 dS$, $\int r^2 ds$; the dimensions of these second moments are [L^5], [L^4], [L^3], respectively.

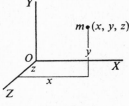

Fig. 103

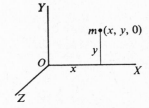

Fig. 104

11·7. Theorems on moments of inertia.

11·71. Parallel axes theorem. Let I be the moment of inertia of a system of particles about any axis, I_G the moment of inertia about the parallel axis through the centre of mass G of the system, M the total mass, and A the distance between the axes. Then

$$I = I_G + MA^2. \tag{50a}$$

To prove this theorem, take a reference frame $OXYZ$ in which OZ coincides with the given axis. Let (x, y, z) be the coordinates of a typical particle of mass m and $(\bar{x}, \bar{y}, \bar{z})$ the coordinates of G, referred to $OXYZ$. Let (x', y', z') be the coordinates of the typical particle referred to a frame with origin at G and axes parallel to OX, OY, OZ.

Then we have (Fig. 103)
$$\begin{aligned} I &= \Sigma[m(x^2 + y^2)] \\ &= \Sigma[m\{(\bar{x} + x')^2 + (\bar{y} + y')^2\}] \\ &= \Sigma\{m(\bar{x}^2 + \bar{y}^2)\} + \Sigma(2mx'\bar{x} + 2my'\bar{y}) + \Sigma\{m(x'^2 + y'^2)\} \\ &= (\Sigma m)(\bar{x}^2 + \bar{y}^2) + 2\{\Sigma(mx')\}\bar{x} + 2\{\Sigma(my')\}\bar{y} + I_G, \end{aligned}$$
since $\bar{x}, \bar{y}$ (but not m, x', y') keep unchanging values through the summations involved.

Now $\Sigma m = M$, and $\bar{x}^2 + \bar{y}^2 = A^2$. Also, by (47b), $\Sigma(mx')$, $\Sigma(my')$ are M times the coordinates of G referred to the frame with G as origin, and are therefore zero.

Hence $\qquad\qquad I = I_G + MA^2$.

11·711. In obvious notation, the corresponding result for radii of gyration, obtained by dividing (50a) through by M, is
$$k^2 = k_G{}^2 + A^2. \tag{50b}$$

11·712.
If I_1, I_2 are the moments of inertia about any two parallel axes, and A_1, A_2 are the respective distances of these axes from the parallel axis through G, we deduce that
$$I_1 - I_2 = M(A_1{}^2 - A_2{}^2).$$
It should be noted that $A_1{}^2 - A_2{}^2$ is not in general equal to the square of the distances between the two axes.

11·72. Perpendicular axes theorem. For a system of particles *in the plane OXY*,
$$I_z = I_x + I_y, \tag{51}$$
where I_x, I_y, I_z are the moments of inertia about OX, OY, OZ, respectively (OX, OY, OZ being as usual mutually orthogonal).

The proof of this theorem is (see Fig. 104) simply:
$$I_z = \Sigma\{m(x^2 + y^2)\} = \Sigma(my^2) + \Sigma(mx^2) = I_x + I_y.$$
The italicised restriction attending this theorem should be noted; the parallel axes theorem is not thus restricted.

11·73.
The above two theorems have been proved for the case of discrete systems of particles. On the basis of the remarks in § 11·2, the theorems hold also for continuous mass-systems, the perpendicular axes theorem being restricted to laminas (or other coplanar distributions).

11·8. Moments of inertia in particular cases.

11·81. Uniform rod, length 2a. (*a*) *About axis through centre perpendicular to rod.* The moment of inertia I about OY (Fig. 105) is, in usual notation, given by

$$I = \int_{-a}^{a} x^2(\lambda dx),$$

where λ is constant. Hence $I = \tfrac{2}{3}\lambda a^3$.

It is usual to express I in the form (49*b*). Since $M = 2a\lambda$, we have

$$I = \tfrac{1}{3}Ma^2; \qquad (k^2 = \tfrac{1}{3}a^2).$$

(*b*) *About axis through an end perpendicular to the rod.* By the parallel axes theorem, we now have

$$I = \tfrac{1}{3}Ma^2 + Ma^2 = \tfrac{4}{3}Ma^2; \qquad (k^2 = \tfrac{4}{3}a^2).$$

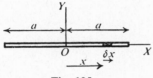
Fig. 105

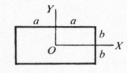
Fig. 106

11·82. Uniform rectangular lamina, side-lengths 2a, 2b. (*a*) *About axis through centre parallel to a side.* Take axes OX, OY as shown in Fig. 106. The formal proof of § 11·81 (*a*) still avails, the mass per unit length parallel to OX being equal to the constant $2b\sigma$, where σ is the surface-density of the lamina; hence $I_y = \tfrac{1}{3}Ma^2$. Similarly, $I_x = \tfrac{1}{3}Mb^2$.

(It similarly follows that the moment of inertia of a uniform parallelogram lamina about a line through its centre parallel to two of the sides, distant 2*a* apart, is $\tfrac{1}{3}Ma^2$. This result is used in Chapter XVI.)

(*b*) *About axis through centre perpendicular to lamina.* By the perpendicular axes theorem, we have $\qquad I_z = I_x + I_y = \tfrac{1}{3}M(a^2 + b^2)$.

11·83. Uniform rectangular block, side-lengths 2a, 2b, 2c. It follows from § 11·82 (*b*) that the moment of inertia about an axis through the centre parallel to the sides of length 2*c* is $\tfrac{1}{3}M(a^2 + b^2)$; and similarly in other cases.

11·84. Thin uniform ring, radius a. (*a*) *About axis through centre perpendicular to plane of ring.* We have $I = \int r^2 dM$, where $r = a$. Hence $I = \int a^2 dM = a^2 \int dM = Ma^2$.

(b) *About a diameter as axis.* By the perpendicular axes theorem, taking OZ to be perpendicular to the plane of the ring, we have $Ma^2 = I_z = I_x + I_y$, where each of I_x, I_y is equal to the required moment of inertia, which is therefore $\frac{1}{2}Ma^2$.

11·85. Uniform disc, radius a. (a) *About axis through centre perpendicular to plane of disc.* Dividing the disc up into elements approxi-

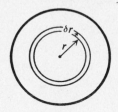

Fig. 107

mating to 'thin rings', of which a typical one is shown in Fig. 107, we have, using § 11·84 (a),

$$I = \int_0^a 2\pi r dr . \sigma . r^2 = \tfrac{1}{2}\pi a^4 \sigma.$$

Also $M = \pi a^2 \sigma$. Hence $I = \tfrac{1}{2}Ma^2$.

(b) *About a diameter as axis.* We deduce, after the manner in § 11·84 (b), that $I = \tfrac{1}{4}Ma^2$.

11·86. Uniform spherical shell, radius a. *About a diameter as axis.* Consider the element of mass cut off between the two planes RS, TU of § 11·36 (Fig. 97). The mass of this element was seen to be $2\pi a \delta x . \sigma$, where σ is the surface-density of the shell. This element may be treated as a 'thin ring' of radius y, where $x^2 + y^2 = a^2$; hence, by § 11·84 (a), its moment of inertia about OX

$$\approx 2\pi a \sigma \delta x . y^2 = 2\pi a \sigma (a^2 - x^2)\delta x.$$

Hence the moment of inertia I of the whole shell about OX

$$= 2\pi a \sigma \int_{-a}^a (a^2 - x^2)dx, \text{ which reduces to } \tfrac{8}{3}\pi a^4 \sigma.$$

But the mass M of the shell $= 4\pi a^2 \sigma$.

Hence $I = \tfrac{2}{3}Ma^2$.

11·87. Uniform solid sphere, radius a. *About a diameter as axis.* Divide the sphere up into elements approximating to thin discs by planes at right angles to the given diameter; and consider the particular element indicated in Fig. 108, bounded by planes RS, TU, distant δx apart.

The mass of the element $\approx \pi y^2 \delta x \cdot \rho$, where ρ is the density of the sphere. By § 11·85 (a), its moment of inertia about OX

$$= \tfrac{1}{2}\pi y^2 \rho \delta x \cdot y^2 = \tfrac{1}{2}\pi(a^2 - x^2)^2 \rho \delta x.$$

Hence the moment of inertia I of the sphere about OX

$$= \tfrac{1}{2}\pi \rho \int_{-a}^{a} (a^4 - 2a^2 x^2 + x^4)dx,$$

which reduces to $\tfrac{8}{15}\pi a^5 \rho$. The mass M of the sphere $= \tfrac{4}{3}\pi a^3 \rho$. Hence $I = \tfrac{2}{5}Ma^2$.

The student should as an exercise deduce this result from the result in § 11·86.

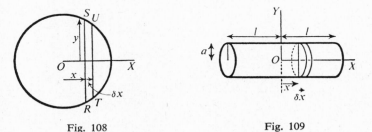

Fig. 108   Fig. 109

11·88. Uniform (right circular) solid cylinder, radius a, length $2l$.
(a) *About the cylinder's axis.* A little consideration (cf. § 11·82 (a)) shows that this result must be the same as for a uniform disc of radius a about its axis; i.e., $I = \tfrac{1}{2}Ma^2$.

(b) *About line through centre of cylinder, perpendicular to cylinder's axis.* Take a disc-like element as indicated in Fig. 109. The mass of this element $= \pi a^2 \delta x \cdot \rho$, and its moment of inertia about its diameter is $(\pi a^2 \rho \delta x)\tfrac{1}{4}a^2$. Its moment of inertia about OY is, by the theorem of parallel axes, equal to $(\pi a^2 \rho \delta x)(\tfrac{1}{4}a^2 + x^2)$.

Hence the moment of inertia I of the cylinder about OY

$$= \int_{-l}^{l} \pi a^2 \rho (\tfrac{1}{4}a^2 + x^2)dx$$

$$= 2\pi a^2 \rho (\tfrac{1}{4}a^2 l + \tfrac{1}{3}l^3).$$

The mass M of the cylinder $= 2\pi a^2 l \rho$. Hence $I = M(\tfrac{1}{4}a^2 + \tfrac{1}{3}l^2)$.

11·89. Uniform (right circular) cylindrical shell, radius a, length $2l$.
By argument similar to that in § 11·88, we find the moments of inertia corresponding to (a), (b) of § 11·88 to be Ma^2, $M(\tfrac{1}{2}a^2 + \tfrac{1}{3}l^2)$, respectively.

11·9. Routh's rule.

We now state (without reference to a theoretical background) a rule which enables certain of the results in §§ 11·81 to 11·87, and some other results, to be immediately written down.

Let B be a body which (i) is of uniform density, (ii) is symmetrical with respect to three mutually orthogonal axes OX, OY, OZ, (iii) has no hollows; ((iii) rules out rings and shells). Let a, b, c be the semi-axes of B parallel to OX, OY, OZ; (in some cases, e.g. lamina and rod, one or two of a, b, c may be zero). Then the radius of gyration of B about one of the axes, OZ say, is given by

$$k_z^2 = \frac{a^2 + b^2}{3, 4 \text{ or } 5}, \qquad (52)$$

where the denominator is 3, 4 or 5 according as the shape of the body is rectangular, elliptical (including circular), or ellipsoidal (including spherical), respectively.

The student should check that this rule gives the main results previously found in §§ 11·81–11·87, excepting the ring and spherical shell and the case of § 11·81 (b). The excepted formulae are, moreover, easily remembered when the others are determined by the rule.

The rule incidentally gives also the radii of gyration about the axes of symmetry of a uniform elliptical disc or uniform ellipsoid.

11·(10). Examples on moments of inertia.
In § 11·8, we have shown how to determine the moments of inertia, about certain axes, of many of the bodies that will be involved in later problems. Other moments of inertia may be deduced from the results in § 11·8, or determined by similar methods. We add seven further examples.

(1) *Find the moment of inertia of a uniform solid hemisphere of mass M and radius a about a diameter of the circular boundary.*

Fig. 110

Consider the hemisphere as part of a complete sphere of the same (uniform) density.

Let I be the required moment of inertia of the hemisphere about the diameter AB (Fig. 110).

Then $2I$ is the moment of inertia of the whole sphere about AB. But the moment of inertia of this sphere about AB is $\frac{2}{5}(2M)a^2$;
thus $\qquad 2I = \frac{2}{5}(2M)a^2$.
Hence $\qquad I = \frac{2}{5}Ma^2$.

(2) *Find the moment of inertia of a uniform square lamina of mass M and side-length 2a about a diagonal.*

Let the diagonals PR, QS intersect in O (Fig. 111).
Take axes OX, OY, OZ along OP, OQ, and perpendicular to $PQRS$.
Then, in the notation of § 11·72, $I_z = I_x + I_y$.
But $I_z = \tfrac{2}{3}Ma^2$, and $I_x = I_y$.
Hence the moment of inertia about a diagonal $= I_x = \tfrac{1}{3}Ma^2$.

Fig. 111   Fig. 112

(3) *Find the radius of gyration, about its axis, of a wheel assumed to consist of a mass M uniformly distributed over the circumference of a circle of radius a, and six thin uniform spokes, each of mass m and length a, radiating from the axis.* (Fig. 112.)

The moment of inertia of the rim (about the axis) $= Ma^2$.
The moment of inertia of each spoke $= \tfrac{4}{3}m(\tfrac{1}{2}a)^2 = \tfrac{1}{3}ma^2$.
Hence the moment of inertia of the wheel $= Ma^2 + 2ma^2$.

Hence the required radius of gyration of the wheel $= a\sqrt{\left(\dfrac{M+2m}{M+6m}\right)}$.

(4) *A system of particles is in the plane OXY. OK is a line in this plane making the angle θ with OX. Prove that the moment of inertia of the system about OK is of the form $A \cos^2 \theta - 2H \cos \theta \sin \theta + B \sin^2 \theta$, where $A = \Sigma(my^2)$, $B = \Sigma(mx^2)$, $H = \Sigma(mxy)$.*

(*A*, *B* are the moments of inertia about OX, OY; *H* is called the *product of inertia* with respect to OX, OY.)

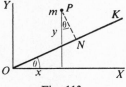

Fig. 113

Let P, of coordinates (x, y), be the position of a typical particle of mass m, and let PN be perpendicular to OK (Fig. 113).
The required moment of inertia $= \Sigma(m.PN^2)$
$= \Sigma\{m(y \cos \theta - x \sin \theta)^2\}$
$= \Sigma(my^2) \cos^2 \theta - 2\Sigma(mxy) \cos \theta \sin \theta + \Sigma(mx^2) \sin^2 \theta$,
$= A \cos^2 \theta - 2H \cos \theta \sin \theta + B \sin^2 \theta$.

(5) *Find the moment of inertia of a uniform rectangular lamina of mass M and side-lengths 2a, 2b, about a diagonal.*

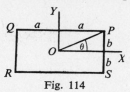

Fig. 114

Let O be the centroid of the rectangle $PQRS$, and take axes OX, OY as in Fig. 114. Let $\angle XOP = \theta$.

In the notation of Ex. (4), $A = \frac{1}{3}Mb^2$, $B = \frac{1}{3}Ma^2$.
Also $H_1 = \Sigma(mxy)$, is zero, by symmetry.

Hence, by Ex. (4), the required moment of inertia
$$= \tfrac{1}{3}Mb^2 \cos^2\theta + \tfrac{1}{3}Ma^2 \sin^2\theta$$
$$= \tfrac{2}{3}Ma^2b^2/(a^2+b^2).$$

(6) *The mass of a uniform solid cone is M, its height is h, and the radius of its base is a. Find its moment of inertia about a line through its vertex perpendicular to its axis of symmetry.*

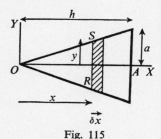

Fig. 115

Take axes as in Fig. 115, OA being the axis of the cone.

Divide the cone up into disc-like elements perpendicular to OA, a typical element corresponding to the shaded area in Fig. 115.

To sufficient accuracy, the moment of inertia of this element about RS is $\tfrac{1}{4}\delta M.y^2$, where δM is its mass, and $y/a = x/h$. Its moment of inertia about OY is, by the parallel axes theorem, $\delta M(\tfrac{1}{4}y^2 + x^2)$.

Also, $\delta M = \pi y^2 \delta x.\rho$, where ρ is the density of the cone.

Hence the moment of inertia I of the cone about OY
$$= \int_0^h \pi\rho y^2 (\tfrac{1}{4}y^2 + x^2)\,dx$$
$$= \frac{\pi a^2 \rho}{h^4} \int_0^h (\tfrac{1}{4}a^2 + h^2)x^4\,dx$$
$$= \tfrac{1}{20}\pi a^2 h \rho (a^2 + 4h^2).$$

But the mass M of the cone $= \tfrac{1}{3}\pi a^2 h \rho$. Hence $I = \tfrac{3}{20}M(a^2 + 4h^2)$.

(7) *Find the radius of gyration of the rod in the Ex. of § 11·22 about a line through A perpendicular to the rod.*
Let k be the required radius of gyration.
Using Fig. 92 and the notation of § 11·22, we then have

$$k^2 = \frac{\int_0^l \lambda x^2 dx}{\int_0^l \lambda dx}$$

$$= \frac{\int_0^l \beta x^{n+2} dx}{\int_0^l \beta x^n dx}$$

$$= \frac{n+1}{n+3} l^2.$$

Since $n = 6$ (§ 11·22), $k = l\sqrt{7}/3.$

S.P.I.C.E. (α) The simple way in which the moment of inertia of a uniform hemisphere about a diameter was deduced in Ex. (1) from the formula for the case of a sphere should be noted. The method can obviously be employed in other similar cases.

(β) Ex. (2) involves a simple use of the perpendicular axes theorem.

(γ) Ex. (3) is included to show that the moment of inertia of a composite body (about any axis) is the sum of the moments of inertia of its parts. (On the other hand, squares of radii of gyration cannot be added in this way.)

(δ) The theorem in Ex. (4) enables us to deduce the moment of inertia of a coplanar mass-distribution about any line in its plane, knowing the moments of inertia about any two lines in its plane which are at right angles and its product of inertia with respect to these axes. Ex. (5) illustrates the use of this theorem.

(ϵ) Exs. (6) and (7) are typical cases where the calculus is needed. In (6), the student should satisfy himself of the falsity of taking the moment of inertia about OY of the element in Fig. 115 as $\delta M.x^2$.

(ζ) Ex. (7) is a case where the density is not constant.

11·(11). Equimomental systems. When two systems of particles have equal moments of inertia about all lines in space, they are said to be *equimomental*.

If two systems are equimomental, it can be shown that they have the same total mass and same centre of mass.

A uniform triangular lamina of mass M is equimomental with the system consisting of three particles, each of mass $\frac{1}{3}M$, situated at the midpoints of the sides of the lamina. This result is used in § 16·314, and the interested student should prove at least the restricted result for the case in which moments of inertia are taken about lines in the plane of the lamina. (See Exs. 23, 25 below.)

EXAMPLES XI

1. Particles of masses 1, 2, 3, 4, 5, 6 units are situated at the points $(-6, 6)$, $(2, -2)$, $(-4, -3)$, $(-1, 0)$, $(0, -7)$, $(3, 0)$, respectively, referred to axes OX, OY. Find the distance of the centre of mass from the origin.

[Answer: 2 units of length.]

2. A number of particles move with uniform velocities. Prove that their centre of mass moves in a straight line.

3. Particles of masses m, m' are attached to the ends of a light inelastic string passing round a fixed smooth pulley of horizontal axis. The system is released so that m, m' move in vertical lines. Find the acceleration of the centre of mass of m, m'.

$$\left[\text{Answer: } \left(\frac{m - m'}{m + m'}\right)^2 g.\right]$$

4. A rod of uniform thickness has half its length composed of one metal and the other half of another metal. The centre of mass is distant one-seventh of the whole length from the centre of the rod. Find the ratio of the densities of the two metals.

[Answer: 11 : 3.]

5. $ABCD$ is a piece of uniform wire, of total length $8a$, in the shape of a square; E and F are the midpoints of BC, CD, respectively. If the portion ECF of the wire is removed, find the distance from A of the centre of mass of the remainder.

[Answer: $3a/2\sqrt{2}$.]

6. A piece of paper in the form of a parallelogram is cut along a straight line joining a corner to the middle point of an opposite side. Find the ratio of the distances of the centroid of the trapezoidal area so formed from its parallel sides.

[Answer: 4 : 5.]

7. $ABCD$ is a uniform trapezoidal lamina in which the angles at A and D are right angles, $AB = 2$ in, $AD = 1$ in, $DC = 1$ in. Find the distance of the centre of mass from A.

(Hint:—First find the distances from AB, AD.) [Answer: 0·896 in.]

8. A uniform solid is made up of a right circular cylinder, of length 1·2 m, and a hemisphere which are joined together so as to have a common base whose centre is A and radius 0·8 m. Find the distance of the centre of mass of the solid from A.

[Answer: 0·323 m.]

9. A hollow right circular cone of height h and semi-vertical angle α is made of uniform thin material and is open at the base. Find the distance of the centre of mass from the vertex.

A frustum of this cone is cut off by two planes U, V perpendicular to the axis, the radii of the circular ends being a, b. Find the distance of the centre of mass of this frustum from the plane midway between U, V.

$$\left[\text{Answers: } \tfrac{2}{3}h; \quad \frac{(a - b)^2 \cot \alpha}{6(a + b)}.\right]$$

10. A frustum of a uniform solid right circular cone is cut off by a plane parallel to the base and bisecting the axis. If the height of the frustum is h, find the distance of its centre of mass from the base. [Answer: $11h/28$.]

11. A hollow vessel of uniform small thickness consists of a cone and hemisphere joined together at their circular rims. The densities of the materials of the cone and the hemisphere are in the ratio 2 : 1, the radius of the common circular rim is a, and the height of the cone is $3a$. Find the distance of the centre of mass of the vessel from the vertex of the cone. [Answer: $2·36a$.]

12. A uniform wire is bent in the form of a triangle ABC. Prove that the line joining A to its centre of mass cuts BC at D, where $BD : DC = (a + b) : (a + c)$, where a, b, c are the side-lengths of the triangle.

13. If the sides BC, CA, AB of a piece of wire as in Ex. (12) are 20, 12, 16 cm, respectively, find the distances of the centre of mass from AB and AC.
[Answer: 4, 6 cm.]
If particles of masses 3·5, 4·5, 4 g are now attached to the wire at A, B, C respectively, prove that the position of the centre of mass is not altered.

14. Find the centre of mass of a tetrahedron $ABCD$ in which the density at any point varies as the distance from the face BCD.
[Answer: In AK, at a distance from A equal to $0·6AK$, where K is the centroid of the face BCD.]

15. A square revolves through 2π radians about a line parallel to one diagonal and passing through the extremity of the other diagonal. Using the theorems of Pappus, find the surface area and volume of the figure formed.
[Answers: $4\pi a^2 \sqrt{2}$, $\pi a^3 \sqrt{2}$, where a is the side of the square.]

16. The line-density of a rod of length l and mass M is proportional to the distance from one end A. Find the distance of the centre of mass from A, and the radius of gyration about an axis through A perpendicular to the rod.
[Answers: $2l/3$; $l/\sqrt{2}$.]

17. An L-shaped piece is cut from a uniform thin sheet of metal. The outer edges of the limbs are 6 ft, 4 ft, respectively; and the width of each is 1 ft. Prove that the radius of gyration about the outer edge of the shorter limb is $2·85$ ft.

18. The length and mass of a uniform rod are $2l$, M, respectively. Prove that the moment of inertia of the rod about an axis through the centre inclined at an angle θ to the rod is $\frac{1}{3}Ml^2 \sin^2 \theta$.
Deduce expressions for the moments of inertia, about its edges, of a uniform lamina in the shape of a parallelogram $ABCD$ of mass M, side-lengths $2a$, $2b$, and with the angle ABC equal to θ. [Answers: $\frac{4}{3}Mb^2 \sin^2 \theta$; $\frac{4}{3}Ma^2 \sin^2 \theta$.]

19. Prove that the moments of inertia about all lines through the centre of a uniform square lamina and in its plane are equal.
(Hint:—Use the perpendicular axes theorem.)

20. A hollow closed vessel of mass M, bounded by a hemispherical surface of radius a and a plane circular area, is made of thin metal of uniform surface density. Prove that its moment of inertia about a diameter of the plane boundary is $19Ma^2/36$.
(Hint:—Use the result for the moment of inertia of a hollow sphere about a diameter.)

21. A hollow circular cylinder with open ends is made out of thin uniform sheet metal. Its mass per unit area is σ, its length is $2a$ and its radius is a. Find its moment of inertia (i) about its axis; (ii) about a line through its centre of mass perpendicular to its axis. [Answers: (i) $4\pi\sigma a^4$; (ii) $10\pi\sigma a^4/3$.]

If the cylinder is closed at both ends with circular discs made of the same sheet metal, find the new moments of inertia corresponding to the cases (i), (ii).

[Answers: (i) $5\pi\sigma a^4$; (ii) $35\pi\sigma a^4/6$.]

22. A hollow sphere of metal of uniform density has internal and external radii a, b, respectively. Find its radius of gyration about any tangent line.

$$\left[\text{Answer: } \sqrt{\left(\frac{7b^5 - 5a^3b^2 - 2a^5}{5(b^3 - a^3)}\right)}.\right]$$

23. Prove that a uniform triangular lamina ABC of mass M, and the system consisting of three particles, each of mass $M/3$, at the midpoints of BC, CA, AB, are equimomental.

(For students not familiar with elementary three-dimensional coordinate geometry, it will be necessary to limit consideration to the plane of the lamina.)

24. Find the moment of inertia of a uniform solid right circular cone of mass M, height h and base-radius a: (*a*) about a diameter of its base; (*b*) about a line through the centre of mass perpendicular to the axis of the cone.

[Answers: $M(3a^2 + 2h^2)/20$; $3M(4a^2 + h^2)/80$.]

25. If two systems are equimomental, prove that they have the same total mass and same centre of mass.

26. A uniform circular plate of radius a and a uniform square plate of side $2b$ are of different surface-densities. Prove that the two plates can be equimomental, and find in this case (*a*) the ratio of a to b; (*b*) the ratio of the surface-densities.

[Answers: $2/\sqrt{3}$; $3/\pi$.]

CHAPTER XII

TWO-DIMENSIONAL DYNAMICS OF A SYSTEM OF PARTICLES

"One, two, three, four, many."
—Common mode of counting of Australian aboriginal.

The term 'two-dimensional' of the chapter heading needs a word of explanation. It will of course include cases where all the particles of a system move in a fixed plane. But the term will also be used in cases where a body is three-dimensional but where each of its particles moves in a fixed plane, all such planes being parallel. An example of 'two-dimensional motion' would be the case of a sphere rolling in simple fashion down a line of greatest slope of an inclined plane.

It is sufficient in a two-dimensional problem to consider all the forces (and other localised vectors) which enter the equations as acting in parallel planes.

(The term 'two-dimensional' could of course also be used to describe the motion of a system of particles over any curved surface. But in this book the use of the term will be limited to the circumstances stated above.)

In the theory to be set up in Chapter XII, it is to be understood, when the contrary is not mentioned or implied, that conditions are coplanar. Subject to the slight modification mentioned in the last paragraph of § 12·11 below, the theory may be applied directly to more general two-dimensional problems.

12·1. Localised vectors. In treating the dynamics of a system of particles it becomes necessary to take cognisance of the *lines of action* of certain vectors, e.g. forces. Hence localised vectors (see § 6·12) have to be discussed. The term *localised vector* in this book will mean a vector localised *in a line*.

12·11. Moment of a localised vector about a given point. Let the magnitude and direction of a localised vector be given by $\mathbf{F}$, and let the line of action be l as in Fig. 116. We shall sometimes denote this localised vector by $(\mathbf{F}, l)$.

The moment of $(\mathbf{F}, l)$ about a given point O is defined to be of magnitude Fp, where F is the magnitude of $\mathbf{F}$, and p is the perpendicular from O to l.

Sign is attached to the moment according to the sense of the curved arrow drawn in Fig. 116. In theoretical two-dimensional work, the moment is often taken to be positive if the sense is anticlockwise.

Fig. 116

(The reader who is familiar with the notion of vector product—Appendix, § 17·1—may note that the moment of $(\mathbf{F}, l)$ about O is, more generally, defined as $\mathbf{r} \times \mathbf{F}$, where $\mathbf{r}$ is the position vector, referred to O, of any point on l. This definition gives moment as a (free) *vector* (perpendicular to the plane containing O and l), and includes the two-dimensional definition given above. Where, as in this chapter, all localised vectors and points such as O are treated as being in one plane, all moment vectors are parallel, being perpendicular to this plane; and so the vector aspect of moment need not here be emphasised—cf. §§ 10·1, 13·221 (iii).)

For general two-dimensional problems, the term 'moment about a point' strictly needs to be replaced by 'moment about an axis', the axis being at right angles to the planes of motion of particles. The moment of a localised vector about such an axis is taken as the moment about the point in which the axis cuts the plane containing the line of the vector.

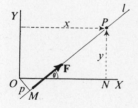 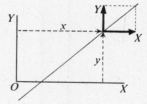

Fig. 117   Fig. 118

12·111. Cartesian formula for moment. Let (x, y) be the coordinates, referred to a pair of axes, OX, OY, of any point P on the line of action l of a localised vector $(\mathbf{F}, l)$.

Let (X, Y) be the components of $\mathbf{F}$ parallel to OX, OY.

Then (see Fig. 117), $p = x \sin \theta - y \cos \theta$.

Hence the moment of $(\mathbf{F}, l)$ about O

$$= Fp$$
$$= F\sin\theta.x - F\cos\theta.y$$
$$= Yx - Xy.$$

This formula may easily be remembered by noting that $Yx - Xy$ is the sum of the results obtained by taking moments about O (in the anticlockwise sense) of vectors of magnitudes X, Y localised as shown in Fig. 118. (This is a particular case of the theorem of moments—§ 12·15.)

12·12. Sum of two localised vectors. Let $(\mathbf{F}_1, l_1)$, $(\mathbf{F}_2, l_2)$ be the given localised vectors, and suppose that l_1, l_2 intersect at K, as shown in Fig. 119.

Then the sum or resultant of $(\mathbf{F}_1, l_1)$, $(\mathbf{F}_2, l_2)$ is *defined* to be the localised vector whose magnitude and direction are given by $\mathbf{F}_1 + \mathbf{F}_2$, and whose line of action, l say, passes through K.

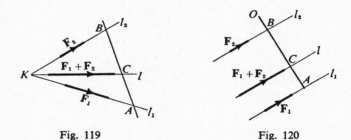

Fig. 119   Fig. 120

12·121. As a preliminary to examining the special case in which l_1, l_2 are parallel, let AB be any transversal cutting l_1, l_2, l in A, B, C, respectively (Fig. 119).

Now $$\mathbf{F}_1 + \mathbf{F}_2 = \left(\frac{F_1}{KA}\mathbf{KA} + \frac{F_2}{KB}\mathbf{KB}\right).$$

Hence, by Ex. (1) of § 6·19 (this Ex. might now be 'promoted' to the status of bookwork), taking $\lambda = F_1/KA$, $\mu = F_2/KB$, we have

$$\mathbf{F}_1 + \mathbf{F}_2 = \left(\frac{F_1}{KA} + \frac{F_2}{KB}\right)\mathbf{KC}, \qquad \text{(i)}$$

where C is given by

$$\left(\frac{F_1}{KA}\right)AC = \left(\frac{F_2}{KB}\right)CB. \qquad \text{(ii)}$$

12·13. Parallel (localised) vectors. Now take the limit of the case in § 12·12 as K recedes to infinity, AB being kept finite. Then (see Fig. 120) l_1, l_2, l become parallel lines, and the ratio of the lengths KA, KB, KC becomes $1:1:1$.

It follows that when l_1, l_2 are parallel, the resultant of $(\mathbf{F}_1, l_1)$, $(\mathbf{F}_2, l_2)$ is a parallel localised vector of magnitude $F_1 + F_2$ (by (i)), passing through the point C of AB, where (by (ii))

$$F_1.AC = F_2.CB. \tag{iii}$$

12·131. 'Like' and 'unlike' parallel vectors. When F_1, F_2, have the same direction, the parallel vectors $(\mathbf{F}_1, l_1)$, $(\mathbf{F}_2, l_2)$ are sometimes said to be 'like'; otherwise 'unlike'.

Except in the case of § 12·132 (below), the results in § 12·13, including (iii), hold for unlike, as well as like, parallel vectors, provided (a) F_1, F_2 are given signs according the directions of $\mathbf{F}_1$, $\mathbf{F}_2$, (b) sense is attached to the order of the letters in the line ACB. With this understanding, the argument in § 12·13 covers both cases.

The student is advised to follow out the argument in § 12·13, drawing separate diagrams for the different cases.

When $\mathbf{F}_1$, $\mathbf{F}_2$ are in the same direction, C divides AB internally in the ratio $F_2 : F_1$. When $\mathbf{F}_1$, $\mathbf{F}_2$ are in opposite directions, C divides AB externally in the ratio $|F_2| : |F_1|$, and lies in AB produced or BA produced according as $|F_1|$ is less than or greater than $|F_2|$.

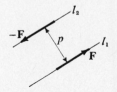

Fig. 121

12·132. Vector couple. The results in § 12·13 are intelligible in all cases except the special case in which: l_1, l_2 are parallel and distinct lines, $\mathbf{F}_1$, $\mathbf{F}_2$ are in opposite directions, and $|F_1| = |F_2|$. There is no single resultant localised vector in this case; and the pair of given localised vectors is said to constitute a *couple*.

The sum of the moments of the pair of vectors constituting a couple, called the *moment of the couple*, is the same about all points, being easily seen to be of magnitude Fp, where F is the magnitude of either vector and p is the distance between the lines of action (Fig. 121).

The sign of the moment is determined as in § 12·11; it is positive in Fig. 121 if the anticlockwise sense is taken as positive.

In the theory to follow, a couple will be taken to be sufficiently specified by its moment. (This suffices for the dynamics of rigid bodies (§ 13·1), but not of deformable bodies. See also § 12·22.) A 'couple L' will mean a couple whose moment is L.

12·14. Resultant of a set of localised vectors. If we are given a set of localised vectors, we may take the sum of any two members (not constituting a couple), add this to any third member (not producing a couple), and so on, by the processes set out in §§ 12·12, 12·13, 12·131.

Finally we shall be left with: (a) a single localised vector; or (b) a couple; or (c) nothing at all. What is left in this way is called the *resultant* of the original set of localised vectors; in case (c) we say that 'the resultant is zero'.

It will be shown in § 12·16 that the resultant as thus defined is uniquely determined (subject to the qualification that in the case (b) it is only the moment of the resultant couple that is uniquely determined).

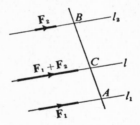

Fig. 122

12·15. Theorem of moments. We now prove the theorem of moments, viz.: the sum of the moments of a set of localised vectors about any point O is equal to the moment of their resultant about O.

Consider first the case in which the given set of vectors consists of just two vectors, $(\mathbf{F}_1, l_1)$, $(\mathbf{F}_2, l_2)$, say.

If l_1, l_2 intersect, say at the point (x, y) referred to axes OX, OY through O, then, by § 12·111, the sum of the moments of the vectors about O

$$= Y_1 x - X_1 y + Y_2 x - X_2 y$$
$$= (Y_1 + Y_2)x - (X_1 + X_2)y,$$

where $\mathbf{F}_1 = (X_1, Y_1)$, $\mathbf{F}_2 = (X_2, Y_2)$; and the last expression is the moment of the resultant of the two vectors about O.

If l_1, l_2 are parallel and the two given vectors do not constitute a couple, then, taking the transversal AB of Fig. 120 to be the line through O at right angles to l_1, l_2, we have (see Fig. 122)

$$F_1.OA + F_2.OB = F_1(OC + CA) + F_2(OC - BC)$$
$$= (F_1 + F_2)OC,$$

by (iii) of § 12·13. This gives the required result in this case.

The foregoing argument holds for unlike as well as for like parallel vectors, the provisos (*a*), (*b*) of § 12·131 being understood; (the student should draw different diagrams to cover representative cases).

If the vectors $(\mathbf{F}_1, l_1)$, $(\mathbf{F}_2, l_2)$ constitute a couple, the result is covered by the definition of 'resultant' in § 12·14.

The theorem is therefore proved in all cases for a set of two vectors; and may obviously be extended to any set of (coplanar) vectors.

12·16. Uniqueness of resultant of a set of localised vectors. By the processes of free vector addition, the magnitude and direction (when relevant) of the resultant of a given set of localised vectors, as defined in § 12·14, are uniquely determined. By the theorem of moments, the moment of the resultant about any point is also uniquely determined, since any sum of moments is independent of the order of addition.

Hence in case (*a*) of § 12·14, the single resultant localised vector is uniquely determined in magnitude, direction and line of action.

In case (*b*), the free vector sum of the given set is zero, and the moment of the resultant couple is uniquely determined. (The separate magnitudes of F and p (Fig. 121) are not uniquely determined, but, as pointed out in § 12·132, it is only the moment of a couple that matters in the applications to be made.)

In case (*c*), both the free vector sum and the sum of the moments of the given set about any point are uniquely determined to be zero.

12·17. 'Equivalence' of sets of localised vectors. It follows from § 12·16 that for separate sets of localised vectors to have the same resultant, it is *necessary and sufficient* that:

(i) the pair of sums of the resolved parts in any two definite non-parallel directions should be the same; and

(ii) the sum of the moments about any one definite point should be the same.

12·171. For the resultant of a set of localised vectors to be zero, it is necessary and sufficient for the three sums involved in (i) and (ii) to be zero.

12·2. Systems of particles in more than one dimension. The theory in Chapter IV will now be generalised from the rectilinear to the coplanar case. In some sections (§§ 12·3-12·35, 12·5-12·53), the argument given covers the three-dimensional case as well.)

The formal proofs in this chapter relate to discrete systems of particles, but the results may, by § 11·2, be applied to continuously distributed mass-systems.

One new concept, that of 'angular momentum', will need to be introduced (§ 12·4); this was not needed in the one-dimensional case.

The *configuration* of a system of particles is now specified by the set of position vectors of the particles referred to some origin.

We have already in Chapter XI generalised § 4·11 on the position, velocity and acceleration of the centre of mass. We now proceed to the further detail required.

12·21. Momentum (linear momentum). Corresponding to § 4·13, the *momentum* of a system is now defined as $\Sigma(m\mathbf{v})$. By (48), this momentum is also equal to $M\bar{\mathbf{v}}$, where M is the total mass Σm, and $\bar{\mathbf{v}}$ is the velocity of the centre of mass.

The 'momentum' is referred to as 'linear momentum' when it is desired to emphasise the distinction between momentum and angular momentum.

12·22. Forces on systems of particles. Forces will now be treated as localised vectors in the sense of § 12·1. This means that account is now taken of the lines of action of forces, and that 'resultants' of forces are formed by the rules summarised in § 12·14.

In the case of forces acting on a single particle, this procedure is no different from that in Chapter VI.

When the forces act on different particles of a system, the 'resultant' is a more artificial concept, being not in general entirely equivalent in effect to the original set of forces.

(For example, the effect on the reader of two simultaneously acting equal parallel forces, one directed towards his head and the other towards his feet, is in some respects not equivalent to a force of double the magnitude directed towards his stomach.)

But in the important model case in which the system of particles is *rigid* (see § 13·1), it may be shown (this is covered in (i) of § 13·(10)2) that the dynamical effect of any set of forces is entirely equivalent to that of its resultant. (The reader of course is not a rigid body.)

On account of this last property and of the extensive application of the subject of Statics—see Chapter XV—to problems on rigid bodies, two sets of forces are said to be *statically equivalent* if they have the same resultant (as defined in § 12·14.)

A necessary and sufficient set of conditions that two sets of forces are statically equivalent is the set (i), (ii) of § 12·17.

12·221. Any given set of forces is statically equivalent to a single force through any assigned point, A say, together with a couple.

To show this for the case (*a*) of § 12·14, let **F** in the line *l* (Fig. 123) be the resultant of the given set of forces. The resultant is obviously not affected if forces **F**, −**F** acting through A are inserted. Hence the given set is statically equivalent to the single force **F** at A together with a couple of moment Fp, where p is the perpendicular from A to l.

In case (*b*) of § 12·14, the single force has the particular value zero.

In case (*c*), both the single force and the couple are zero. The set of forces is then said to be *in equilibrium*.

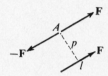

Fig. 123

12·23. Weight; centre of gravity. Consider a system of particles in a vertical plane, and let OX, OY be horizontal and vertical axes in this plane. Then (taking the acceleration of gravity, g, to be constant, as usual) we have by (47*b*), in usual notation,

$$(\Sigma mg)\bar{x} = \Sigma(mgx). \tag{i}$$

By § 12·22, the resultant of the weights of the individual particles has magnitude equal to the sum $\Sigma(mg)$ of the magnitudes of the weights of the particles. Also, by (i) and the theorem of moments (§ 12·15) for the case of parallel forces, the resultant acts vertically downwards through the centre of mass G of the system.

(This result holds generally in three dimensions. To show this, it is necessary to extend the theorem of moments to the case of a set of parallel localised vectors not necessarily in one plane.)

This resultant is called the *weight* of the system of particles, and G, the point through which it acts, is commonly called the *centre of gravity*.

It should be noted that the term 'centre of gravity' is not relevant unless the system of particles is in a constant gravitational field (or in certain other special circumstances that we shall not consider in this book). On the other hand, the term 'centre of mass' is not restricted in this way. Moreover, whenever a centre of gravity exists, its position coincides with that of the centre of mass. Hence there is no real need for the term 'centre of gravity' in mechanics, and we shall not refer to it again.

(In ordinary speech, the term 'centre of gravity' is very often used where what is meant is really 'centre of mass', or 'centroid'.)

12·3. Principle of (linear) momentum.

12·31. Internal and external forces. We add to the postulate of § 4·2 that action, reaction forces have the same line of action.

We then classify forces on systems of particles into internal and external forces, as in § 4·31.

Since internal forces occur in pairs which have equal magnitude, opposite senses and the same line of action, it follows that:

(i) The free vector sum of all the internal forces on a system of particles is zero.

(ii) The sum of the moments of all the internal forces about any point is zero.

12·32. Principle of momentum. At time t, let **v**, **f** be the velocity and acceleration of a typical particle P (of mass m) of a given system of particles. Let $\mathbf{F}_i$, $\mathbf{F}_e$, respectively, be the resultants of the internal and external forces acting on P, so that $\mathbf{F}_i + \mathbf{F}_e$, $= \mathbf{F}$ say, is the resultant of all forces on P.

By (26a), $\qquad m\mathbf{f} = \mathbf{F} = \mathbf{F}_i + \mathbf{F}_e.$

Summing for all particles of the system, we have
$$\Sigma(m\mathbf{f}) = \Sigma \mathbf{F}_e,$$
since, by § 12·31 (i), $\qquad \Sigma \mathbf{F}_i = 0.$

Hence $$\mathbf{R} = \frac{d}{dt} \Sigma(m\mathbf{v}), \qquad (53a)$$

where **R** is the resultant of all the *external* forces acting on the system, treated as non-localised vectors.

(53a) embodies the principle of momentum, viz.: The rate of change of the momentum of a system of particles is equal to the resultant of all the external forces acting.

(Lines of action are not involved with the principle of momentum.)

12·33. Motion of centre of mass. An exceedingly useful form of the principle of momentum is given by using (48) to write (53a) in the form

$$\mathbf{R} = \frac{d}{dt}(M\bar{\mathbf{v}}), \qquad (53b)$$

or $\qquad \mathbf{R} = M\bar{\mathbf{f}}, \qquad (53c)$

where M is the total mass Σm, and $\bar{\mathbf{v}}$, $\bar{\mathbf{f}}$ are the velocity and acceleration of the centre of mass G.

According to (53b) or (53c), G moves as if it were a particle of mass M acted on by forces whose magnitudes and directions are those of the *external* forces on the system.

12·331. If $\bar{x}$, $\bar{y}$ are the coordinates of G, referred to *fixed* axes OX, OY, the principle of momentum may also be expressed in either of the following forms:

$$\Sigma X = M\ddot{\bar{x}}, \qquad \Sigma Y = M\ddot{\bar{y}}, \tag{53d}$$

$$\mathbf{R} = M\ddot{\bar{\mathbf{r}}}, \tag{53e}$$

where (X, Y) are the components of a typical external force on the system, and $\bar{\mathbf{r}} (= \bar{x}\mathbf{i} + \bar{y}\mathbf{j})$ is the position vector of G.

12·34. Conservation of momentum.

The momentum $\Sigma(m\mathbf{v})$, or $M\bar{\mathbf{v}}$, is *conserved* so long, but only so long, as the resultant $\mathbf{R}$ is zero. The centre of mass of the system then moves (in a straight line) with constant velocity.

12·35. Influence of impulsive forces.

The principle of momentum may be expressed in integrated form by equations analogous to $(21c, f)$—§§ 4·32, 4·33—obtained by replacing $R, u, v, \bar{u}, \bar{v}$ in $(21c, f)$ by corresponding vectorial symbols.

The statements on impulsive forces in § 4·35 then continue to hold.

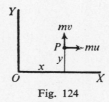

Fig. 124

12·4. Principle of angular momentum.

12·41. Angular momentum of a particle about a point. The angular momentum (A.M.) of a single particle P about a given point (or *axis* in the case of general two-dimensional motion) is defined as the moment of the momentum of P about the point (or axis), the momentum of P being now regarded as a localised vector through the position of P.

The dimensions of angular momentum are $[ML^2T^{-1}]$, those of (linear) momentum being $[MLT^{-1}]$.

Let (x, y) be the coordinates of P at time t, and (mu, mv) the components of its momentum, referred to axes OX, OY through O. Then, by § 12·111, the angular momentum of P about O is $mvx - muy$.

12·42. Angular momentum of a system of particles about a point. The angular momentum of a system of particles about a point O is defined as the sum of the angular momenta of the individual particles about O, i.e. as $\Sigma(mvx - muy)$.

12·43. Principle of angular momentum. Consider the rate of change of the angular momentum H of a system of particles about O, where O is a *fixed* point.

Since O is fixed, we may with the above notation (Fig. 124) put $u = \dot{x}$, $v = \dot{y}$. (If O were not fixed, $(\dot{x}, \dot{y})$ would give not the absolute velocity, but the velocity *relative to O*.)

Thus the rate of change of angular momentum, i.e. dH/dt,

$$= \frac{d}{dt} \Sigma(m\dot{y}x - m\dot{x}y)$$
$$= \Sigma(m\ddot{y}x) - \Sigma(m\ddot{x}y), \qquad (i)$$

(the terms $\Sigma m\dot{x}\dot{y}$, $-\Sigma m\dot{x}\dot{y}$ cancelling out).

Next consider the sum of the moments of the external forces about O. By § 12·31 (ii), this is equal to the sum of the moments of all the forces (internal and external) about O; and so, by § 12·111, $= \Sigma(Yx - Xy)$, where (X, Y) are here the components of the resultant force on the particle at (x, y); i.e. by (26a), $= \Sigma(m\ddot{y}x - m\ddot{x}y)$, which is the same as (i).

This gives the principle of angular momentum, viz.:

$$\left.\begin{array}{l}\textit{Rate of change of A.M. of a}\\ \textit{system of particles about a}\\ \textit{fixed point O}\end{array}\right\} = \left\{\begin{array}{l}\textit{Sum of moments of the}\\ \textit{external forces about O.}\end{array}\right. \qquad (54a)$$

12·44. Conservation of angular momentum.

If, in particular, the resultant of the external forces is (*a*) zero, *or* (*b*) a single force through O, then the rate of change of angular momentum about O is zero. So long as this holds, the angular momentum is *conserved*.

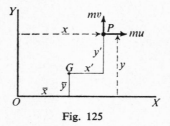

Fig. 125

12·45. Extended form of angular momentum principle. We now consider the rate of increase of the angular momentum, H_G say, *about the centre of mass G* of the system, where G may or may not be fixed.

We take axes OX, OY through the fixed point O as before. Let $(\bar{x}, \bar{y})$ be the coordinates of G referred to these axes, and let (x', y') be the coordinates of the typical particle P referred to parallel axes through G.

The required rate of change of H_G
$$= \frac{d}{dt} \Sigma(m\dot{y}x' - m\dot{x}y')$$
$$= \Sigma(m\ddot{y}x') - \Sigma(m\ddot{x}y') + \Sigma(m\dot{y}\dot{x}') - \Sigma(m\dot{x}\dot{y}').$$
The last two terms are together equal to zero, for
$$\Sigma(m\dot{y}\dot{x}') - \Sigma(m\dot{x}\dot{y}') = \Sigma\{m\dot{y}(\dot{x} - \dot{\bar{x}})\} - \Sigma\{m\dot{x}(\dot{y} - \dot{\bar{y}})\}$$
$$= -\dot{\bar{x}}\Sigma(m\dot{y}) + \dot{\bar{y}}\Sigma(m\dot{x})$$
$$= -\dot{\bar{x}}(\Sigma m)\dot{\bar{y}} + \dot{\bar{y}}(\Sigma m)\dot{\bar{x}}, \quad \text{by (48),}$$
$$= 0.$$

The remaining expression, $\Sigma(m\ddot{y}x') - \Sigma(m\ddot{x}y')$, is equal to the sum of the moments of all the external forces about G. This follows by argument similar to that in the fourth paragraph of § 12·43.

Hence we have the extended form of the principle, viz.:

$$\left.\begin{array}{l}\text{Rate of change of A.M. about} \\ \text{the C.M. } G \text{ (even if } G \text{ is} \\ \text{moving)}\end{array}\right\} = \left\{\begin{array}{l}\text{Sum of moments of the} \\ \text{external forces about } G.\end{array}\right. \quad (54b)$$

12·451. Alternative extended form. In § 12·45, it is the *absolute* angular momentum about G that is involved.

It happens, however, that the absolute angular momentum H_G about G is equal to the angular momentum relative to G about G. By the latter, which we denote by H_G', we mean the angular momentum about G as calculated using velocities relative to G.

To show the equality of H_G and H_G', we have
$$H_G' = \Sigma(m\dot{y}'x' - m\dot{x}'y')$$
$$= \Sigma\{m(\dot{y} - \dot{\bar{y}})x'\} - \Sigma\{m(\dot{x} - \dot{\bar{x}})y'\}$$
$$= H_G - \dot{\bar{y}}\Sigma(mx') + \dot{\bar{x}}\Sigma(my');$$
and $\Sigma(mx')$, $\Sigma(my')$ are zero, being, respectively, (Σm) times the coordinates of G referred to axes with G as origin. Hence $H_G' = H_G$.

Hence in using the extended form of the principle, we may take either the absolute angular momentum (about G) or the angular momentum relative to G. This statement will be called . . . . (54c).

12·46.

Note that the proofs in subsections of § 12·4 could be shortened with the use of vector products. (In vector form, the angular momentum **H** about O would be $\Sigma\{\mathbf{r} \times (m\mathbf{v})\}$; and the sum of the moments of the external forces about O would be $\Sigma(\mathbf{r} \times \mathbf{F})$, in obvious notation; etc.)

Moreover, the results in vector form hold in the general three-dimensional case.

12·5. Energy.
In order to complete the generalisation of Chapter IV, we have yet to generalise the subsections of § 4·5 on energy.

12·51. Kinetic energy.
The kinetic energy, T say, of a system of particles continues to be defined as in § 4·51, i.e. as $\Sigma(\tfrac{1}{2}mv^2)$, in usual notation.

12·511. Translational and internal kinetic energy.
Let T_0 be the kinetic energy of a particle whose mass is equal to the total mass M of the system, and whose velocity is equal to that of the centre of mass G. Let T_1 be the kinetic energy as computed for the system using velocities of the particles relative to G, instead of their absolute velocities.

We show that
$$T = T_0 + T_1. \tag{55}$$

The terms T_0, T_1 are sometimes called the 'translational' and 'internal' kinetic energy, respectively, of the system of particles.

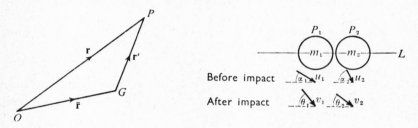

Fig. 126   Fig. 127

To prove (55), let $\mathbf{r}$ and $\mathbf{v}$ be the position vector and velocity of a typical particle P referred to a fixed origin O, let $\bar{\mathbf{r}}$ be the position vector of G referred to O, and let $\mathbf{r}'$ be the position vector of P relative to G (Fig. 126). Then the kinetic energy T of the system

$$= \Sigma(\tfrac{1}{2}mv^2)$$
$$= \Sigma(\tfrac{1}{2}m\dot{\mathbf{r}}\cdot\dot{\mathbf{r}})$$
$$= \Sigma\{\tfrac{1}{2}m(\dot{\bar{\mathbf{r}}} + \dot{\mathbf{r}}')\cdot(\dot{\bar{\mathbf{r}}} + \dot{\mathbf{r}}')\}$$
$$= \Sigma(\tfrac{1}{2}m\dot{\bar{\mathbf{r}}}\cdot\dot{\bar{\mathbf{r}}}) + \Sigma(m\dot{\bar{\mathbf{r}}}\cdot\dot{\mathbf{r}}') + \Sigma(\tfrac{1}{2}m\dot{\mathbf{r}}'\cdot\dot{\mathbf{r}}').$$

The first term in the last expression $= \tfrac{1}{2}(\Sigma m)\dot{\bar{\mathbf{r}}}^2 = T_0$. The second term $= \dot{\bar{\mathbf{r}}}\cdot\Sigma(m\dot{\mathbf{r}}')$, which is zero since $\Sigma(m\dot{\mathbf{r}}')/\Sigma m$ is zero, being equal to the velocity of G relative to G. The third term is equal to T_1.

Thus the theorem is proved.

12·52. Potential energy. If the system is subject to a conservative field of force, the corresponding potential energy is (as in § 4·52) defined as the sum of the potential energies of the constituent particles.

12·521. Potential energy associated with weight. In the particular case of a system of particles under constant gravity, the corresponding potential energy (see § 6·831) is $\Sigma(mgz)$, where z is the height of a typical particle above some reference level.

By (47b), this potential energy $= Mg\bar{z}$, where $M = \Sigma m$, and $\bar{z}$ is the height of the centre of mass above the reference level.

12·53. Principle of energy. The statements in §§ 4·53, 4·54, continue to hold without further qualification.

Thus we still have:

$$\left. \begin{array}{l} \text{Sum of works by all forces} \\ \text{(internal and external)} \end{array} \right\} = \left\{ \begin{array}{l} \text{Increase in K.E.} \\ \text{of system.} \end{array} \right. \quad (22a)$$

or

$$\left. \begin{array}{l} \text{Sum of works by extraneous} \\ \text{forces (internal and external)} \end{array} \right\} = \left\{ \begin{array}{l} \text{Increase in (K.E. +} \\ \text{P.E.) of system.} \end{array} \right. \quad (22b)$$

12·6. Applications. The theory of Chapter XII will now be applied to problems on systems consisting of a pair or small number of particles. Further applications will be made in Chapter XIII.

12·61. Elastic impacts of spheres. Let m_1, m_2 be the masses of two spherical bodies P_1, P_2 which collide. We shall treat the bodies as particles when we apply dynamical theory, but regard them as spheres in referring to the 'line of centres' at collision.

Let the velocity of P_1 just before the impact be u_1 in a direction inclined at α_1 to the line, L say, joining the centres of P_1, P_2 at the instant of impact, as indicated in Fig. 127; let (v_1, θ_1), (u_2, α_2), (v_2, θ_2) apply similarly to the velocity of P_1 just after, and of P_2 just before and just after the impact, respectively.

The momentum of the system consisting of the pair of particles is conserved at the instant of impact (§ 12·35); and we have, resolving along and at right angles to the line L,

$$m_1 u_1 \cos \alpha_1 + m_2 u_2 \cos \alpha_2 = m_1 v_1 \cos \theta_1 + m_2 v_2 \cos \theta_2, \quad \text{(i)}$$

$$m_1 u_1 \sin \alpha_1 + m_2 u_2 \sin \alpha_2 = m_1 v_1 \sin \theta_1 + m_2 v_2 \sin \theta_2. \quad \text{(ii)}$$

A third equation comes from a generalised form of the postulate in § 4·8, viz.,

$$v_2 \cos \theta_2 - v_1 \cos \theta_1 = -e(u_2 \cos \alpha_2 - u_1 \cos \alpha_1); \quad (23a)$$

i.e. we take Newton's 'law of impact' as given in § 4·8 by (23), with the qualification that each velocity is resolved parallel to L.

Suppose the bodies are smooth. Then considering P_1 alone, we have, by § 6·33, taking resolved parts at right angles to L,

$$m_1 v_1 \sin \theta_1 - m_1 u_1 \sin \alpha_1 = 0. \tag{iii}$$

From (i), (ii), (iii) and (23a), we can determine v_1, v_2, θ_1, θ_2, when m_1, m_2, u_1, u_2, α_1, α_2, e are given.

(A fifth equation, $m_2 v_2 \sin \theta_2 - m_2 u_2 \sin \alpha_2 = 0$, comes from considering P_2 alone; this equation adds nothing to the solution, being deducible from (ii) and (iii).)

12·611. Remarks.

(i) If the spheres were rough, the equations (i), (ii), (23a) of § 12·61 would still hold; but in equation (iii) we should need to include a frictional component of impulse. The student who is interested may obtain equations sufficient to give the solution, assuming that limiting friction (involving a coefficient μ) holds. The problem is more involved if this last assumption does not hold.

(ii) For the case of a sphere colliding with a fixed wall, the equations (i) and (ii) of § 12·61 are not longer relevant. The equation (23a) degenerates to

$$v_1 \cos \theta_1 = -e u_1 \cos \alpha_1, \tag{23b}$$

where P_1 is taken to be the given sphere, and P_2 is replaced by the wall, which is at right angles to L in Fig. 127. The equation (iii) continues to hold. The equations (23b) and (iii) are sufficient to determine v_1, θ_1 in terms of u_1, α_1, e.

(iii) We postulate that (23a) is relevant to problems such as Ex. (2) of § 12·612 below, where there is a constraint due to a taut string.

(iv) Some problems involving collisions between more than two spheres simultaneously can be solved by extending the theory in § 12·61 along the lines indicated in § 4·81 (v) and (vi), and § 4·82, Ex. (2).

12·612. Examples.
(1) *A smooth billiard ball P is projected on a billiard table towards an equal ball Q at rest, the path of the centre of P before impact being tangential to the surface of Q. The coefficient of restitution for the impact is e. Find the change in the direction of motion of the centre of P which takes place at the impact.*

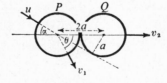

Fig. 128

Let a be the radius of each ball and m the mass.

Let the velocity of P before the impact be given by u, α, and after the impact by v_1, θ, as indicated in Fig. 128.

Due to the impact, Q receives an impulse along the line of centres and must therefore move in this direction after the impact. Let v_2 be the speed imparted to Q.

By the principle of momentum, resolving along and at right angles to the line of centres, we have

$$mu \cos \alpha = mv_1 \cos \theta + mv_2,$$
$$mu \sin \alpha = mv_1 \sin \theta.$$

By Newton's law of impact, (23a),

$$v_2 - v_1 \cos \theta = -e(0 - u \cos \alpha).$$

From these three equations, we readily derive

$$\tan \theta = 2 \tan \alpha / (1 - e).$$

But by the data we have (see Fig. 128) $\sin \alpha = a/2a$. Hence $\alpha = \pi/6$. The required deflection in the motion of P

$$= \theta - \alpha$$
$$= \tan^{-1}\left\{\frac{2}{(1-e)\sqrt{3}}\right\} - \frac{\pi}{6}.$$

(2) *A smooth sphere Q of mass 10 g is tied to a fixed point O by a light inelastic thread. Initially Q is at rest and the thread is just taut. A sphere P of mass 5 g impinges directly against Q with a velocity of 6 cm/sec in a direction inclined at 45° to the thread, the component of this velocity along the thread being in the sense away from O. The coefficient of restitution is 0·5. Find the speed of Q just after the impact.*

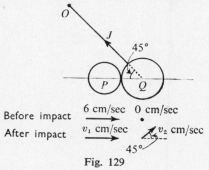

Fig. 129

This problem is complicated by the presence of an *external impulsive force*, of impulse J say, acting along the thread on the system consisting of the two spheres.

Let the velocities of P, Q just after the impact be as indicated in Fig. 129.

The direction of the velocity of P just after the impact is along the line of centres, because the velocity of P before impact and the only impulse on P (the reaction impulse from Q) are along this line.

The direction of the velocity of Q just after impact is at right angles to the thread since the thread is of fixed length.

We neither know nor want J. Hence, in using the principle of momentum for the pair of particles, we take resolved parts at right angles to the thread, obtaining
$$5 \times 6 \cos 45° = 5v_1 \cos 45° + 10v_2.$$
By Newton's law, (23a),
$$v_2 \cos 45° - v_1 = -0{\cdot}5(0 - 6).$$
These equations yield $\quad v_2 = 9\sqrt{2}/5$ cm/sec.

(3) *For the collision discussed in* § 12·61 (*see Fig.* 127), *show that there is a loss of kinetic energy at the impact equal to*
$$\tfrac{1}{2}\frac{m_1 m_2}{m_1 + m_2}(1 - e^2)(u_2 \cos \alpha_2 - u_1 \cos \alpha_1)^2.$$

We shall use the formula (55) on translational and internal K.E.

By § 12·33, the velocity of the C.M. G of a pair of particles is not changed at the impact, the impulsive forces between the particles being internal forces for the whole system.

Since the mass of the system is of course also unchanged, the translational K.E. of the system is not changed at the impact.

Hence the loss in K.E. is the loss of internal K.E. alone.

Let $\mathbf{u}_1, \mathbf{u}_2$ be the velocities of P_1, P_2 just before impact.

The velocity, $\mathbf{u}$ say, of G just before (also just after) impact
$$= (m_1\mathbf{u}_1 + m_2\mathbf{u}_2)/(m_1 + m_2).$$

The velocities of P_1, P_2, relative to G, just before impact
$$= \mathbf{u}_1 - \mathbf{u}, \qquad \mathbf{u}_2 - \mathbf{u}$$
$$= \frac{m_2(\mathbf{u}_1 - \mathbf{u}_2)}{m_1 + m_2}, \quad \frac{m_1(\mathbf{u}_2 - \mathbf{u}_1)}{m_1 + m_2}.$$

Hence the internal K.E. before impact
$$= \frac{\tfrac{1}{2}m_1 m_2^2(\mathbf{u}_1 - \mathbf{u}_2)^2}{(m_1 + m_2)^2} + \frac{\tfrac{1}{2}m_2 m_1^2(\mathbf{u}_2 - \mathbf{u}_1)^2}{(m_1 + m_2)^2}$$
$$= \frac{\tfrac{1}{2}m_1 m_2}{m_1 + m_2}(\mathbf{u}_2 - \mathbf{u}_1)^2.$$

Similarly, the internal K.E. after impact
$$= \frac{\tfrac{1}{2}m_1 m_2}{m_1 + m_2}(\mathbf{v}_2 - \mathbf{v}_1)^2,$$
where $\mathbf{v}_1, \mathbf{v}_2$ are the velocities of P_1, P_2 just after impact.

Hence the loss of K.E. due to the impact
$$= \frac{\tfrac{1}{2}m_1 m_2}{m_1 + m_2}\{(\mathbf{u}_2 - \mathbf{u}_1)^2 - (\mathbf{v}_2 - \mathbf{v}_1)^2\}$$
$$= \frac{\tfrac{1}{2}m_1 m_2}{m_1 + m_2}\{(U^2 + U'^2) - (V^2 + V'^2)\}, \qquad \text{(i)}$$
where U, U' are the components, along and at right angles to L, of $\mathbf{u}_2 - \mathbf{u}_1$, i.e. of the relative velocity before impact; and V, V' are similar components after impact.

By Newton's law of impact, (23a), $V = -eU$. Also $U' = V'$, since velocity components at right angles to L are unaltered by the impact. Hence, by (i), the loss of K.E. $= \frac{1}{2}(1-e^2)\, m_1 m_2 U^2/(m_1+m_2)$.

Since $U = u_2 \cos \alpha_2 - u_1 \cos \alpha_1$, we have the required answer.

S.P.I.C.E. (α) It may be noted that for a billiard ball, e is of order 0·9. Hence the deflection as given in Ex. (1) will be a little less than 60°.

(β) It should be noted that, in Ex. (1), the rolling of the balls was ignored. A more complete treatment would take this and other circumstances into consideration.

(γ) In Ex. (2), the way in which we obtained an equation not involving the unknown impulse J should be noted.

(δ) Ex. (3) shows an interesting use of the formula (55). The reader may be interested to see how much more algebra is needed to solve Ex. (3) when this formula is not used; cf. § 4·81 (i).

12·62. Further examples involving a small number of particles. (1) *A particle of mass m moves down the inclined face (inclination α) of a smooth wedge of mass M, the wedge being on a smooth fixed horizontal table. Find the acceleration of the wedge.*

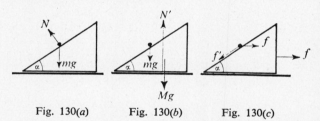

Fig. 130(a)   Fig. 130(b)   Fig. 130(c)

This problem has previously been solved in § 6·921. We shall use the same notation as before. Figs. 130 (a), (c) are the same as Figs. 62 (a), (c), and show: (a) the forces on the particle; (c) the acceleration f of the wedge, and the acceleration of the particle (which is the resultant of f and the acceleration f' relative to the wedge).

Considering the particle alone, we obtain as before (resolving at right angles to N)
$$mg \sin \alpha = m(f' - f \cos \alpha). \qquad \text{(i)}$$

Now, however, we shall consider the system consisting of the particle and wedge (instead of considering the wedge alone, as we did previously). Fig. 130 (b) shows the external forces acting on this system.

Since there is no horizontal component of the resultant of the external forces on this system, the rate of change of the horizontal component of the momentum of the whole system is zero.

Hence
$$m(f - f' \cos \alpha) + Mf = 0. \qquad \text{(ii)}$$

From (i) and (ii), we obtain, as in § 6·921,
$$f = \frac{mg \cos \alpha \sin \alpha}{M + m \sin^2 \alpha}.$$

(2) *Two equal particles P, Q are joined by a light inelastic string which passes through a small hole O in a smooth fixed horizontal table. P moves on the table and Q hangs vertically below O. Initially the string is taut, Q is at rest, and P is 1 ft from O and moving with a velocity of 12 ft/sec at right angles to OP. Find the distance of P from O and the speed of P when P is next moving at right angles to OP.*

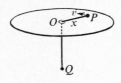

Fig. 131

Let m lb be the mass of each particle. Let x ft be the required distance of P from O when P is moving at right angles to OP, with speed v ft/sec, say.

The forces on P are its weight, the vertical support of the table, and the tension of the string. The first two of these forces are seen to be equal and opposite. The tension force passes through O.

Hence the resultant force on P passes through O.

Hence the angular momentum of P about O is constant.

Hence $$mvx = m \times 12 \times 1. \qquad \text{(i)}$$

The mechanical energy of the whole system is conserved since, of all the forces present, only the weight of Q does work.

The K.E. of Q is zero both initially and when $OP = x$; for, when P is moving at right angles to OP, then OP, and therefore OQ, is stationary, i.e. the velocity of Q is zero.

The P.E. of P does not change.

Hence the loss in K.E. of P is equal to the gain in P.E. of Q.

Hence $$\tfrac{1}{2}m(12^2 - v^2) = mg(x - 1). \qquad \text{(ii)}$$

From (i), (ii), we derive
$$\frac{12^2(x^2 - 1)}{x^2} = 2g(x - 1) ;$$

i.e. $$(x - 1)(4x^2 - 9x - 9) = 0;$$

i.e. $$(x - 1)(x - 3)(4x + 3) = 0.$$

The result $x = 1$ corresponds to the initial circumstances; the result $x = -\tfrac{3}{4}$ is irrelevant (the reader might show this in two ways).

The required distance is therefore 3 ft; and the required speed of P is, by (i), 4 ft/sec.

(3) *Two particles of masses m_1, m_2, connected by a light string of length b are in motion in a vertical plane, the string being taut. Prove that the angular velocity ω of the string is constant, and find the tension in terms of m_1, m_2, b, ω.*

Fig. 132   Fig. 133

Let G be the C.M. of the system. Then the distances of the particles from G are r_1, r_2, where

$$m_1 r_1 = m_2 r_2, \quad r_1 + r_2 = b. \tag{i}$$

The resultant external force on the system is the weight $(m_1 + m_2)g$, which acts through G, and hence has zero moment about G. Hence, by the extended principle of angular momentum (54c), the rate of change of angular momentum about G of the motion of the system relative to G is zero.

Relative to G, m_1 moves in a circle of radius r_1 with speed $r_1\omega$.

Hence the angular momentum (moment of momentum) H_1 of m_1 about G is given by

$$H_1 = m_1(r_1\omega)r_1 = m_1 r_1^2 \omega;$$

and similarly with m_2.

Hence $\dfrac{d}{dt}(m_1 r_1^2 \omega + m_2 r_2^2 \omega) = 0.$

Since $m_1 r_1^2 + m_2 r_2^2$ is constant, it follows that $d\omega/dt = 0$, i.e. that ω is constant.

To determine the tension T, it is necessary to consider a single particle, say m_1, separately from the rest of the system. This particle is acted on by the forces shown in Fig. 133.

The acceleration of m_1 is the resultant of: the acceleration of G, i.e. g vertically downwards (by the principle of linear momentum applied to the whole system); and the acceleration of m_1 relative to G, i.e. $r_1\omega^2$ towards G.

By the fundamental equation (26a), we have the vector equation:

$m_1 g$ downwards $+ T$ towards G
$\qquad\qquad = m_1(g$ downwards $+ r_1\omega^2$ towards $G)$.

Hence $\quad T$ towards $G = m_1 r_1 \omega^2$ towards G.

By (i), we then obtain

$$T = m_1 m_2 \omega^2 b/(m_1 + m_2).$$

(4) *A shell of mass M, moving in any manner, bursts into three equal pieces, each of mass $\tfrac{1}{3}M$. If v_{12}, v_{23}, v_{31} are the magnitudes of the mutual relative velocities of the pieces after the explosion, prove that the increase in kinetic energy due to the explosion is $M(v_{23}{}^2 + v_{31}{}^2 + v_{12}{}^2)/18$.*

Before explosion   After explosion

Fig. 134

Let $\mathbf{V}$ be the velocity of M just before the explosion, and $\mathbf{v}_1, \mathbf{v}_2, \mathbf{v}_3$ the velocities of the pieces (P, Q, R say) just afterwards (Fig. 134).

There are no external impulsive forces during the explosion.

Hence by the principle of linear momentum,
$$\tfrac{1}{3}M\mathbf{v}_1 + \tfrac{1}{3}M\mathbf{v}_2 + \tfrac{1}{3}M\mathbf{v}_3 - M\mathbf{V} = 0;$$
thus $\qquad \mathbf{V} = \tfrac{1}{3}(\mathbf{v}_1 + \mathbf{v}_2 + \mathbf{v}_3).$ \hfill (i)

The required gain in K.E.
$$= \tfrac{1}{2}(\tfrac{1}{3}M)\mathbf{v}_1{}^2 + \tfrac{1}{2}(\tfrac{1}{3}M)\mathbf{v}_2{}^2 + \tfrac{1}{2}(\tfrac{1}{3}M)\mathbf{v}_3{}^2 - \tfrac{1}{2}M\mathbf{V}^2$$
$$= \frac{M}{18}\{3(\mathbf{v}_1{}^2 + \mathbf{v}_2{}^2 + \mathbf{v}_3{}^2) - (\mathbf{v}_1 + \mathbf{v}_2 + \mathbf{v}_3)^2\}, \qquad \text{by (i),}$$
$$= M(2\mathbf{v}_1{}^2 + 2\mathbf{v}_2{}^2 + 2\mathbf{v}_3{}^2 - 2\mathbf{v}_2\cdot\mathbf{v}_3 - 2\mathbf{v}_3\cdot\mathbf{v}_1 - 2\mathbf{v}_1\cdot\mathbf{v}_2)/18$$
$$= M\{(\mathbf{v}_2 - \mathbf{v}_3)^2 + (\mathbf{v}_3 - \mathbf{v}_1)^2 + (\mathbf{v}_1 - \mathbf{v}_2)^2\}/18.$$

The result follows immediately, since the velocity $\mathbf{v}_{23}$ of Q relative to R is equal to $\mathbf{v}_2 - \mathbf{v}_3$, and $\mathbf{v}_{23}{}^2 = v_{23}{}^2$; etc.

S.P.I.C.E. (α) An advantage of the method used in Ex. (1) over that in § 6·921 is that the acceleration f is determined without introducing the reaction N into the equations. Of course if N is also required, this advantage is lost.

(β) Ex. (2) shows how the principles of energy and angular momentum can help in a problem whose solution might otherwise be difficult.

(γ) In Ex. (3), where a required force is *internal* to the whole system of particles, it is necessary to consider separately a part of the system for which the force is *external*.

(δ) Ex. (4) is a case in which vector methods are used to advantage. The student should satisfy himself of this advantage by trying to work Ex. (4) out independently of vector methods.

EXAMPLES XII

1. D, E, F are midpoints of the sides BC, CA, AB of an equilateral triangle. Forces of magnitudes 1, 2, 3, 4, 5, 6 units act along BC, CA, AB, FE, ED, DF, respectively. Prove that the line of action of the resultant divides each of AB, AC in the ratio $5:1$.

2. A force of components X, Y, parallel to the axes OX, OY, acts through the point (x, y). Prove that its moment about (a, b) is $(x - a)Y - (y - b)X$.

Forces of magnitudes 4, 5, 6 units act along the sides OA, BA, CB, of a square $OABC$ of side-length l in the senses indicated by the order of the letters. Taking OA, OC as the x-, y-axes, find the equation of the line of action of the resultant.
[Answer: $5x + 10y = 11l$.]

3. A, B, C are the points $(3, 2)$, $(5, -4)$, $(-1, 7)$, respectively. Forces whose components parallel to OX, OY, are $(8, -5)$, $(-3, 4)$, $(7, 6)$ units, act at A, B, C, respectively. Prove that the distance from O of the line of action of the resultant is 6 units.

4. The diagonals AC, BD of a quadrilateral $ABCD$ intersect at O, and the parallelograms $AODE$, $COBF$ are completed. Prove that the resultant of forces represented completely by DC, AB is represented completely by EF.
(Hint:—Complete the parallelogram whose sides are along EA, ED, FB, FC.)
(Notice that the solution incidentally proves the concurrency of AB, CD, EF.)

5. Four forces act from a point O to the corners of a convex quadrilateral $ABCD$, their magnitudes being proportional to OA, OB, OC, OD. If the forces are in equilibrium, find the position of O. (Hint:—Use Ex. (1) of § 6·19.)
[Answer: O is the intersection of KM, NL, where K, L, M, N are the mid-points of AB, BC, CD, DA.]

6. Two perfectly elastic smooth spheres A, B of equal mass are on a table. A is at rest initially, and is then struck obliquely by B. Prove that after the impact the directions of motion are at right angles.

7. Just before two equal smooth balls impinge, their velocities are U_1, U_2, at angles α_1, α_2 with the line of impact. Prove that just after the impact their relative velocity makes an angle
$$\cot^{-1}\left\{\frac{e(U_2 \cos \alpha_2 - U_1 \cos \alpha_1)}{U_1 \sin \alpha_1 - U_2 \sin \alpha_2}\right\}$$
with the line of impact, where e is the coefficient of restitution.

8. Two equal smooth balls, each of radius 1 in, are moving in parallel lines 1 in apart, each with speed U and in opposite directions. The balls subsequently collide, the coefficient of restitution being $\frac{3}{4}$. Prove that, after the impact, their speeds are $U\sqrt{43}/8$, and that their directions of motion are inclined at $\tan^{-1}(4/3\sqrt{3})$ with the line of centres.

9. A red ball R is placed on a spot 6 in from the cushion AB, and 3 ft from each of the cushions CA, DB of a billiard table $ABDC$, where $AB = 6$ ft, $AC = 12$ ft. The coefficient of restitution between the ball and the cushion is $\frac{3}{4}$. Find the point on the cushion AC from which a white ball, of mass equal to that of R, should be sent in order to strike R directly and send it after two reflections into the pocket at C, all friction being neglected. [Answer: 7 ft from C.]

10. A particle is projected with speed V at an inclination α, and rebounds from the ground which is smooth and horizontal. The coefficient of restitution is e. Prove that the particle covers a horizontal distance of $V^2 \sin 2\alpha/\{g(1 - e)\}$ before ceasing to bounce, and that the time occupied is $2V \sin \alpha/\{g(1 - e)\}$.

11. A, B are fixed points at the same level, distant $2a$ apart. Initially, two equal ivory balls, P, Q of radii a hang vertically from A, B by equal light inelastic strings AC, BD. P is then raised so that AC is kept taut and turned away from BD through $90°$ in the vertical plane through AB. P is now released from rest and strikes Q which rises until the string BD has turned through $80°$. Find the coefficient of restitution for the balls. [Answer: 0·82.]

12. A ball is projected from the point O with speed $\sqrt{(gh)}$ at an inclination α, strikes a fixed smooth vertical wall whose plane is perpendicular to the plane of the motion, and returns through O. The distance from O to the wall is d. Find the coefficient of restitution. [Answer: $d/(h \sin 2\alpha - d)$.]

*13. Two equal smooth balls lie in contact on a fixed table, and a third equal ball strikes them simultaneously and symmetrically. If the coefficient of restitution is 0·8, prove that 0·216 of the kinetic energy is dissipated at the impact.

14. A shell moving through the air is broken into two portions of masses m_1, m_2, by an internal explosion which imparts kinetic energy E to the system. Prove that the relative speed of the portions after the explosion is
$$\sqrt{\{2(m_1 + m_2)E/m_1 m_2\}}.$$

15. A mass M, which can slide on a smooth fixed horizontal rod, carries a light smooth pulley P round which passes a light inelastic cord OPA, fixed to the rod at one end O and carrying a mass m at the other end A. The cord is taut and the system is moving in a vertical plane so that the inclination θ of PA to the vertical is constant. Prove that the acceleration f of M is constant, and show that f, θ are given by $\tan \theta = f/g$; $m/M = \sin \theta/(1 - \sin \theta)^2$.

16. Two particles of masses m_1, m_2 are joined by a light inelastic string and are initially close together on a smooth table. A smooth ring of mass M is threaded by the string and hangs over the edge of the table, and the portions of the string on the table are at right angles to the edge. Find the acceleration of the ring.
[Answer: $(m_1 + m_2)Mg/\{(m_1 + m_2)M + 4m_1 m_2\}$.]

*17. Two equal particles A, B are joined by a light rod AB of length $2a$. Initially the system is at rest with AB vertical, A above B, and B in contact with a smooth table. The system is then given a slight angular displacement and begins to fall on to the table. Prove that the midpoint of AB moves in a straight line, and find the speed of B when the inclination of the rod to the vertical is $60°$. [Answer: $\sqrt{(ag/7)}$.]

18. Two particles of masses m, m', connected by a light taut inelastic string, lie on a smooth table, and a blow of impulse J is applied to m in a direction that makes an angle α with the line of the particles and keeps the string taut. Find the impulsive tension produced in the string. [Answer: $Jm' \cos \alpha/(m + m')$.]

*19. Three particles A, B, C, each of mass m, are placed at the angular points of an equilateral triangle; and A, B are connected to C by light inelastic strings that are just taut. C is then projected at right angles to and towards the line of A, B with speed v. Find the loss in kinetic energy at the instant when A, B are jerked into motion.
[Answer: $3mv^2/10$.]

20. Two particles, each of mass 0·05 kg, are connected by a light perfectly elastic string of natural length 30 cm and modulus 0·4 Kg, and the system moves on a smooth table. Initially the particles are close together, A is at rest, and B is moving with a speed of 1 m/s directly away from A. Find the greatest length of the string.

[Answer: 34·4 cm.]

*21. Two particles A, B of masses m, m' are connected by a light perfectly elastic string of natural length a, which is in motion on a smooth table. Initially A is fixed, and B is rotating steadily about A with angular speed Ω, the string being stretched to a length b. If A is now released, prove that the angular speed of the string if and when the string becomes slack is $b^2\Omega/a^2$.

Prove that the string will become just slack during the motion if
$$(m + m')a^2 = mb(a + b).$$

22. A uniform chain of length l slides over the edge of a smooth fixed table, the portion on the top of the table being at right angles to the edge. Prove that when a length x of chain is over the edge, where $x < l$, the acceleration is gx/l. If the table is rough, with coefficient of friction μ, prove that the acceleration is $g\{x(1 + \mu) - l\mu\}/l$.

23. A uniform chain of length l and weight W is hanging vertically from its two ends A, B which are close together, when the end B is suddenly released. Find the tension at A when B has fallen a distance x (where $x < l$). [Answer: $\frac{1}{4}W(1 + 3x/l)$.]

*24. A and B are two equal particles, joined by a light inelastic string. The system is in motion on a smooth table, the string being taut and A being constrained to move in a smooth straight groove on the table. Prove that the force exerted by the groove on A at any time is proportional to $\sin\theta/(1 + \cos^2\theta)^2$, where θ is the inclination of AB to the groove.

*25. Three equal particles, A, B, C are lying on a smooth table in a straight line, and are joined together by light taut inelastic strings AB, BC, where $AB = BC = a$. A and C are now simultaneously projected in the same direction with the same speed V at right angles to ABC. When either string has turned through an angle θ, prove that its angular speed is $V/\sqrt{\{a^2(2 - \cos 2\theta)\}}$.

CHAPTER XIII

TWO-DIMENSIONAL DYNAMICS OF A RIGID BODY

"We are not divided,
All one body we."
—Hymns A. & M.

13·1. Rigid body. A *rigid body* is a system of particles which has the special property that the distance between each pair of the particles is permanently fixed.

A pair of particles joined by a light rod of fixed length would be a very simple example of a rigid body on this definition. Continuously distributed mass-systems may also be rigid bodies.

By virtue of the remarks in the second paragraph of § 11·2, all the theorems proved in Chapter XII for a general discrete system of particles may be applied to any rigid body or system of rigid bodies as particular cases. For the same reason it will be permissible for us in the present chapter (e.g. in §§ 13·41, 13·61, 13·81) to set up further theory for any rigid body by considering only discrete cases.

It should be noted that the rigid body, like the particle, is a mathematical model. All 'real' bodies are deformable to greater or less degrees. In introducing the concept of rigid body we take one step further in our attempts to describe our observations of the natural world.

13·2. Kinematical considerations.

13·21. Degrees of freedom. If k is the minimum number of coordinates (the coordinates may be distances from a fixed point, or angular coordinates, or may be more general) needed to determine the configuration of a system of particles, the system is said to have k degrees of freedom.

A single particle free to move in two-dimensional space thus has 2 degrees of freedom; in three-dimensional space, 3.

A system of n entirely free particles in three-dimensional space has $3n$ degrees of freedom. If, however, the particles belong to a rigid body, the coordinates have to satisfy equations expressing the fixity of the distances between the particles, and the number of degrees of freedom is accordingly reduced.

A rigid body restricted to two-dimensional motion, but otherwise unrestricted, has 3 degrees of freedom. (The configuration is determined if we know two coordinates giving the position of any particular particle P of the body, together with a single angular coordinate.)

If the body in the last paragraph is further restricted to rotation about a fixed axis, it has just 1 degree of freedom, since its configuration can then be specified by a single angular coordinate.

In the theory and examples to follow in this chapter, it is to be understood without special mention that conditions are either coplanar or two-dimensional (the context will make it clear which of these is the case). When, as e.g. in § 13·22, the coplanar case is taken, the application to general two-dimensional motion will be obvious to the reader.

13·22. Definition of angular velocity of a rigid body. Let l, m be any pair of lines fixed to a moving rigid body, and L, M a pair of lines fixed in space.

Let $\alpha, \beta, \theta, \phi, \eta$ be as in Fig. 135.

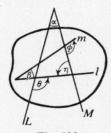

Fig. 135

Then $\theta - \alpha = \eta = \phi - \beta$.

As the time t changes, θ and ϕ in general change, but α, β are fixed. Hence, differentiating with respect to t, we have $\dot{\theta} = \dot{\phi}$.

It follows that at any instant the rate of change of the angle which any line fixed in the body makes with a line fixed in space is independent of the particular lines taken. (It is understood that the angle is measured in a definite sense—cf. the directions of the arrows round θ, ϕ in Fig. 135.)

This rate of change is called the *angular velocity* of the body at the instant.

13·221. Remarks.

(i) The angular velocity of a point (or a particle) is defined (§ 7·11) only in relation to some assigned reference point. In contrast, there is no reference point involved in specifying the angular velocity of a rigid body.

(ii) The existence of the angular velocity of a rigid body is a special consequence of its being rigid.

(iii) A much more elaborate discussion is needed before the angular velocity of a rigid body in three-dimensional motion can be defined (see Appendix, § 17·2). In that case, the angular velocity must be treated as a vector; in the two-dimensional treatment, to which we limit ourselves in this chapter, the vector aspect need not be stressed—cf. the third paragraph of § 12·11.

13·23. Angular acceleration of a rigid body. The angular acceleration of a rigid body is defined as the rate of change of the angular velocity (sign conventions being understood as usual).

The angular acceleration of the body in Fig. 135 is thus $\ddot{\theta}$.

13·231. As in § 7·12, we may express the angular acceleration in the form (36), viz.

$$\ddot{\theta} = \frac{d}{d\theta}(\tfrac{1}{2}\dot{\theta}^2). \tag{36}$$

13·24. Velocity of a point of a rigid body. At time t, let $\mathbf{v}$, $\mathbf{v}_0$ be the velocities of any two points P, P_0 of a moving rigid body (Fig. 136), and let ω be the angular velocity of the body. Let $P_0P = r$.

Relative to P_0, P is describing a circle of radius r with angular velocity ω about P_0. Hence $\mathbf{v}_{P\,\text{rel}\,P_0} = r\omega\mathbf{t}$, where $\mathbf{t}$ is a unit vector at right angles to P_0P (drawn in the appropriate sense). Hence, by the relative velocity formula (§ 6·91),

$$\mathbf{v} = \mathbf{v}_0 + r\omega\mathbf{t}.$$

Fig. 136   Fig. 137

13·241. Now suppose the body is rolling without sliding over a fixed surface, and at time t take P_0 in particular at the point of contact A (Fig. 137). Since there is no sliding, P_0 is now instantaneously at rest. Hence, from the formula of § 13·24, the velocity of P at time t is given by

$$v = r\omega,$$

the direction being at right angles to AP.

13·25. Examples. (1) *A cylinder A of radius a is rolling without slipping, with its axis always horizontal, over a fixed cylinder B of radius b and axis parallel to that of A. When the plane containing the axes of the cylinders is inclined at θ to the upward vertical, prove that the angular velocity of A is $\dot\theta(a+b)/a$.*

In Fig. 138, which shows a cross section, P is the point of contact corresponding to the given θ, G and O are the centres of A and B, OD and HG are vertical, and C is the point of A which would make contact with the highest point D of B.

Since GH, being vertical, is in a direction fixed in space, and since GC is a line fixed in the moving cylinder A, the angular velocity of A is $\dot\phi$ in the sense indicated.

We proceed to connect ϕ with θ. We have
$$\phi = \theta + \angle OGC$$
$$= \theta + (\text{arc } CP)/a$$
$$= \theta + (\text{arc } DP)/a$$
$$= \theta + \theta b/a$$
$$= \theta(a+b)/a.$$

Differentiating with respect to t, we have
$$\dot\phi = \dot\theta(a+b)/a.$$

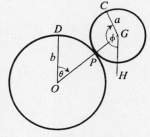

Fig. 138

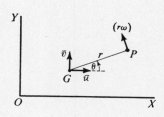

Fig. 139

(2) *At a given time t, the components of the velocity of the centre of mass of a moving rigid body are $(\bar{u}, \bar{v})$ and the angular velocity is ω. P is a point of the body distant r from G; and at the time t, GP is inclined at θ to the x-axis. Find the components of the velocity of P at this time.*

The velocity of G is $\bar u \mathbf{i} + \bar v \mathbf{j}$.

Since P is describing a circle relatively to G, the velocity of P relative to G is $r\omega$ at right angles to GP, as indicated in Fig. 139; i.e., is equal to
$$-(r\omega \sin\theta)\mathbf{i} + (r\omega \cos\theta)\mathbf{j}.$$
The velocity, $u\mathbf{i} + v\mathbf{j}$ say, of P is therefore given by
$$u\mathbf{i} + v\mathbf{j} = (\bar u - r\omega \sin\theta)\mathbf{i} + (\bar v + r\omega \cos\theta)\mathbf{j}.$$
Hence
$$u = \bar u - r\omega \sin\theta, \qquad v = \bar v + r\omega \cos\theta.$$

S.P.I.C.E. (α) It will be noticed that, in Ex. (1), θ is not the angular velocity of the moving cylinder. In determining this angular velocity, it was necessary to refer to the definition of § 13·22.

(β) Ex. (2) is a particular case of § 13·24, and gives the velocity of any given point of a rigid body in terms of the three quantities $\bar{u}, \bar{v}, \omega$.

13·3. Pin-joints. In many of the problems to follow, a rigid body will be rotating about a cylindrical pin which pierces the body.

Unless the contrary is stated, we shall work in terms of a mathematical model pin which is so small that we may neglect the space it occupies when calculating the moment of inertia of the body; but not so small that we may in general treat all the forces (in any plane at right angles to the axis of the pin) exerted by the pin on the body, as acting through a single point.

By § 12·221, the forces just referred to are statically equivalent to a single force passing through the axis of the pin, together with a couple. The single force is in practice usually represented by its components X, Y in two directions at right angles, and the couple by the magnitude L of its moment; these features are commonly represented on a diagram as in Fig. 140.

Friction at the joint produces a couple on the body which acts against the direction of rotation, and is referred to as the *friction couple L*. (Fig. 140 would thus correspond to anticlockwise rotation of the body about the pin.)

If in particular the pin is smooth, the friction couple L is zero.

In the theory to follow, the term 'rotation about an axis' will be understood to mean rotation about a pin-joint of the above character.

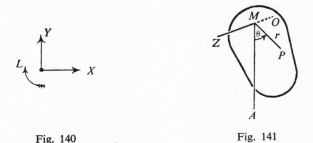

Fig. 140   Fig. 141

13·4. Kinetic energy of a rigid body.

13·41. Case of fixed axis of rotation. Let OZ be a given fixed axis about which a rigid body rotates, and let ω be the angular velocity at any instant. Let P, of mass m, be any particle of the body, and let

PM, $= r$ say, be perpendicular to OZ. Let MA be a line fixed in space and at right angles to OZ, and let the angular coordinate θ be as in Fig. 141.

Then the angular velocity of P about $M = \dot\theta = \omega$, by § 13·22. Hence the speed of $P = r\omega$. Thus the kinetic energy of $P = \frac{1}{2}m(r\omega)^2$.

Hence, by § 12·51, the kinetic energy of the body
$$= \Sigma(\tfrac{1}{2}mr^2\omega^2)$$
$$= \tfrac{1}{2}(\Sigma mr^2)\omega^2$$
$$= \tfrac{1}{2}I\omega^2, \tag{56a}$$
where I here denotes the moment of inertia of the body *about the axis of rotation* OZ.

13·42. General two-dimensional case.
By § 12·511 and § 13·41, the kinetic energy of a body in general (two-dimensional) motion
$$= \tfrac{1}{2}M\bar{v}^2 + \tfrac{1}{2}I\omega^2, \tag{56b}$$
where M is the total mass, $\bar{v}$ is the speed of the centre of mass G, I is the moment of inertia *about G**, and ω is the angular velocity of the body; (the student should satisfy himself that ω is relevant to the motion relative to G).

The 'internal' kinetic energy in this case, viz. $\tfrac{1}{2}I\omega^2$, is sometimes referred to as the 'rotational' kinetic energy.

As an exercise, the student should use (56b) to derive the formula (56a) for the special case of rotation about a fixed axis. (Note that the 'I' in (56a) is not the same as the 'I' in (56b).)

13·5. Potential energy.
The statement on potential energy in § 12·52 covers the case of a rigid body.

13·51.
In a field of constant gravity (see § 12·521), the corresponding potential energy of a rigid body is $Mg\bar{z}$, where M is the mass of the body and $\bar{z}$ is its height above a reference level.

13·6. Use of the principle of energy.
Many problems on the two-dimensional motion of a rigid body can be solved (and are best solved) by using only the principle of energy. For this reason, we shall in the present chapter consider the principle of energy before the principles of linear and angular momentum.

The main theory has already been given in Chapter XII; §§ 13·61–13·64 contain results special to the case of a rigid body.

*In two-dimensional problems, we sometimes speak of the moment of inertia 'about a point', meaning 'about a line through the point perpendicular to the plane of motion'.

13·61. Work of internal forces in a rigid body. We first prove that during any displacement of a *rigid* body the total rate of work of the set of internal (action, reaction) forces is zero. (This result does *not* hold in general for a system of particles that is not rigid.)

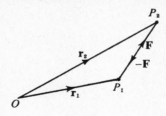

Fig. 142

At time t, let $\mathbf{r}_1$, $\mathbf{r}_2$ be the position vectors, referred to a fixed origin O, of any two particles P_1, P_2 of the body, and let $-\mathbf{F}$, $\mathbf{F}$ give the magnitude and direction of the action, reaction forces between them, as shown in Fig. 142.

The sum of the rates of work of the pair of forces in Fig. 142

$$= (-\mathbf{F}) \cdot \dot{\mathbf{r}}_1 + \mathbf{F} \cdot \dot{\mathbf{r}}_2$$
$$= \mathbf{F} \cdot (\dot{\mathbf{r}}_2 - \dot{\mathbf{r}}_1)$$
$$= \mathbf{F} \cdot \frac{d}{dt} \mathbf{P}_1\mathbf{P}_2. \qquad (i)$$

The expression (i) is zero if $\dfrac{d}{dt} \mathbf{P}_1\mathbf{P}_2$ is either zero or perpendicular to $\mathbf{F}$, i.e. to $\mathbf{P}_1\mathbf{P}_2$.

Now, since the body is rigid, $P_1P_2^2 =$ constant; i.e. $\mathbf{P}_1\mathbf{P}_2 \cdot \mathbf{P}_1\mathbf{P}_2 =$ constant; hence, differentiating with respect to t, (using § 6·171 (xiv)), we have $2\mathbf{P}_1\mathbf{P}_2 \cdot \dfrac{d}{dt} \mathbf{P}_1\mathbf{P}_2 = 0$. Since $\mathbf{P}_1\mathbf{P}_2 \neq 0$, it follows that $\dfrac{d}{dt} \mathbf{P}_1\mathbf{P}_2$ is either zero or perpendicular to $\mathbf{P}_1\mathbf{P}_2$, and hence that the expression (i) is zero.

Summing for all pairs of particles of the body, we then have that the total rate of work of the whole set of internal forces is zero; and this holds at any and hence every instant.

From this the theorem follows.

This proof incidentally holds for the general three-dimensional case.

13·611.

The theorem of § 13·61 applies also to an *inelastic* string (assumed as usual to be flexible) in certain cases where the string is not kept straight, e.g. in problems of the type solved in § 6·85.

13·62. Form of energy principle for a rigid body. For a *rigid* body, the total work done by the *external* forces is equal to the increase in the kinetic energy of the body.

This follows from the more general energy principle (22a) as given in § 12·53 for a system of particles, in view of the result proved in § 13·61; and is valid in three as well as in two dimensions.

13·63. Case of conservative forces. In place of the works done by forces which belong to conservative fields, we may in using the energy equation substitute corresponding decreases in potential energy.

In particular, if all the external forces which do non-zero work on a rigid body are conservative, the sum of the kinetic and potential energies is constant.

13·64. Work done by forces at contacts.

(i) If a body rotates about a fixed smooth axis, the work done by the force at the axis is zero, since there is no displacement of its point of action.
The case in which there is a friction couple at the axis will be discussed in § 13·831.

(ii) If a body A is rolling without sliding on a second body B which is fixed (the contacts may be rough or smooth), the work done on A by the forces at the points of contact is zero.
To show this, let $\mathbf{F}$ be the force at any particular instant on A at its point of contact, P say, with B. Since there is no sliding, the velocity, $\mathbf{v}$ say, of P at this instant is zero. Hence the rate of work of $\mathbf{F}$, namely $\mathbf{F} \cdot \mathbf{v}$, is zero. This gives the result.

(iii) Suppose that the conditions are as in (ii) except that B is now not fixed. Thus at the instant under consideration, the velocity $\mathbf{v}$ of P is not necessarily zero. The rate of work of the contact force $\mathbf{F}$ at P on A is equal to $\mathbf{F} \cdot \mathbf{v}$, while that by the reaction force $-\mathbf{F}$ on B is equal to $(-\mathbf{F}) \cdot \mathbf{v}$. The sum of these rates of work is zero. Hence, in forming the energy equation *for the system consisting of A and B together*, the work due to the action, reaction forces at P may be disregarded.

(iv) The result in (iii) may be extended. For example, if a string passes round the groove of a pulley sufficiently rough to prevent slipping, the work done at the contacts with the groove may be disregarded if we form the energy equation *for a system which includes the string and the pulley*.

(v) If a body is slipping against a *smooth* fixed surface, the work done by the contact forces is zero; for at every instant the contact force is at right angles to the velocity of the point of the body on which it acts.

(vi) If the surface in (v) is rough, negative work will be done on the body by the contact forces.

13·65. Examples. (1) *The shaft of a wheel can rotate about a fixed smooth axis. A light inelastic cord is tightly coiled round the shaft and is pulled with a constant force of 25 lb.wt. When 10 ft of the cord have been uncoiled, the wheel, which was initially at rest, is rotating at 5 revolutions per second. Find the moment of inertia, about the axis of rotation, of the system consisting of the wheel and shaft.*

The forces acting on the moving system include: (a) the weight of the system; (b) the force exerted at the axis on the system; (c) the pull of $25g$ pdl; (d) internal forces. (See Fig. 143.)

Of these forces, only (c) does work during the motion; the work $= 25g \times 10$ ft-pdl.

By the principle of energy, this work $= \frac{1}{2}I\Omega^2$ ft-pdl, where I lb-ft^2 is the required moment of inertia, and Ω rad/sec, the angular speed acquired, $= 5 \times 2\pi$ rad/sec.

Hence $\qquad\qquad 250g = 50\pi^2 I.$

Hence the required moment of inertia

$\qquad\qquad = 5g/\pi^2$ lb-ft^2.

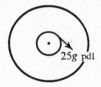

Fig. 143

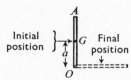

Fig. 144

(2) *A uniform rod OA of length 2a can rotate smoothly about the end O which is fixed. It is slightly displaced from the equilibrium position in which A is above O. Find its angular speed when horizontal.*

Let M be the mass of the rod.

By Routh's rule, its moment of inertia about the centre of mass G is $\frac{1}{3}Ma^2$.

Hence, by the theorem of parallel axes, its moment of inertia about O is $\frac{4}{3}Ma^2$.

The only force which does work during the displacement is the weight. Hence the mechanical energy is conserved.

Hence, if Ω is the required angular speed, we have

$\qquad \frac{1}{2} \cdot \frac{4}{3} M a^2 \Omega^2 =$ increase in K.E.

$\qquad\qquad\qquad = $ decrease in P.E.

$\qquad\qquad\qquad = Mga.$

Hence $\qquad\qquad \Omega = (3g/2a)^{\frac{1}{2}}.$

(3) *A uniform solid sphere rolls without slipping down a line of greatest slope of a rough fixed inclined plane of inclination α. Find the acceleration of its centre.*

The forces on the sphere are: (a) its weight; (b) the force at the point of contact with the inclined plane; (c) internal forces.

By §§ 13·61, 13·64, (b) and (c) do no work during the motion. Hence the mechanical energy is conserved.

Let G be the position of the C.M., v the speed of the C.M., and ω the angular speed of the sphere, at any time t.

At the particular time $t = 0$, let the C.M. be at G_0, and let $v = v_0$. Let $G_0G = x$.

By § 13·42, the K.E. at time t is $\frac{1}{2}Mv^2 + \frac{1}{2}I\omega^2$, where M is the mass of the sphere, I (the moment of inertia *about* G) is $\frac{2}{5}Ma^2$ (by Routh's rule), a being the radius of the sphere, and $\omega = v/a$ (by § 13·241).

This K.E. $= \frac{1}{2}Mv^2 + \frac{1}{5}Ma^2(v/a)^2 = 7Mv^2/10$.

In particular, the initial K.E. $= 7Mv_0^2/10$.

The loss of P.E. $= Mgx \sin \alpha$.

Hence $\qquad 7Mv^2 - 7Mv_0^2 = 10Mgx \sin \alpha$.

Differentiating with respect to x, we have by (3), (p. 12),
$$7M\ddot{x} = 5Mg \sin \alpha.$$
Thus the acceleration $\ddot{x}$ of the C.M. is constant and equal to $(5g/7) \sin \alpha$.

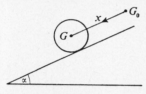
Fig. 145

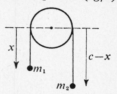
Fig. 146

(4) *Masses m_1, m_2 are attached to the ends of a light inelastic string which passes round a pulley of mass M and radius a. The pulley can rotate about a frictionless horizontal fixed axis, but its groove is sufficiently rough to prevent the string from slipping round it. The radius of gyration of the pulley about its axis is k. Find the acceleration of either mass.*

In the light of §§ 13·61, 13·611, 13·63, 13·64, it is seen that the mechanical energy of the system consisting of the pulley, the string and the two masses is conserved.

At time t, let the speeds of m_1, m_2, be v; and let the distance of m_1 below the axis of the pulley be x.

The K.E. of the system is then $\frac{1}{2}m_1v^2 + \frac{1}{2}m_2v^2 + \frac{1}{2}Mk^2\omega^2$, where $\omega = v/a$; and the P.E. (taking the level of the axis as standard) is $-m_1gx - m_2g(c - x)$, where c is constant.

Hence $\qquad \frac{1}{2}(m_1 + m_2)v^2 + \frac{1}{2}Mk^2v^2/a^2 - (m_1 - m_2)gx =$ constant.

Differentiating with respect to x, we have
$$(m_1 + m_2 + Mk^2/a^2)\ddot{x} = (m_1 - m_2)g,$$
where $\ddot{x}$ is the downward acceleration of m_1.

Hence $\qquad \ddot{x} = (m_1 - m_2)g/(m_1 + m_2 + Mk^2/a^2)$.

S.P.I.C.E. (α) It will be noticed that in each of the above examples there is just one degree of freedom; i.e., one coordinate is sufficient to specify the configuration. The equation given by the energy principle is sufficient to show how this coordinate changes with the time.

(β) Before the energy equation is formed, exhaustive consideration of the possible work by the various forces present is seen to be necessary. Sometimes, but by no means always, this consideration shows that the mechanical energy is conserved.

(γ) The sphere in Ex. (3) is a three-dimensional body. But it is evident that the solution of this problem must be the same as that of the corresponding two-dimensional problem in which a circular disc of appropriate radius, mass and moment of inertia rolls down the inclined plane. (See the opening remarks in Chapter XII.)

(δ) The reader should check that if the sphere in Ex. (3) is replaced by the more general case of any body symmetrical about an axis rolling with this axis horizontal down the inclined plane, then the acceleration of the axis is $a^2 g \sin \alpha/(a^2 + k^2)$ where a is the distance of the axis from the inclined plane, and k is the radius of gyration about the axis.

(ε) The 'slight' displacement referred to in the enunciation of Ex. (2) means that the displacement is sufficient to upset the equilibrium, but not sufficient to need taking into account in the ensuing equations. Other words, e.g. 'just', are used in analogous ways in dynamics.

(ζ) If, in particular, we put M equal to zero in the result of Ex. (4), we get the acceleration formula given in § 4·71 as a particular case.

13·7. Use of the principle of (linear) momentum.

The energy principle is sufficient to determine the velocities, accelerations and positions of the rigid bodies in the examples of § 13·6, but does not enable all the acting forces to be determined.

(Indeed, a special charm of the energy principle is the number of forces that it enables us to avoid introducing into the equations used.)

If we seek to determine the resultant force acting, we need to use the principle of (linear) momentum. The most useful form for a rigid body is that stated in § 12·33, viz.: The centre of mass moves as if it were a particle acted on by forces whose magnitudes and directions are those of the external forces present.

By § 12·331, we then have

$$\Sigma X = M\ddot{\bar{x}}, \qquad \Sigma Y = M\ddot{\bar{y}}, \qquad (53d)$$

where M is the mass of the body, $\bar{x}, \bar{y}$ are the coordinates of G referred to fixed axes OX, OY, and X, Y are components of a typical external force acting on the body.

13·71. *Suppose for example we wish to determine the force at the axis on the moving system in Ex. (1) of § 13·65, with the additional data that the mass of the system is 10 lb and the pull of 25 lb.wt. acts vertically downward.*

Let X, Y pdl be the horizontal and vertical components of the force due to the axis on the system, as indicated in Fig. 147.

Fig. 147

Then by the principle of momentum, taking horizontal and vertical resolved parts, we have
$$X = 10f_x,$$
$$Y - 10g - 25g = 10f_y,$$
where f_x, f_y are components of acceleration of the centre of mass of the system; f_x, f_y are clearly zero in this problem.
Hence $X = 0$, $Y = 35g$.

Thus the force exerted at the axis on the system is 35 lb.wt., vertically upward throughout the motion.

13·72.

It should be noted that, in general, the force at an axis of rotation cannot be determined as easily as in the case of § 13·71.

For example, suppose it is required to determine the force at the axis in any position of the rod in Ex. (2) of § 13·65 (the mass of the rod being now given). It is now necessary to use the principle of linear momentum and, also, either the principle of energy or the principle of angular momentum. The process of solution will be indicated in § 13·931.

13·8. Use of the principle of angular momentum; case of fixed axis of rotation.
For a rigid body rotating about a fixed axis, the angular momentum principle gives an equation alternative to that given by the principle of energy.

13·81. Angular momentum of a rigid body rotating about a fixed axis.
We take circumstances and notation as in § 13·41 (Fig. 141). The angular momentum of the typical particle P about OZ (in the sense indicated by the arrow in Fig. 141) is its moment of momentum $m(r\omega)r$ about OZ. Hence, by § 12·42, the angular momentum of the body about OZ
$$= \Sigma(mr^2\omega)$$
$$= I\omega, \qquad (57)$$
where ω is the angular velocity of the body, and I is its moment of inertia about OZ.

13·82. Equation expressing the angular momentum principle. At time t, let N be the moment about OZ of a typical external force acting on the body, the arrow in Fig. 141 indicating the positive sense.

(In the present context, N is sometimes called by engineers the *torque* of the force.)

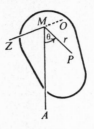

Fig. 141

Then the angular momentum principle of § 12·43 and (57) give

$$\Sigma N = \frac{d}{dt}(I\omega). \qquad (58a)$$

Since I is constant, we may write (58a) in the form

$$\Sigma N = I\ddot{\theta}, \qquad (58b)$$

where θ is the angle AMP of Fig. 141.

In the use of (58b), the arm MP of the angle θ may be taken to pass through any particular point P of the body. For many purposes, it is convenient to take it as passing through the centre of mass G.

13·83. Expression for work done during rotation. By §§ 13·41, 13·62, the sum of the rates of work of the external forces

$$= \frac{d}{dt}(\tfrac{1}{2}I\dot{\theta}^2)$$
$$= I\ddot{\theta}\dot{\theta}$$
$$= (\Sigma N)\dot{\theta},$$

by (58b). Hence the work done by the external forces during a finite rotation of the body about the axis OZ

$$= \Sigma \int N d\theta, \qquad (59)$$

the integral being taken over the relevant range of values of θ.

13·831. Work of friction couple. As a particular case, the work done on the body, during a motion in which θ continually increases, by a friction couple L at the axis is

$$-\int L d\theta. \qquad (59a)$$

If, further, L is constant, the work done by the friction couple is $-L\theta$, where θ is the angle turned through.

13·84. Examples.

Exs. (1), (2), (3) below will each be solved by two methods: (*a*) using the principle of angular momentum; (*b*) using the principle of energy.

(1) *A uniform thin magnet of length 2a and mass M can rotate about a fixed smooth vertical axis through its centre. There is a constant force of magnitude P acting each end, one towards the North and the other towards the South. The magnet is released at rest when pointing to the East. Find its angular speed when it points to the North.*

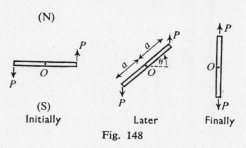

Fig. 148

(*a*) Introduce the angular coordinate θ (Fig. 148) to specify the general position of the magnet.

The external forces on the magnet are its weight, the two forces P and the force at the centre O.

The resultant of the moments of these forces about $O = 2Pa \cos \theta$.
The angular momentum about $O = I\dot\theta$, where $I = \tfrac{1}{3}Ma^2$.
By the principle of angular momentum, we then have

$$2Pa \cos \theta = \frac{d}{dt}(\tfrac{1}{3}Ma^2\dot\theta);$$

i.e. $\qquad Ma\ddot\theta = 6P \cos \theta.$

Integrating with respect to θ, we have

$$\tfrac{1}{2}Ma\dot\theta^2 = 6P \sin \theta + C,$$

where the constant C is zero, since $\dot\theta = 0$ when $\theta = 0$.
Hence the angular speed Ω when $\theta = \pi/2$ is given by

$$\tfrac{1}{2}Ma\Omega^2 = 6P;$$

i.e. $\qquad \Omega = \sqrt{(12P/Ma)}.$

(*b*) During the whole motion, the total work done on the magnet
$$= Pa + Pa = 2Pa.$$
The corresponding increase in K.E. $= \tfrac{1}{2}I\Omega^2,$
where $\qquad I = \tfrac{1}{3}Ma^2.$
By the principle of energy,
$$2Pa = \tfrac{1}{6}Ma^2\Omega^2,$$
giving $\qquad \Omega = \sqrt{(12P/Ma)},$ as before.

(2) *A wheel, whose moment of inertia about its axis is* 2000 *lb-ft²*, *initially rotating at an angular speed of* 100 *rev/min, is brought to rest in* 5 *min by a constant friction couple at its axis, which is horizontal. Find the magnitude of this couple, and the number of revolutions made before the wheel comes to rest.*

(*a*) Let OA be the vertical through the axis, B a fixed point of the wheel, θ the angle between OA and OB at time t, and let the motion be in the direction of increasing θ.

The external forces acting on the wheel are: the weight Mg; and the forces at the axis, represented by the components X, Y and the friction couple, L say. (See Fig. 149.)

By the principle of angular momentum,

$$-L = \frac{d}{dt}(2000\dot\theta). \qquad (i)$$

Integrating (i) with respect to t, we have
$$-Lt = 2000\dot\theta + C,$$
where C is constant.

The initial conditions give
$$\dot\theta = 100 \times 2\pi/60 \text{ when } t = 0.$$
Hence C is determined by
$$0 = 2000 \times 10\pi/3 + C.$$
The final conditions give $\dot\theta = 0$ when $t = 0$.
Hence $\qquad -300L = 0 - 20{,}000\pi/3.$
Hence the required couple $= 200\pi/9$ pdl-ft
$\qquad\qquad\qquad\qquad\quad = 2{\cdot}18$ lb.wt.-ft.

Integrating (i) with respect to θ, we have
$$-L\theta = 1000\dot\theta^2 + C',$$
where the constant C' is determined (taking $\theta = 0$ initially) by
$$0 = 1000(100 \times 2\pi/60)^2 + C'.$$
If n is the number of revolutions required, we then have
$$-L \times 2\pi n = 0 - 1000(10\pi/3)^2,$$
which gives $\qquad n = 250.$

(*b*) During the motion, the only work done on the wheel is the work done by the friction couple L.

At time t sec (where $t < 300$), this work $= -L\theta$ ft-pdl, where θ is the angle turned through.

Hence, by the principle of energy,
$$-L\theta = \tfrac{1}{2}I\dot\theta^2 - \tfrac{1}{2}I\Omega^2, \qquad (ii)$$
where $I = 2000$ (lb-ft²) and Ω, the initial angular speed in rad/sec, is $10\pi/3$.

Differentiating (ii) with respect to θ gives (i), and L may then be determined as in (*a*).

The number of revolutions n may then be determined directly from (ii), putting in particular $\dot\theta = 0$ and $\theta = 2\pi n$.

(3) *A gyroscope which is dynamically equivalent to a uniform disc of radius 3 in and mass 5 lb, is set spinning by coiling a light inelastic string round its axle of radius $1\frac{1}{2}$ in, attaching a mass of 2 lb to its end, and letting the mass fall until 12 ft of the string have become uncoiled. The axis, assumed to be smooth, is kept fixed during the process, and the string does not slip during the uncoiling. Find the angular velocity generated.*

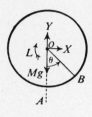

Fig. 149

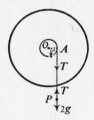

Fig. 150

(a) Let T pdl be the tension of the string, θ radians the angle turned through by the gyroscope, and x ft the downward displacement of the mass P after time t sec. (See Fig. 150.)

For the gyroscope, we have by the principle of angular momentum
$$T \times \tfrac{1}{8} = I\ddot{\theta}, \text{ where } I = \tfrac{1}{2} \times 5 \times (\tfrac{1}{4})^2 = 5/32.$$
For the mass P, we have
$$2g - T = 2f,$$
where f is the acceleration of P. Now $x = \tfrac{1}{8}\theta$; hence, differentiating twice, we have $f = \tfrac{1}{8}\ddot{\theta}$.

Eliminating T, we obtain
$$2g - \tfrac{5}{4}\ddot{\theta} = \tfrac{1}{4}\ddot{\theta}, \qquad \text{i.e. } \ddot{\theta} = \tfrac{4}{3}g.$$
Integrating with respect to θ, we have
$$\tfrac{1}{2}\dot{\theta}^2 = \tfrac{4}{3}g\theta + C,$$
where the constant C is zero since $\dot{\theta} = 0$ when $\theta = 0$.

The required angular velocity, Ω rad/sec, is given by the value of $\dot{\theta}$ when $x = 12$, i.e. when $\theta = 96$ (rad).

Hence $\tfrac{1}{2}\Omega^2 = \tfrac{4}{3}g \times 96$; i.e. the required angular velocity is $64\sqrt{2}$ rad/sec.

(b) For the system consisting of the gyroscope, the string and the mass P, the mechanical energy is conserved.

The loss in P.E. for the system $= 2 \times g \times 12$ ft-pdl.
The gain in K.E. for the system $= \tfrac{1}{2}I\Omega^2 + \tfrac{1}{2} \times 2 \times V^2$ ft-pdl, where I, Ω are as defined in (a), and $V = \tfrac{1}{8}\Omega$.
Hence $\qquad 24g = \tfrac{1}{2} \times \tfrac{5}{32}\Omega^2 + \tfrac{1}{2} \times 2(\tfrac{1}{8}\Omega)^2,$
which gives $\Omega = 64\sqrt{2}$, as before.

(4) *A mass A of* 12 *lb is attached to the outer string (passing round the wheel) of a 'wheel and axle', which can rotate about a fixed smooth horizontal axis, in order to raise a mass B of* 40 *lb attached to the inner string (passing round the axle). The radii of the wheel, axle are* 5, 1 *in, respectively, and the masses of the wheel and axle and of the strings (which are inelastic) are to be neglected. Find the tensions in the strings, the accelerations of A, B, the supporting force at the axis, and the time taken to raise B from rest through* 32 *ft.*

The external forces on the wheel and axle are T, T' pdl and the force (X, Y) pdl at the axis, as shown in Fig. 151. By the principle of angular momentum, taking moments about the axis, we have
$$T' - 5T = 0, \quad \text{(i)}$$
the right-hand side being zero since the mass, and hence the moment of inertia, of the wheel and axle are taken as zero.

Let f, f' ft/sec² be the downward accelerations of A, B. It is easy to show that
$$f + 5f' = 0. \quad \text{(ii)}$$
For the motions of A, B, we have, respectively,
$$12g - T = 12f, \quad \text{(iii)}$$
$$40g - T' = 40f'. \quad \text{(iv)}$$
From (i), (ii), (iii), (iv), we derive
$$f = 5g/17; \quad f' = -g/17; \quad T = 8\cdot47g; \quad T' = 42\cdot4g;$$
these give four of the required answers.

Let t sec be the time taken to raise B through 32 ft from rest.
Then since f' is constant, we have
$$32 = \tfrac{1}{2}gt^2/17; \quad \text{i.e.} \quad t = \sqrt{34} \text{ (sec)}.$$
By the principle of linear momentum, we have for the wheel and axle
$$X = 0; \quad Y - T - T' = 0.$$
Hence the support at the axis $= (8\cdot5 + 42\cdot4)g$ pdl $= 50\cdot9$ lb.wt.

S.P.I.C.E. (α) It will be noticed that, in Exs. (1) and (3), the energy method is the faster. When, however, friction forces are present and do work, the angular momentum method is often easier to use than the energy method. In Ex. (2), there is no special gain in using the energy principle.

(β) In problems where the angular acceleration is constant, the angular analogues of the formulae of § 2·5 are relevant. In Ex. (2), for example, n could be determined much more rapidly using an analogue of the formula '$s = \tfrac{1}{2}(u + v)t$'; the student should check this. Nevertheless, it is well that the student should rely basically on the more general principles which are less restricted in application.

(γ) A common error in attempts at solving problems of the type of Ex. (3), using the method (*a*), is to say that 'the rate of change of angular momentum of the whole system (about the axis of rotation) is $mg.OA$, where m is the mass of P'. This is invalid because the *whole* system is not a single rotating rigid body. The error has the effect of taking $T = mg$, which is obviously not the case.

It may be remarked that if, in using the method (*b*), the kinetic energy of P is overlooked, the same (invalid) answer is obtained as when the above error in using the method (*a*) is made.

(δ) The energy method, if used in Ex. (4), would enable the accelerations to be rapidly determined, but the method actually followed leads most rapidly to the determination of the tensions.

(ε) In Ex. (4), the relation $T' = 5T$ would not hold if the moment of inertia of the wheel and axle were not zero.

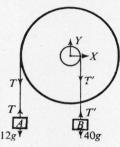

Fig. 151

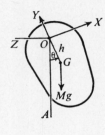

Fig. 152

13·9. The rigid-body pendulum. This consists simply of a rigid body oscillating about a fixed horizontal axis. It is sometimes called a *compound pendulum*, and corresponds somewhat more closely to a 'real' pendulum than does a simple pendulum. We shall take frictional resistances (including the frictional couple at the axis) to be zero, except where we indicate the contrary.

13·91. Equation of motion. Let G be the position of the centre of mass at time t; let GO, $= h$ say, be the perpendicular from G to the axis OZ of rotation; let OA be vertical; and let the angle $AOG = \theta$, as in Fig. 152. Let M be the mass of the body, and k its radius of gyration about OZ.

We first derive the equation of motion using the principle of angular momentum. The forces on the body are equivalent to the weight Mg at G and the forces at the axis. Since the friction couple is being neglected, the weight is the only force which has a non-zero moment about OZ. Hence, by (58b),
$$-Mgh \sin \theta = Mk^2 \ddot{\theta};$$
i.e.
$$\ddot{\theta} + (gh/k^2) \sin \theta = 0. \qquad \text{(i)}$$

The same equation may be derived by the energy principle. Only the weight does work during a displacement, so that the mechanical energy is conserved. Hence
$$\tfrac{1}{2} M k^2 \dot{\theta}^2 - Mgh \cos \theta = \text{constant}. \qquad \text{(ii)}$$
Differentiating (ii) with respect to θ and dividing by Mk^2, we obtain (i) again.

13·92. Period of small oscillations. If θ is always small, so that $\sin \theta \approx \theta$, the equation of motion (i) becomes approximately the simple harmonic equation

$$\ddot{\theta} + (gh/k^2)\theta = 0.$$

Hence, by § 8·9, θ changes simple harmonically, the period T being given by

$$T = 2\pi \sqrt{\frac{k^2}{gh}}. \tag{60}$$

The length of the equivalent simple pendulum is k^2/h.

It needs to be remarked that small oscillations can occur only about a stable equilibrium configuration. (See § 15·8.)

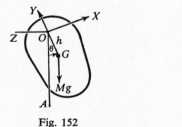

Fig. 152   Fig. 153

13·921. Example. *A uniform solid cone of mass M, height h and base-radius a, swings as a rigid-body pendulum about a diameter of its base. Find the period of small oscillations.*

Let AB be the diameter about which oscillations occur (Fig. 153).

We can show, as in Ex. (6) of § 11·(10), that the moment of inertia of the cone about a line through the vertex O parallel to AB is $3M(a^2 + 4h^2)/20$.

Since $OG = \tfrac{3}{4}h$, the moment of inertia about a parallel axis through $G = 3M(a^2 + 4h^2)/20 - M(3h/4)^2$.

Hence the radius of gyration about AB, k say, is given by
$$Mk^2 = 3M(a^2 + 4h^2)/20 - M(3h/4)^2 + M(h/4)^2$$
$$= M(3a^2 + 2h^2)/20.$$

The distance 'h' of the C.M. from $AB = h/4$.

Hence the length of the E.S.P. = 'k^2/h' = $(3a^2 + 2h^2)/5h$.

Hence the period of small oscillations = $2\pi \sqrt{\left(\dfrac{3a^2 + 2h^2}{5gh}\right)}$.

13·93. Force at the axis. Let the force at the axis on the pendulum be represented by the components X, Y as in Fig. 152; (we are still neglecting the friction couple).

By the principle of linear momentum, G moves as a particle of mass M acted on by forces X, Y and Mg in the directions shown in Fig. 152. By § 7·3, since G moves in a circle of radius h, the components of its acceleration parallel to X, Y, respectively, are $h\ddot\theta$, $h\dot\theta^2$. Hence

$$X - Mg \sin \theta = Mh\ddot\theta, \qquad \text{(i)}$$
$$Y - Mg \cos \theta = Mh\dot\theta^2. \qquad \text{(ii)}$$

By the principle of angular momentum (see § 13·91)

$$-Mgh \sin \theta = Mk^2\ddot\theta. \qquad \text{(iii)}$$

By the principle of energy, or alternatively integrating (iii) with respect to θ, we have

$$Mgh(\cos \theta - \cos \alpha) = \tfrac{1}{2}Mk^2(\dot\theta^2 - \Omega^2), \qquad \text{(iv)}$$

where Ω is the angular speed in any particular position $\theta = \alpha$.

For any position θ, X is then given by a simple elimination of $\ddot\theta$ between (i) and (iii), and Y by elimination of $\dot\theta^2$ between (ii) and (iv).

For a definite answer, Ω and α must in general be given, in addition to M, h, k.

13·931. Example. *In Ex. (2) of § 13·65, find the magnitude of the reaction at the axis when the rod has turned through an angle θ.*

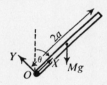

Fig. 154

The external forces on the rod are equivalent to the components X, Y at the axis and the weight Mg, as shown in Fig. 154.

By the principle of linear momentum,

$$X - Mg \cos \theta = -Ma\dot\theta^2, \qquad \text{(i)}$$
$$Mg \sin \theta - Y = Ma\ddot\theta. \qquad \text{(ii)}$$

By the principle of energy,

$$Mga(1 - \cos \theta) = \tfrac{1}{2} \cdot \tfrac{4}{3} Ma^2\dot\theta^2. \qquad \text{(iii)}$$

By (iii), or alternatively by the principle of angular momentum,

$$Mga \sin \theta = \tfrac{4}{3} Ma^2 \ddot\theta. \qquad \text{(iv)}$$

By (i) and (iii), $\qquad X = \tfrac{5}{2} Mg \cos \theta - \tfrac{3}{2} Mg.$

By (ii) and (iv), $\qquad Y = \tfrac{1}{4} Mg \sin \theta.$

The required magnitude $= \sqrt{(X^2 + Y^2)}$
$$= \frac{Mg}{4} \sqrt{(99 \cos^2 \theta - 120 \cos \theta + 37)}.$$

13·94. It may be noted that the general equations obtained for a rigid-body pendulum are relevant to more general problems than cases of ordinary pendulum motion. The following is an example.

A uniform rod of length 2a is held horizontally on a fixed rough table at right angles to the edge of the table and with its centre at a distance b beyond the edge. If the rod is released from rest in this position, prove that it will begin to slide off the table after it has turned through an angle of $\tan^{-1}\{\mu a^2/(a^2+9b^2)\}$, where μ is the coefficient of friction at the table's edge.

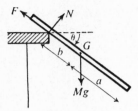

Fig. 155

Until the rod starts to slide, it is turning about the edge of the table as about a fixed axis. (The reader should by using the principle of linear momentum, taking components along the length of the rod, convince himself that if $\mu \neq 0$ the rod will not immediately start to slide.)

Let M be the mass of the rod, and N, F the normal and frictional components (as in Fig. 155) of the force on the rod when the angular displacement is θ, where θ is not greater than the angular displacement at which sliding starts.

By the principle of momentum,
$$Mg \cos \theta - N = Mb\dot\theta, \qquad \text{(i)}$$
$$F - Mg \sin \theta = Mb\dot\theta^2, \qquad \text{(ii)}$$
since the C.M. G is moving in a circle of radius b.

By the principle of energy,
$$Mgb \sin \theta = \tfrac{1}{2}M(\tfrac{1}{3}a^2 + b^2)\dot\theta^2, \qquad \text{(iii)}$$
and hence
$$Mgb \cos \theta = M(\tfrac{1}{3}a^2 + b^2)\ddot\theta. \qquad \text{(iv)}$$

By (i) and (iv), and by (ii) and (iii), we derive
$$N = \frac{Mga^2}{a^2 + 3b^2} \cos \theta; \qquad F = \frac{Mg(a^2 + 9b^2)}{a^2 + 3b^2} \sin \theta.$$

Since F is zero when $\theta = 0$, then (provided $\mu \neq 0$) $F < \mu N$, initially. Sliding commences when F reaches the value μN, i.e. when
$$(a^2 + 9b^2) \sin \theta = \mu a^2 \cos \theta,$$
i.e. when
$$\theta = \tan^{-1}\{\mu a^2/(a^2 + 9b^2)\}.$$

13·95. Case where friction couple is taken into account. Suppose now that a friction couple L acts at the axis. The equations (i) and (ii) of § 13·93 still hold. The equation (iii) needs the addition of the term $-L$ or $+L$ on the left-hand side, according as the angular velocity is positive or negative. The equation (iv) correspondingly needs the addition of the term $-\int_\alpha^\theta L d\theta$ or $+\int_\alpha^\theta L d\theta$ on the left-hand side; (the equation thus derived is relevant only to a motion in which $\dot\theta$ does not change sign).

13·951. Example. *A wheel of diameter 5 ft can rotate in a vertical plane about a fixed horizontal axis through its centre O, and carries a particle of mass 4 oz at a point Q of its rim. Motion of the wheel is resisted by a constant friction couple at the axis. The system is released from rest with OQ at 30° above the horizontal, and comes instantaneously to rest when OQ has been carried 45° beyond the vertical. Prove that when the system comes finally to rest, OQ will not have returned to the vertical position again.*

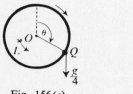

Fig. 156(a)   Fig. 156(b)

Fig. 156 (a) corresponds to circumstances during the first swing. In the general position θ, the forces on the system which have moments about O are the weight, $g/4$ pdl, and the friction couple, L pdl-ft.

By the principle of angular momentum, we have
$$\frac{g}{4} \times \frac{5}{2} \sin\theta - L = I\ddot\theta. \qquad (i)$$
(The reader should check that this holds for the whole range of values of θ, from $\pi/3$ to $5\pi/4$.)

Integrating between $\theta = \pi/3$ and $\theta = 5\pi/4$, we have
$$\left[-\frac{g}{4} \times \frac{5}{2}\cos\theta\right]_{\pi/3}^{5\pi/4} - L\left(\frac{5\pi}{4} - \frac{\pi}{3}\right) = 0,$$
the right-hand side being zero since $\dot\theta$, and therefore $\tfrac{1}{2}I\dot\theta^2$, is zero initially and finally.

Hence $L = 15g(\sqrt{2} + 1)/44\pi$.

During the second swing, the friction couple is reversed in direction, but (i) still applies if we interpret θ as in Fig. 156 (b).

The initial value of θ is now $\tfrac{3}{4}\pi$, and since for this value the left-hand side of (i) is found (on carrying out the arithmetic) to be positive, it follows that there will be a second swing.

Let Ω be the value of $\dot\theta$ when OQ next passes through the vertical. On integrating (i) from $\theta = \tfrac{3}{4}\pi$ to $\theta = \pi$, we now obtain

$$\left[-\frac{g}{4} \times \frac{5}{2} \cos\theta\right]_{3\pi/4}^{\pi} - L(\pi - \tfrac{3}{4}\pi) = \tfrac{1}{2}I\Omega^2.$$

On substituting the value already found for L, we find the left-hand side of this equation to be negative. Hence there is no real value of Ω at which OQ can be vertical.

Hence the wheel must come to rest before OQ is vertical.

Since the friction couple always acts against the direction of rotation, it is evident that the wheel will now remain at rest.

S.P.I.C.E. (α) The problem could be neatly solved by equating the gain in the mechanical energy to the work done by the extraneous forces; (both these quantities are of course negative in this problem).

(β) The data do not give the moment of inertia I of the wheel about its axis. But it transpires that the particular value of I is not needed in getting the answer.

(γ) The problem could have been solved by finding the value of θ at which the system comes to rest in the second swing. But the equation which gives this value of θ is a transcendental equation of the form $A\cos\theta + B\theta = C$, whose solution involves recourse to tables. The method we have followed gives the solution with less labour.

13·(10). Motion of a rigid body when there is no fixed axis of rotation.

If there is no fixed axis of rotation, the body (unless otherwise restricted) has three degrees of freedom. In these circumstances, the configuration at time t is often specified in terms of the coordinates $\bar{x}$, $\bar{y}$ of the centre of mass G ($\bar{x}$, $\bar{y}$ are referred to fixed axes), and an angular coordinate θ (referred to a direction fixed in space).

The principle of momentum as stated in § 13·7 will supply the two equations (53d), involving the acceleration components $\ddot{\bar{x}}$, $\ddot{\bar{y}}$.

A third equation, involving $\ddot{\theta}$, may often be obtained using the principle of energy (see e.g. Ex. (3) of § 13·65; in this Ex., $\ddot{\theta} = \ddot{x}/a$).

When the principle of energy cannot be conveniently used, recourse must be had to a form of the angular momentum principle for a third equation. The theory under § 13·8 is not relevant for this purpose, but the *extended form* (§ 12·45) of the angular momentum principle can be used; we now discuss this.

13·(10)1. Use of extended principle of angular momentum.

At time t, let N be the moment *about* G of a typical external force acting on the rigid body. By argument similar to that in § 13·81, it follows that the angular momentum of the body about G is $I\omega$, where ω is the angular velocity of the body, and I is its moment of inertia *about G*. Then, by § 12·45, we have

$$\Sigma N = \frac{d}{dt}(I\omega) = I\ddot{\theta}. \tag{61}$$

13·(10)2. Remarks.

(i) It follows from (53*d*) and (61) that the acceleration of G and the angular acceleration of the rigid body will be the same under all statically equivalent sets of forces (see § 12·23). For ΣX, ΣY, ΣN are the same for all such sets.

(ii) The three equations given by (53*d*) and (61) are together equivalent to the three equations given by (53*d*) and the energy principle.

(iii) There is often a temptation to apply the principle of angular momentum about an 'instantaneous axis' (e.g. the line of contact in the case of a cylinder rolling without slipping down an inclined plane) as if this were a fixed axis. This procedure would in general be invalid. (The brighter student may be interested to show that the procedure is valid *if* the distance from G to the instantaneous axis remains constant. The ordinary student is advised at this stage to 'take moments about G' in all cases, unless there is a permanently fixed axis of rotation.)

(iv) The kinetic energy formula (56*a*), i.e. $\frac{1}{2}I\omega^2$, may, however, be used with impunity in the case of an instantaneous axis of rotation. (The reader may check that the proof of § 13·41 still holds if OZ is an instantaneous axis not necessarily permanently fixed.) It needs to be appreciated that the moment of inertia I of a body about an instantaneous axis *does not in general remain constant* as time goes on, since in general an instantaneous axis shifts relatively to the body as time goes on.

13·(10)3. Examples.
(1) *A rigid body is in motion as a projectile under constant gravity, resistances being neglected. Prove that its centre of mass describes a parabola, and that its angular velocity remains constant.*

The first result follows from the principle of momentum (§ 13·7), according to which the C.M. G moves as if it were a particle of mass M acted on by only the weight Mg, where M is the mass of the body.

The second result follows from the extended principle of angular momentum (§ 13·(10)1). Thus if I is the moment of inertia of the body about G, and ω is the angular velocity, we have, by (61), $0 = I\dot{\omega}$; hence $\dot{\omega}$ is zero and ω is constant.

(2) *Solve Ex. (3) of § 13·65 (on the motion of a sphere rolling without slipping down a rough fixed inclined plane) independently of the energy principle.*

The external forces on the sphere are its weight Mg, and the normal and frictional components N, F of the force at the point of contact, as shown in Fig. 157.

Taking moments about G, we have

$$Fa = d(I\omega)/dt, \tag{i}$$

where a, I, ω are as in Ex. (3) of § 13·65; I, the moment of inertia about G, $= 0·4Ma^2$.

The equation (i) involves F which can be eliminated only if we set down another equation containing F. By the principle of linear momentum, resolving parallel to F, we have
$$Mg \sin \alpha - F = Mf, \qquad \text{(ii)}$$
where f is the acceleration of G.

Eliminating F between (i) and (ii), we have
$$Mg \sin \alpha - 0\cdot 4 Ma\dot\omega = Mf.$$
Since $v = a\omega$ (§ 13·241), where v is the velocity of G, we have
$$f = a\dot\omega,$$
and hence $\qquad g \sin \alpha - 0\cdot 4 f = f;$
i.e., $\qquad f = (5g/7) \sin \alpha.$

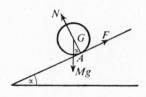　　　　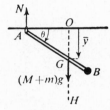

Fig. 157　　　　　　　　Fig. 158

(3) *A small ring A of mass m can slide on a fixed smooth horizontal wire, and has attached to it a light rod AB of length l to which is attached a particle B of mass M. Initially the system is at rest with B held close to the wire. If B is now released, find the angular velocity of the rod when it is vertical.*

The system consisting of the ring, the rod and the particle may be treated as a single rigid body.

Its C.M., G, is given by
$$AG = Ml/(M+m), \quad \text{or} \quad BG = ml/(M+m). \qquad \text{(i)}$$
Its moment of inertia, I, about G is given by
$$I = m\left(\frac{Ml}{M+m}\right)^2 + M\left(\frac{ml}{M+m}\right)^2$$
$$= \frac{Mml^2}{M+m}. \qquad \text{(ii)}$$

The external forces on the system are the weight $(M+m)g$ at G, and the (smooth) support N at A, as shown in Fig. 158.

By the principle of linear momentum, resolving horizontally, we have $0 = (M+m)\ddot{x}$, where $\ddot{x}$ is the horizontal component of the acceleration of G. Hence $\dot{x}$ is constant. But initially $\dot{x}$ was zero; hence $\dot{x}$ is always zero; i.e. G moves in a fixed vertical line, OH say.

Hence the velocity of G is $\dot{y}$, where $\bar{y}$ is as in Fig. 158.

In this problem the only force which does work is the weight. Hence the mechanical energy is conserved. When the configuration is as in Fig. 158, we thus have

$$\tfrac{1}{2}(M+m)\dot{\bar{y}}^2 + \tfrac{1}{2}I\dot{\theta}^2 = \text{gain in K.E.}$$
$$= \text{loss in P.E.}$$
$$= (M+m)g\bar{y}.$$

When the rod is vertical, $\dot{\bar{y}} = 0$ and $\bar{y} = AG$; hence the required angular velocity, Ω say, is given by

$$\tfrac{1}{2}I\Omega^2 = (M+m)g.AG.$$

On substituting from (i), (ii), we derive

$$\Omega = \sqrt{\left\{\frac{2(M+m)g}{ml}\right\}}.$$

(4) *A thin uniform hoop of radius a can move on a fixed rough horizontal surface (coefficient of friction μ), and smooth constraints confine the motion to a vertical plane. The hoop is set in motion with ω_0 as its initial angular velocity, and v_0 as the initial velocity of its centre G, in the senses shown in Fig. 159 (a); and $\omega_0 > v_0/a$. Prove that after a certain time the hoop will be rolling without slipping, the velocity of G being then opposite to the direction of v_0. Find this time, the final velocity of G, and the final angular velocity.*

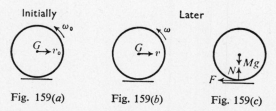

Fig. 159(a)   Fig. 159(b)   Fig. 159(c)

Let t_1 be the required time, and let Figs. 159 (b) and (c) refer to circumstances at any time t, where $0 \leqslant t \leqslant t_1$.

The hoop will slip so long as v is unequal to $a(-\omega)$.

Let v_1, ω_1 be the values of v, ω (Fig. 159 (b)) when slipping ceases, i.e. when $t = t_1$; thus

$$v_1 = -a\omega_1. \qquad (i)$$

Until slipping ceases, the external forces on the hoop are (in usual notation) as shown in Fig. 159 (c).

The friction force F is positive since the hoop slips (at the contact) to the right, by virtue of the data $\omega_0 > v_0/a$.

Also, until slipping ceases,

$$F = \mu N. \qquad (ii)$$

By the principle of momentum,

$$-F = M\dot{v}; \qquad N - Mg = 0. \qquad (iii)$$

By the extended principle of angular momentum,
$$-Fa = Ma^2\dot\omega. \qquad (iv)$$
By (ii), (iii), (iv), so long as there is slipping, we have
$$\dot v = -\mu g; \qquad a\dot\omega = -\mu g. \qquad (v)$$
Integrating (v) from $t = 0$ (when $v = v_0$ and $\omega = \omega_0$) to $t = t_1$ (when $v = v_1$ and $\omega = \omega_1$), we have
$$v_1 - v_0 = -\mu g t_1, \qquad (vi)$$
$$a\omega_1 - a\omega_0 = -\mu g t_1. \qquad (vii)$$
From (i), (vi), (vii), we derive
$$t_1 = (a\omega_0 + v_0)/2\mu g, \qquad (viii)$$
$$v_1 = -a\omega_1 = -\tfrac{1}{2}(a\omega_0 - v_0). \qquad (ix)$$
(viii) gives the required time, and (ix) the final velocity of G and the final angular velocity. Since $a\omega_0 > v_0$, v_1 is negative and ω_1 is positive; thus finally G is moving to the left, and the hoop is spinning in the anticlockwise sense.

S.P.I.C.E. (α) The result in Ex. (1) is of considerable generality.

(β) In Ex. (2) we did not resolve at right angles to the plane. This would give an equation which would fulfil the purpose of determining N; but it happens that the value of N is not needed in the solution.

(γ) Remark (ii) of § 13·(10)2 is relevant to Ex. (2).

(δ) To get the answer in Ex. (2), it is not necessary to examine whether F is positive or not in Fig. 157. (It is easy to deduce from the working in Ex. (2) that F is positive.) On the other hand, in Ex. (4), where there is slipping, it is necessary to examine the direction in which the friction force F acts.

(ϵ) Ex. (3) could be solved by forming equations for the ring and the particle separately. But by treating the ring, rod and particle as one system, it was possible to save labour by avoiding the introduction of the tension or thrust of the rod into the equations.

(ζ) In Ex. (3), a notable simplification arises from perceiving that G moves in a straight line. Had there been friction at A, this would of course not have been the case.

(η) Ex. (2) was at least equally well solved earlier using the energy principle. On the other hand, the energy principle would be much more difficult to apply in Ex. (4) on account of the difficulty in estimating the work done by the unknown force F. (In Ex. (2), F does no work.)

(θ) In Ex. (4), $F = \mu N$ since there is slipping. In Ex. (2) on the other hand we may not write $F = \mu N$.

13·(11). Work done by external forces during displacement of a rigid body. Let the configuration of the body at time t be specified in terms of $\bar x, \bar y, \theta$ as in § 13·(10). Then $\dot{\bar x}^2 + \dot{\bar y}^2 = \bar v^2$, where $\bar v$ is the speed of the centre of mass G; and $\dot\theta$ is the angular velocity.

Let M be the mass of the body, and I its moment of inertia about G.

Then by §§ 13·42, 13·62, the rate of work of the external forces

$$= \frac{d}{dt}\{\tfrac{1}{2}M(\dot{\bar{x}}^2 + \dot{\bar{y}}^2) + \tfrac{1}{2}I\dot{\theta}^2\}$$
$$= M\ddot{\bar{x}}\dot{\bar{x}} + M\ddot{\bar{y}}\dot{\bar{y}} + I\ddot{\theta}\dot{\theta}$$
$$= (\Sigma X)\dot{\bar{x}} + (\Sigma Y)\dot{\bar{y}} + (\Sigma N)\dot{\theta}, \qquad (62a)$$

by (53d) and (61), where X, Y, N are the x-component, the y-component, and the moment about G, of a typical external force.

Hence the work done by the external forces during a general two-dimensional displacement of a rigid body

$$= \Sigma\{\int X d\bar{x} + \int Y d\bar{y} + \int N d\theta\}, \qquad (62b)$$

the integrals being taken over the relevant ranges of $\bar{x}$, $\bar{y}$, θ.

13·(12). Effect of impulsive forces.

13·(12)1. Case of rigid body free to rotate about fixed axis. Take the fixed axis of rotation as OZ. At time t, let $\mathbf{F}$ ($= X\mathbf{i} + Y\mathbf{j}$) be a typical external force, (x, y) any point on the line of action of this force, and N ($= Yx - Xy$, see § 12·111) its moment about OZ. Then by the principles of linear and angular momentum,

$$\Sigma\mathbf{F} = M\dot{\bar{\mathbf{v}}}, \qquad (i)$$
$$\Sigma(Yx - Xy) = I\dot{\omega}, \qquad (ii)$$

where M is the mass of the rigid body, I its moment of inertia about OZ, $\bar{\mathbf{v}}$ the velocity of the centre of mass G and ω the angular velocity.

Now integrate (i), (ii) over a time-interval from t to $t + \tau$, and take the limit $\tau \to 0$. If impulsive forces are present, $\bar{\mathbf{v}}$, ω will in general suffer finite changes, say from $\bar{\mathbf{v}}_1$, ω_1 to $\bar{\mathbf{v}}_2$, ω_2, although the configuration is not changed during the action of these forces—see § 3·541; (thus x, y are treated as constants in the integration). We obtain

$$\Sigma\mathbf{J} = M\bar{\mathbf{v}}_2 - M\bar{\mathbf{v}}_1, \qquad (63)$$
$$\Sigma(J_y x - J_x y) = I\omega_2 - I\omega_1, \qquad (64)$$

where $\mathbf{J} = \lim_{\tau \to 0} \int_t^{t+\tau} \mathbf{F} dt$, and J_x, J_y are the resolved parts of $\mathbf{J}$; i.e. $\mathbf{J} = J_x\mathbf{i} + J_y\mathbf{j}$. ($\mathbf{J}$ is of course zero unless $\mathbf{F}$ is impulsive.)

Hence, by (63), the increase in the velocity of G at time t is equal to that which would be given by treating G as a particle of mass M subjected to the external impulsive forces acting on the body at time t.

By (64), the increase in the angular momentum of the body about OZ at time t is equal to the sum of the moments about OZ of the external instantaneous impulses at time t.

Contributing to the left-hand side of (64), there may be moments of *impulsive couples*, of dimensions $[ML^2T^{-1}]$, analogous to force-couples and specified by the moments of their impulses.

13·(12)2. Rigid body with no fixed axis of rotation. In this case, (63) still holds without reservation.

Also, by the *extended* principle of angular momentum, the form (64) holds, *if OZ is now taken to pass through G*; $(x, y, I$ now of course refer to this new axis through G).

Alternatively, we may replace (64) by the more general form

$$\Sigma(J_y x - J_x y) = H_2 - H_1, \tag{64a}$$

where OZ does not necessarily contain G, and H_1, H_2 are the initial and final angular momenta about OZ. The proof of this rests on our ignoring configuration changes, so that any axis may be treated as fixed, during the action of an impulsive force. (The use of (64a) in practice may be complicated by difficulties in calculating H_1 or H_2; Ex. 28 of Set XIV gives a clue to these calculations.)

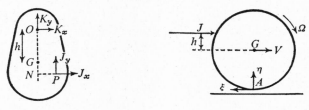

Fig. 160   Fig. 161

13·(12)3. Examples. (1) *A rigid body pendulum initially at rest receives an instantaneous impulse* **J**. *Write down equations which determine the impulsive reaction at the axis, and hence find the conditions that this impulsive reaction should be zero.*

Let O in Fig. 160 be on the axis of rotation, and let G be the C.M. Let P be the point at which the impulse **J** acts, let PN be perpendicular to OG, and let J_x, J_y be the components of **J** parallel to NP, NO. Let K_x, K_y be the components of the reaction impulse at O on the body.

Let Ω be the angular velocity of the body just after the impact; the velocity of G will then be $h\Omega$ in the direction NP, where $h = OG$.

By (63), taking resolved parts, and (64), we have

$$J_x + K_x = Mh\Omega, \tag{i}$$
$$J_y + K_y = 0, \tag{ii}$$
$$J_x \cdot ON + J_y \cdot NP = Mk^2 \Omega, \tag{iii}$$

where M, k are the body's mass and radius of gyration about O.

(i), (ii) and (iii) are sufficient to determine K_x, K_y and Ω. From K_x and K_y, the impulsive reaction at the axis is determined.

If there is to be *no* impulsive reaction at the axis, K_x and K_y must both be zero; i.e., by (ii), $J_y = 0$, and, by (i) and (iii),

$$J_x = Mh\Omega, \quad \text{and} \quad J_x \cdot ON = Mk^2 \Omega,$$

and hence $ON = k^2/h$.

Thus we require that the line of action of the applied impulse should be at right angles to OG, and should pass through a point N on OG such that ON is equal to the length of the equivalent simple pendulum.

In this context, the point N is called the *centre of percussion* corresponding to the given axis through O.

(2) *A billiard ball is at rest on a table. It is struck horizontally by a cue, which moves in a vertical plane through the centre G of the ball, at a vertical height h above G. Assuming the ball to be homogeneous, find its motion just after the impact.*

Let J be the magnitude of the applied impulse; M and a the mass and radius of the ball; Ω the angular velocity, and V the velocity of G, just after the impact; the senses being as in Fig. 161.

Let ξ, η be the frictional and normal components of the reaction impulse on the ball at the point of contact A with the table.

By (63), resolving vertically, we have $\eta = 0$. Also, if μ is the coefficient of friction at A, $|\xi| \leqslant \mu\eta$ (by the definition of impulse, and the relation between normal and frictional components of forces at contacts). Hence $\xi = 0$. Thus there is no impulsive reaction at A.

By (63), resolving horizontally, we then have $J = MV$; and, by (64),
$$Jh = 0{\cdot}4Ma^2\Omega.$$
Hence
$$V = 2a^2\Omega/5h. \tag{i}$$

The ball will start with pure rolling if $a\Omega = V$, i.e. by (i) if $h = 2a/5$.

If $h > 2a/5$, it follows from (i) that $a\Omega > V$; i.e. the ball is 'topped'.

If $h < 2a/5$, then $a\Omega < V$, and the ball drags.

(3) *A uniform spherical ball moving in a vertical plane and spinning about a diameter at right angles to the plane of motion with angular velocity Ω, impinges on the ground with velocity components u, v as shown in Fig. 162 (a). The ground is horizontal and sufficiently rough to prevent slipping; and e is the coefficient of restitution at the impact. Investigate the motion just after the impact.*

Let M be the mass of the ball and a its radius. Let the components of impulse at the point of contact be (ξ, η) as in Fig. 162 (b); and let u', v' be the velocity components of the centre and Ω' the angular velocity just after the impact, as shown in Fig. 162 (c).

By the principle of linear momentum,
$$M(u' - u) = -\xi, \tag{i}$$
$$M(v' + v) = \eta. \tag{ii}$$

By the principle of angular momentum,
$$\tfrac{2}{5}Ma^2(\Omega' - \Omega) = \xi a. \tag{iii}$$

Since the ground is so rough as to prevent sliding, the point A of the ball which meets the ground must be reduced to rest at the impact. Hence
$$u' = a\Omega'. \qquad \text{(iv)}$$
By Newton's law of impact (§ 12·611, equation (23b)),
$$v' = ev. \qquad \text{(v)}$$
From (i), (iii), (iv), we deduce
$$u' = a\Omega' = (5u + 2a\Omega)/7, \qquad \text{(vi)}$$
$$\xi = 2M(u - a\Omega)/7. \qquad \text{(vii)}$$

The equations (v), (vi), (vii) determine the motion just after the impact.

(a) Just before impact   (b) At the impact   (c) Just after impact

Fig. 162(a)   Fig. 162(b)   Fig. 162(c)

Particular cases.

(*A*) If $\Omega = u/a$, we have $\xi = 0$, $u' = u$, $\Omega' = \Omega$. Thus u and Ω are unaltered by the impact.

(*B*) If $\Omega > u/a$ (the case of 'overspin'), then ξ is negative, $u' > u$ and $\Omega' < \Omega$. Thus the spin is reduced and the forward velocity increased by the impact.

(*C*) If $\Omega < u/a$, then ξ is positive, the spin is increased and the forward velocity reduced by the impact.

(*D*) If $\Omega = -5u/2a$, then u' and Ω' are zero; i.e. the ball rises vertically without spin, after the impact.

(*E*) If $\Omega < -5u/2a$, then u' and Ω' are both negative. Thus if before impact the ball has an 'underspin' of magnitude exceeding $5u/2a$, the ball will bounce backwards.

S.P.I.C.E. (α) It will be noticed that no reference was made in the above solutions to non-impulsive forces such as weight forces. These forces are not involved in (63), (64).

(β) Ex. (1) introduces the term *centre of percussion*.

(γ) In Ex. (2), detailed argument was needed to show that the impulsive reaction at the point of contact A was zero. On the other hand, in Ex. (3), both normal and frictional components of the impulsive reaction at A are in general not zero.

(δ) Exs. (2) and (3) have some bearing on certain ball games. The results are not quite the last word, however, since a number of practical factors are ignored; e.g., a tennis ball is not a uniform solid, nor is it rigid.

EXAMPLES XIII

1. A cylinder A of radius a is rolling without slipping on the interior surface of radius b of a fixed hollow cylinder, the axes of the cylinders being horizontal and parallel. When the plane containing the axes of the cylinder makes an angle θ with the vertical, prove that the angular velocity of A is $\dot\theta(b-a)/a$.

*2. A rod AB of fixed length moves with constant angular velocity so that its ends A, B describe two fixed straight lines which intersect at right angles at O. Prove that the sum of the squares of the speeds of A, B is constant, and that the acceleration of any given point P of the rod is directed towards O and is proportional to PO.

3. Find the kinetic energy of a flywheel of mass 2 tons making 100 rev/min: (a) assuming the mass to be concentrated in a circular rim of diameter 6 ft; (b) assuming the wheel to be a uniform disc of diameter 6 ft.
[Answers: 69,000, 35,000 ft-lb.wt., approx.]

4. A uniform rod AB of length 6 ft is smoothly hinged at A, which is fixed, and is initially at rest with B vertically below A. Find what speed must be given to the lower end B in order that B should just reach a point vertically above A.
[Answer: $24\sqrt{2}$ ft/sec.]

5. A uniform solid circular cylinder rolls without slipping, with its generators horizontal, down an inclined plane of inclination α. Using the principle of energy, prove that the acceleration of any point of its axis is $(2g/3)\sin\alpha$.

6. A rectangular door 3 ft wide is of mass 100 lb (which may be assumed to be uniformly distributed). Neglecting friction at the hinges, find in lb.wt. what constant force, always perpendicular to the door and applied at a point 30 inches from the line of hinges, would turn the door from rest through a right angle in 2 sec. Find also in ft-pdl the kinetic energy thus generated.   [Answers: 2·95 lb.wt.; 370 ft-pdl.]

7. A pulley in the form of a uniform circular disc of mass $2m$ can rotate smoothly about its axis, which is fixed and horizontal. A light inelastic string passing over the pulley carries masses $2m$, m at its ends. There is no slip between the string and the pulley, and the hanging parts of the string are vertical. Find the acceleration of either particle, and the tensions in the hanging parts of the string.
[Answers: $0{\cdot}25g$; $1{\cdot}5mg$; $1{\cdot}25mg$.]

8. Two spheres A, B, made of uniform materials (of different densities) have the same mass, the same diameter and same outward appearance. It is known that A is solid and that B has a concentric spherical cavity. Give a simple dynamical test to determine which is A and which is B.
[Answer: The solid sphere will roll with greater acceleration down an inclined plane.]

If the internal and external radii of B are a, b, respectively, and both spheres are left to roll (without sliding) from rest down an inclined plane, prove that the ratio of the distances described in equal times is $(7b^5 - 5a^3b^2 - 2a^5)/(7b^5 - 7a^3b^2)$.

9. The mass of a flywheel is 120 lb and may be regarded as concentrated round the rim, of circumference 66 in. It is initially spinning at 2,400 rev/min, and is then by suitable gearing made to drive machinery that absorbs energy at a steady rate of 1 h.p. Assuming that no energy is lost in transmission, find how long the machinery will be driven before the angular velocity of the flywheel is reduced by one-half.

[Answer: 124 sec.]

*10. A uniform solid sphere A of radius a and mass M rolls without slipping inside a fixed hollow sphere B of centre O and radius b. The angle which the line joining the point of contact of A with B to O makes with the downward vertical is denoted by θ. Initially A is at rest and θ is equal to α. (i) Find the angular velocity of A in terms of a, b, $\dot\theta$; and find in terms of M, a, b, α, θ: (ii) the kinetic energy of A; (iii) $\dot\theta$; (iv) the angular acceleration of A.

[Answers: (i) $(b-a)\dot\theta/a$;

(ii) $Mg(b-a)(\cos\theta - \cos\alpha)$;

(iii) $\dot\theta^2 = 10g(\cos\theta - \cos\alpha)/7(b-a)$;

(iv) $-(5g/7a)\sin\theta$.]

11. Find the period of small oscillations of the sphere A in Ex. 10 about its lowest position. (Hint:—Differentiate with respect to θ the result (iii) in Ex. 10.)

[Answer: $2\pi\sqrt{\{7(b-a)/5g\}}$.]

12. A uniform circular disc of mass M and radius a has a thin uniform rod AB of mass αM and length $4a$ rigidly attached at B normally to a point on its circumference. The pendulum thus formed is suspended from A and allowed to vibrate freely in a vertical plane containing the disc. Find the length of the equivalent simple pendulum. [Answer: $(153 + 32\alpha)a/(30 + 12\alpha)$.]

13. Four uniform rods, each of the same mass and length $2a$, are rigidly joined together to form a square frame $ABCD$. The frame can turn smoothly, in a vertical plane, about the point A which is fixed. The frame is released from rest with DB vertical. Find its angular speed when DB is horizontal.

[Answer: $\{3g\sqrt{2}/5a\}^{\frac{1}{2}}$.]

14. The motion of a flywheel, whose moment of inertia (about its axis) is I, is retarded by a constant frictional couple L. The wheel is set spinning by a constant couple G which acts for a time t and then ceases. Show that the angle through which the wheel turns between starting and stopping is $\tfrac{1}{2}G(G-L)t^2/IL$ radians.

15. A uniform solid sphere and a thin uniform spherical shell of the same diameter and mass are each set spinning about a diameter with the same initial angular speed. There is a resisting couple of magnitude proportional to the angular speed, the constant of proportionality being the same in both cases. Prove that the shell will make 5/3 times as many revolutions as the solid sphere before coming to rest.

16. The L-shaped piece, described in Exs. XI, No. 17, can rotate smoothly about the outer edge of the shorter limb, which is fixed horizontally. Prove that the period of small oscillations is about 2·15 sec.

17. A uniform piece of wire ABC is bent so as to form two straight portions at right angles, AB of length a, and BC of length b. It swings in a vertical plane freely about B which is fixed. Find the length of the equivalent simple pendulum.

[Answer: $\frac{2}{3}(a^3 + b^3)/(a^4 + b^4)^{\frac{1}{2}}$.]

18. A uniform circular lamina of weight W can turn smoothly about a horizontal axis which passes through a point O of its circumference and is perpendicular to its plane. It is released from rest when the diameter OA through O is vertical, with A above O. Prove that when the lamina has turned through an angle θ, the components of force at the axis along and perpendicular to OA are $\frac{1}{3}W(7\cos\theta - 4)$ and $\frac{1}{3}W\sin\theta$, respectively.

19. A uniform rod of weight W, which can turn smoothly about a pivot at one end, is released from rest when horizontal. When the angle between the rod and the vertical is θ, prove that the force on the pivot is $\frac{1}{4}W\sqrt{(1 + 99\cos^2\theta)}$.

20. A uniform rod AB of weight W can turn smoothly about one end A which is fixed. It is released from rest when horizontal. Prove that the initial reaction at A is $\frac{1}{4}W$, and that the initial acceleration of B is $3g/2$.

21. A flywheel of mass 100 kg and radius of gyration (about its axis) 20 cm is mounted on a fixed light horizontal axle of radius 2 cm, and can rotate about frictionless bearings. A light string wound round the axle carries at its free end a mass of 5 kg. If the system is released from rest with the 5 kg mass hanging vertically, find the acceleration of this mass, (i) using the principle of energy; (ii) independently of the principle of energy.

If string slips off the axle after the mass has descended 2 metres, find in Kg-m the moment of the constant couple which must be applied to bring the flywheel to rest in 5 revolutions. [Answers: 0·49 cm/sec²; 0·318 Kg-m.]

22. Energy is transmitted from an engine, which causes a shaft A to rotate about a fixed axis, to a second shaft B which also rotates about a fixed axis, by means of a uniform endless belt moving steadily at a speed of v ft/sec. If T, T' lb.wt. are the tensions in the two straight parts of the belt, where $T > T'$, prove that the transmitted horse-power is $(T - T')v/550$. (Method:—Consider the rate of work done on the shaft B.)

23. Assuming the Earth to be symmetrical about the line NS joining the North and South Poles, prove that if as a result of cooling the radius of gyration of the Earth about NS had diminished by the fraction $1/n$, the length of a day would correspondingly have diminished by the fraction $2/n$, approximately.

24. Solve Ex. 5 above independently of the principle of energy.

25. A uniform rod AOB is let fall from a horizontal position with an initial spin Ω about a horizontal axis through the centre O perpendicular to AB. It makes its first contact with the ground when vertical. If the initial height of the rod above the ground is h, prove that the length of the rod must be

$$2h - g\pi^2(n + \tfrac{1}{2})^2/\Omega^2,$$

where n is zero or a positive integer.

26. A homogeneous circular cylinder of radius a is placed horizontally in contact along its length with a horizontal plane (coefficient of friction μ), and a vertical wall (coefficient of friction μ'). The cylinder is given an initial spin Ω about its axis in the sense required to maintain contact with the wall. Show that it comes to rest in time
$$\frac{\Omega a(1 + \mu\mu')}{2g(\mu + \mu\mu')}.$$

27. A uniform rod of length $2l$ is constrained so that its ends A, B move without friction in fixed grooves along the lines OX, OY, respectively, where OX is horizontal and OY is vertical. Initially the angle OAB is α and the rod is at rest. Find the angular velocity of AB when the angle OAB is θ, (i) using the principle of energy; (ii) independently of the principle of energy. [Answer: $\sqrt{\{3g(\sin\alpha - \sin\theta)/2l\}}$.]

28. A cyclist rides steadily round the inside of an upright fixed cylinder of radius 17 ft. The centre of mass of the cyclist and his machine is always 2 ft from the wall of the cylinder and describes a horizontal circle at 20 m.p.h. Find the minimum value of the coefficient of friction between the tyres and the wall.

[Answer: 0·56.]

29. A railway carriage is travelling on a horizontal curve of radius r with speed v; $2a$ is the distance between the rails, which are in a horizontal plane; and h is the height of the centre of mass of the carriage above the rails. Find the ratio in which the weight of the carriage is divided between the rails, and show that the carriage will start to upset if $v > \sqrt{(gar/h)}$. [Answer to first part: $(gar - hv^2)/(gar + hv^2)$.]

30. A uniform solid cylinder of mass M rolls, with its axis always horizontal, down a smooth plane of inclination α, unwrapping, as it goes, a light inelastic string attached to the highest point A of the plane. The winding is in the sense that results in the string being in contact with the plane at all points between A and the point of contact of the cylinder with the plane. Find the acceleration of the centre of the cylinder and the tension of the string at any instant. [Answers: $\tfrac{2}{3}g \sin\alpha$, $\tfrac{1}{3}Mg \sin\alpha$.]

*31. A uniform rod of length $2a$ is placed with one end in contact with a smooth table, and is then allowed to fall from rest. If θ is the inclination of the rod to the vertical at time t, and α is the initial value of θ, prove that
$$\dot\theta^2 = \frac{6g}{a}\left(\frac{\cos\alpha - \cos\theta}{1 + 3\sin^2\theta}\right).$$

*32. A rough uniform solid circular cylinder of mass M can rotate without friction about its axis which is fixed and horizontal. A particle of mass m is placed on it vertically above the axis, and the system slightly disturbed from rest in this configuration. Prove that the particle will start to slip on the cylinder when the cylinder has turned through an angle θ given by
$$\mu(M + 6m)\cos\theta - M\sin\theta = 4m\mu,$$
where μ is the coefficient of friction between the particle and the cylinder.

*33. A uniform plank AB of length $2a$ is poised vertically with the lower end B at the edge of a platform fixed at a height h above the (horizontal) ground, and is allowed to fall outwards. The end B is maintained in contact with the edge of the platform until the plank is horizontal, and is then released. Neglecting any friction couple at B, prove that the plank will strike the ground horizontally if $h = n\pi a(1+n\pi/3)$, where n is a positive integer.

34. A girder of mass 2000 lb rolls lengthwise down a line of greatest slope on an inclined plane of 1 in 50 on three equal uniform solid rollers, each of mass 200 lb. The rollers roll between the girder and the incline with their axes horizontal; and there is no slipping. Find the acceleration of the girder.
[Answer: About 8 in/sec².]

35. A hollow ball of mass M and radius a is made of uniform thin metal. A fixed point of its surface is smoothly attached to a point O fixed in space, and the ball is hanging at rest under gravity from this point. Find the depth of the centre of percussion below O.
[Answer: $5a/3$.]
If J is the magnitude of the impulse of a blow applied horizontally through the centre of percussion, prove that the energy given to the ball is $5J^2/6M$.

36. A uniform circular disc of radius a is hinged at a fixed point O of its circumference and can rotate smoothly about a horizontal axis (containing O) in its plane. Find the depth of the centre of percussion below O.

Find the depth below O at which a horizontal impulsive blow should be applied at right angles to the plane of the disc, in order that the impulsive reaction on the disc at the hinge should be in the same sense as the applied impulse and of half its magnitude.
[Answers: $5a/4$; $15a/8$.]

37. The ends of a uniform rod AB of mass M can move smoothly in two fixed straight grooves which meet at right angles at O. Initially the rod is at rest and equally inclined to the two grooves. The rod then receives an impulse J at A in the direction AO. Prove that the kinetic energy generated is $\frac{3}{4}J^2/M$.

*38. A uniform circular disc of radius a is rolling without slipping along a smooth fixed horizontal plane, the speed of its centre being v. If the highest point O of the disc be suddenly fixed, prove that if the disc can now rotate smoothly about O, it will make a complete revolution if $v^2 > 24ag$.

39. A uniform rod AB is initially rotating about the end A which is fixed. If the end B is impulsively stopped by an impulse at right angles to AB, prove that the magnitude of the impulse at B is twice that at A.

40. A uniform rod AB resting on a smooth table receives a horizontal blow at right angles to AB at a point of trisection of AB. Prove that the kinetic energy generated is four times as much as if the rod had been smoothly pivoted about a vertical axis through its middle point.

(The K.E. generated is $2J^2/3M$, where J is the impulse of the blow, and M is the mass of the rod.)

CHAPTER XIV

INTERLUDE

"If only I could wake in the morning
And find I had learned the solution,
Wake with the knack of knowledge
Who as yet have only an inkling."
—Louis MacNiece, *Leaving Barra*.

14·1. The programme outlined in § 10·3 has now been carried out.

14·11. From the point of view of dynamical theory, the main features in Chapter XI are the definitions of centre of mass and moment of inertia (and radius of gyration), and immediate deductions as indicated in the equations (47a, b, c), (48), 49a, b, c).

The remarks in § 11·2 on continuously distributed mass-systems and density are of general importance.

Most of the remainder of Chapter XI was devoted to estimating centres of mass and moments of inertia in particular cases; it is helpful if the student can remember the chief results or at least be able to find them rapidly. The theorems of parallel axes and perpendicular axes—equations (50a, b), (51)—are important in calculations of moments of inertia. Routh's rule, (52), is a useful aid to memory.

14·12. In Chapter XII, §§ 12·1-12·171 were concerned with a treatment, mainly two-dimensional, of vectors localised in lines. In these sections, the term *moment* was introduced, and a formal definition was given of the *resultant* of a set of localised vectors. The theory in these sections, and in §§ 12·22, 12·221, includes theory commonly given under the heading of 'statical equivalence of sets of coplanar forces', and includes the 'theorem of moments'.

The main theoretical content of the remainder of Chapter XII is the development of the principles of linear momentum (equations (53a, b, c, d, or e)), angular momentum (equations (54a, b, c) and energy (for which the equations (22a, b) are still relevant). The kinetic energy formula (55) is important.

Other features of Chapter XII are a reference to the term 'centre of gravity', and applications of the theory to particular problems involving small numbers of particles. Newton's law of impact (23), § 4·8, was generalised to the form (23a), § 12·61.

14·13. In Chapter XIII, the theory of Chapter XII was applied to simple problems on rigid bodies in two-dimensional motion.

The definition of a rigid body enabled us: (i) to speak unequivocally of its angular velocity and angular acceleration (§§ 13·22, 13·23); (ii) to ignore the work done by internal forces in the body (§ 13·61). The property (ii) makes it specially expedient to use the energy principle in many problems. In later sections of Chapter XIII the use of the principles of linear and angular momentum in problems was discussed, the extended form of the angular momentum principle being needed when there is no fixed axis of rotation. Auxiliary formulae most frequently needed in the problems of Chapter XIII are (36), (56a, b), (57), (58a, b), (61), and sometimes the kinematical formulae of §§ 13·24, 13·241. In applications of the formulae (56), (57), (58) and (61), formulae for moments of inertia are needed as well.

Other features of Chapter XIII include: a description of 'pin-joints'; the formulae (59), (59a) for work done during the rotation of a rigid body; and the formulae (62a, b) for the work done during a general displacement including translation and rotation.

A special problem of some importance is that of the rigid-body pendulum, the period being given by equation (60).

The effect of impulsive forces was considered in § 13·(12), the essential equations being (63) and (64). The term 'centre of percussion' was introduced in § 13·(12)3, Ex. (1).

14·2. Chapter XV on *Statics* involves comparatively little new theory (except for students who have omitted the major part of Chapters XII and XIII—see § 10·4), and is largely a side-issue from the point of view of the main current of dynamical theory. The basic equations needed are equations (65). Chapter XV is largely devoted to theoretical devices for solving the simpler statical problems. The question of stability of equilibrium is introduced.

From the point of view of the main theory of mechanics, the important features of Chapter XVI on *Hydrostatics* are the introduction of the notion of a deformable body; the concept of *stress* and its relation to *pressure* in the particular case of a fluid at rest; and the definition of *incompressibility* which leads to a discrimination between 'liquids' and 'gases'. The remainder of the chapter is devoted to setting up auxiliary material for the purpose of solving simpler problems in hydrostatics. The numbered equations, (66) to (74), indicate the main formulae that should be remembered.

14·3.
The examples immediately following are on the topics in Chapters XI to XIII, but are, in the main, rather harder or more advanced than Exs. XI to XIII. Several have been deliberately inserted to give the student of talent something to think about.

EXAMPLES XIV

1. Write improved statements in place of the following:

   (a) The centre of gravity of the Earth is at a distance of about 6357 km from the North Pole.

   (b) If a body is at rest, the forces acting on it are in equilibrium.

   (c) The kinetic energy of a system of particles is $\frac{1}{2}Mv^2$, where M is the total mass, and v is the speed of the centre of mass.

   (d) It is a principle of dynamics that the linear and angular momenta of a system of particles are conserved.

   (e) In any motion of a rigid body, $T = IA$, where T is the torque acting, I is the moment of inertia, and A is the angular acceleration.

   (f) The rate of change of angular momentum of a rigid body about the instantaneous centre I of the motion is equal to the sum of the moments of the external forces about I. (Note.—Cf. Ex. 30 below.)

   (g) If a wheel is rotating about its axis, which is fixed, then the angular velocity of the wheel about its axis is equal to the angular velocity of any point on the rim.

   (h) When a wheel rolls, without slipping, down an inclined plane, no work is done by the contact-force, which must therefore act at right angles to the plane.

2. A circular hole of radius $a\sqrt{3}/2$ is bored through a uniform solid hemisphere of radius a, the axis of the hole coinciding with the axis of symmetry of the hemisphere. Find the distance, from the base, of the centre of mass of the remaining solid.

   [Answer: $3a/16$.]

3. Grains of wheat, taken to be indefinitely small, are flowing steadily through an opening A into a receiver whose top B is at a depth h below A. The velocity of the grains at A may be taken as zero. Prove that the centre of mass of the falling grains between A and B is at a distance $h/3$ below A.

4. The boundary of a uniform lamina is the ellipse $(x^2/a^2) + (y^2/b^2) = 1$. Prove that the radii of gyration about the major and minor axes are $\frac{1}{2}b$, $\frac{1}{2}a$, respectively; and about a line through the centre perpendicular to the plane of the lamina, $\frac{1}{2}\sqrt{(a^2+b^2)}$.

5. The moment of inertia of the Earth (assumed symmetrical about its centre) is $0.334Ma^2$, where M, the mass, and a, the radius, are 5.98×10^{27} g, 6370 km, respectively. The Earth consists of a central core of radius 3470 km surrounded by a rocky mantle of thickness 2900 km. Assuming the densities of the mantle and core to be uniform, find these densities. [Answer: 4.2, 12.2 g/cm^3.]

6. The surface-density at any point of a circular disc of radius a is k times the distance of the point from the centre, where k is constant. Find the radius of gyration about a tangent line to the circumference of the disc. [Answer: $a\sqrt{(13/10)}$.]

7. A set of coplanar forces is statically equivalent to forces X, Y acting along OX, OY, and a couple G. Prove that the line of action of the resultant cuts OX, OY at distances G/Y, $-G/X$ from O.

*8. O is any point within a triangle ABC. Forces act along BC, CA, AB, OA, OB, OC, the magnitudes being a, b, c, d, e, f, respectively, times these lengths. Prove that a necessary and sufficient set of conditions that the set of forces is equivalent to a couple is
$$d + b - c = e + c - a = f + a - b = 0.$$
(Hint.—First take resolved parts parallel to BC, drawing perpendiculars from A and O to BC.)

*9. A, B, C, D are four given coplanar points; E is the point of intersection of AB, CD; and P is any point, not necessarily in the plane $ABCD$. Prove that the resultant of vectors represented completely by **PA**, **PB**, **PC**, **PD**, **EP**, passes through a point fixed independently of P.

(Let M be the midpoint of KL, where K, L are the midpoints of AB, CD. Then the required fixed point, R say, is on EM produced, where $ER = 4EM/3$.)

10. A and B are two similar smooth vertical parallel fixed walls. A particle is projected from the base of A between the walls in a plane at right angles to the walls. After three reflections with the walls, the particle returns exactly to the original point of projection. Also the third impact is at right angles to B. Prove that the coefficient of restitution between the particle and a wall satisfies the equation $x^3 + x^2 + x - 1 = 0$.

11. Two inelastic spheres A, B, of masses m, m', are in contact on a smooth table. A receives a blow in a line passing through its centre making an angle α with the line of centres, the direction being such that B as well as A is jerked into motion. Prove that the kinetic energy generated is less than if B had been absent, in the ratio $m + m' \sin^2 \alpha$ to $m + m'$.

*12. Two particles, P, Q, of masses $2m$, m, are lying on a smooth table at a distance a apart, and are connected by a light perfectly elastic string of natural length a and modulus $\tfrac{2}{3}mg$. An instantaneous impulse J is applied to Q in the direction PQ. Prove that the string will first resume its natural length after time $\pi\sqrt{(a/g)}$. And find the distance from the initial position of P of the point at which the particles first meet. [Answer: $\{\pi J\sqrt{(a/g)} + 2ma\}/3m$.]

13. A uniform string in the form of a closed circle is rotating on a smooth fixed horizontal plane about a fixed vertical axis through its centre. If v is the speed at any time, prove that the tension is λv^2, where λ is the line-density.

(Hint.—See the solution of Ex. (4), § 15·(11).)

*14. A uniform chain is sliding over the edge of a fixed table, the portion on the table being coiled up close to the edge. If v, f are the speed and acceleration of the lower end A when a length x of the chain is over the edge, prove that $gx - v^2 = xf$. Hence show that

$$x^2 \frac{dv^2}{dx} + 2xv^2 = 2gx^2;$$

and deduce that if v is zero when x is zero, then A moves with a constant acceleration $g/3$.

*15. If circumstances are as in Ex. 14 except that the chain instead of being uniform has a line-density kx, where k is constant, prove that the acceleration is $g/5$.

16. The masses of a system of particles are $m_1, m_2, \ldots$, and v_{rs} denotes the magnitude of the velocity of m_r relative to m_s. Prove that the internal kinetic energy of the system is $\frac{1}{2}\Sigma(m_r m_s v_{rs}^2)/(\Sigma m)$, where the summation in the numerator applies to each pair of particles once only.

17. Use the result of Ex. 16 to solve Ex. (4) of § 12·62.

18. A heavy flywheel, rotating about its fixed axis of symmetry, is slowing down under the influence of a friction couple at its bearings. During a certain minute, the angular speed is observed to drop by 5%. Find the percentage drop in angular velocity during the next minute on the hypotheses that the magnitude of the friction couple is: (i) constant, (ii) proportional to the angular speed ω, (iii) proportional to ω^2.

[Answers: (i) 5·26%; (ii) 5%; (iii) 4·76%.]

19. I is the moment of inertia of a wheel-and-axle about its axis which is fixed and horizontal. A load of mass M is suspended from the axle, which is of radius a, by a light inelastic string wrapped round the axle. Rotation is opposed by a constant friction couple G. If the system is released from rest with M hanging vertically, find the angular speed of the wheel when the load has descended a distance h; (i) using the principle of energy; (ii) otherwise.   [Answer: $\sqrt{\{2(h/a)(Mga - G)/(I + Ma^2)\}}$.]

20. AB is a diameter of a uniform circular disc of mass M. The disc can turn smoothly about a fixed horizontal axis through A perpendicular to the plane of the disc. The disc is released from rest when AB is horizontal. Find: (a) the magnitude of the reaction at A when the disc has turned through $45°$; (b) the initial acceleration of B.   [Answers: $5Mg/3$; $4g/3$.]

21. A uniform rod of length $2a$ stands vertically on a smooth horizontal table. If the rod is just displaced from rest, prove that:

(i) its centre moves in a vertical line;

(ii) when it has turned through an angle θ, the speed of the centre is $2 \sin \frac{1}{2}\theta \sin \theta \sqrt{\{3ga/(1 + 3 \sin^2 \theta)\}}$;

(iii) the lower end of the rod remains in contact with the table until the rod becomes horizontal;

(iv) just before the rod becomes horizontal, the force on the table is one-quarter of the weight of the rod.

22. A uniform circular lamina of radius a and mass M can turn smoothly in a vertical plane about a fixed horizontal axis perpendicular to its plane at a distance $\frac{1}{3}a$ from its centre. If the lamina just performs complete revolutions, find the greatest and least magnitudes of the reaction at the axis. [Answers: $19Mg/11$; $6\sqrt{2}Mg/11$.]

*23. A light inelastic string AB of length l is fixed at its upper end A, and is attached at its lower end B to a uniform rod BC of length a whose lower end C is supported by a smooth table (whose depth below A lies between l and $l + a$). The system ABC is initially at rest in equilibrium, and C is then slightly displaced along the table in the plane ABC. Find the length of the equivalent simple pendulum in the ensuing small oscillations. [Answer: $2l$.]

*24. A uniform solid circular cylinder of mass M rolls without slipping, with its axis horizontal, down a fixed plane of inclination α. A light inelastic string, which winds up round the cylinder as the cylinder rolls, passes over a small smooth fixed pulley above the level of the cylinder and raises a mass m hanging vertically at the other end. The portion of the string between the pulley and the cylinder is parallel to a line of greatest slope of the plane. Find the acceleration of the mass and the tension of the string.

[Answers: $(4M \sin \alpha - 8m)g/(3M + 8m)$; $(3 + 4 \sin \alpha)Mmg/(3M + 8m)$.]

25. A uniform circular cylinder, with plane ends perpendicular to its axis, rolls without slipping down a fixed plane whose inclination to the horizontal is α, the axis of the cylinder making an angle β with the lines of greatest slope of the plane. Prove that the axis moves with acceleration $\frac{2}{3}g \sin \alpha \sin \beta$.

*26. A uniform bar of length $2a$ and centre O is supported horizontally in equilibrium by two light vertical inelastic strings AB, CD, each of length l, attached to the rod at B, D, where $BO = OD = b$, the upper ends A, C of the strings being attached to fixed points at the same level. The bar is now given a slight angular displacement about the vertical through O, and released from rest. Prove that in the ensuing small oscillations, the length of the equivalent simple pendulum is $la^2/3b^2$.

27. A rough lamina, initially horizontal, can turn smoothly about a fixed horizontal axis through its centre of mass, its moment of inertia about this axis being I. A particle of mass m is gently placed on the lamina at a distance c from the axis, and the system is initially at rest. In the ensuing motion, find the angular speed and angular acceleration of the lamina after it has turned through an angle θ, (where θ is less than the angle at which the particle starts to slip).

[Answers: $\dot{\theta}^2 = 2cmg \sin \theta/(I + mc^2)$; $\ddot{\theta} = cmg \cos \theta/(I + mc^2)$.]

Hence show that the particle starts to slip when the angle turned through is $\tan^{-1}\{\mu I/(I + 3mc^2)\}$, where μ is the coefficient of friction.

28. Prove that the angular momentum of a system of particles about any axis OX is equal to $H_0 + H_1$, where H_0 is the angular momentum of the motion relative to the centre of mass G about a parallel axis through G, and H_1 is the angular momentum about OX of the whole mass supposed collected at G and moving with the velocity of G.

*29. A lamina of mass M and centre of mass G is moving in its plane. Prove that its angular momentum about a point O (not necessarily fixed) in its plane is equal to the angular momentum about O of the motion relative to O, minus the angular momentum about G of a particle of mass M at O moving with the velocity of O.

*30. The angular velocity at time t of a rigid body moving in two dimensions is ω. Prove that $I d\omega/dt + \tfrac{1}{2}\omega dI/dt = L$, where I is the moment of inertia and L the moment of the external forces about the instantaneous centre of rotation.

(Hint.—Consider the rate of increase of the kinetic energy $\tfrac{1}{2}I\omega^2$.)

*31. The cross section of a uniform solid cylindrical body of mass M is a semicircle of radius a. The body is placed with its curved surface in contact with a fixed horizontal plane rough enough to prevent slipping; and makes oscillations about its equilibrium position. Prove that when the angular displacement is θ, the moment of inertia about the line of contact with the plane is $(9\pi - 16\cos\theta)Ma^2/6\pi$.

Hence prove that the length of the equivalent simple pendulum for small oscillations is $(9\pi - 16)a/8$. (Hint.—Use the result of Ex. 30.)

32. A uniform rod AB, of mass M and length l, smoothly hinged at A to a fixed point, is allowed to fall from rest in a horizontal position. Find the angular speed of the rod when its inclination to the horizontal is θ.

[Answer: $\dot\theta^2 = 3g \sin \theta/l$.]

Prove that when the rod is vertical, the tension at a point P of the rod, where $AP = \xi$, is $Mg(5l^2 - 2l\xi - 3\xi^2)/2l^2$. (Hint.—Consider the motion of the portion PB as if it were a separate body.)

*33. A man of mass m stands at A on a horizontal platform which can rotate smoothly about a fixed vertical axis through the point O of the platform. The moment of inertia of the platform about the axis is I, and $OA = a$. Initially, both man and platform are at rest. The man then proceeds to walk on the platform describing (relative to the platform) a complete circle of centre O and radius a. Find the angle through which the platform turns (relative to the ground).

[Answer: $2\pi ma^2/(I + ma^2)$ radians.]

*34. A uniform sphere of mass M and radius a, rolls without sliding on a rough horizontal plane, the initial speed of the centre being U. If the resistance of the air be represented by a horizontal force at the centre of the sphere equal to $M\lambda v^2/a$ and a couple of moment $M\mu v^2$, where v is the speed, and λ, μ are constants, prove that the distance described after time t is $k^{-1}\log_e(1 + kUt)$, where $k = 5(\lambda + \mu)/7a$.

35. A uniform rod AB of length 4 ft and mass 12 lb can turn smoothly about a fixed horizontal axis which is distant 1 ft from the end A. Initially the rod is hanging at rest with B vertically below A. The end B of the rod is then given a horizontal blow of impulse 36 lb-ft/sec at right angles to the axis. Find: (a) the angular velocity of the rod just after the impact; (b) the impulsive action at the axis; (c) the greatest angular displacement of the rod.

[Answers: 3·86 rad/sec; 10·3 lb-ft/sec; 62°·8.]

*36. A uniform plank, of length $2l$ and thickness $2h$, rests symmetrically across the top of a fixed cylinder of radius a whose axis is horizontal, and is then set rolling (without slipping) in a plane perpendicular to the axis. Prove that

$$(4h^2 + l^2 + 3a^2\theta^2)\dot{\theta}^2 + 6g\{(a + h)\cos\theta + a\theta\sin\theta\} = \text{constant},$$

where θ is the inclination of the plank to the horizontal.

Deduce that small oscillations are possible if $h < a$, and find their period.

[Answer: $2\pi\sqrt{[(4h^2 + l^2)/\{3g(a - h)\}]}$.]

Prove also that the greatest angular amplitude for finite oscillations (these will not be simple harmonic) is given as a root of the equation $a\theta = h\tan\theta$. (Hint.—Find the condition that $\dot{\theta} \leqslant 0$ when $\theta = 0$.)

37. A uniform rod of length 0·3 m and mass 1 kg rests on a smooth table and is smoothly pivoted at a distance of 0·05 m from its centre. It receives an impulsive couple of 0·01 N-s-m in the plane of the table. Find the angular speed generated and the impulsive reaction at the pivot.

[Answers: 1 rad/sec; 0·05 N-s.]

38. A bucket of mass 2 lb hangs by a light inelastic string which is coiled round a uniform solid cylinder of radius 6 in and mass 40 lb. The cylinder can be turned smoothly about its axis, which is fixed and horizontal, by means of a light handle distant 2 ft from the axis. Initially the system is held in equilibrium by a force at the handle. If the handle is suddenly released, prove that the tension in the string is suddenly reduced in the ratio 11 : 10.

If after making one complete revolution the handle is suddenly stopped, find the impulsive tension in the string and the impulse at the handle.

[Answers: 8·6, 23·5 lb-ft/sec.]

39. A bar AB of mass M rests on a smooth table. Its centre of mass is distant h from A, and its radius of gyration about a line through A perpendicular to its length is k. It receives a horizontal blow ξ at right angles to AB, at the point P, where $AP = x$. Find the speed with which the point Q of the rod, where $AQ = y$, starts to move.

[Answer: $(\xi/M)\{k^2 - h^2 + (x - h)(y - h)\}/(k^2 - h^2)$.]

Show that if the same blow ξ were applied instead at Q, then P would start to move with the speed just found.

*40. A given lamina initially at rest on a smooth horizontal table is struck by a given blow in its own plane. If E is the energy generated if the lamina is entirely free to move on the table, and E' the energy generated if a point of the lamina is fixed, prove that $E \geqslant E'$.

CHAPTER XV

TWO-DIMENSIONAL STATICS

"The skipper spat disconsolately down the engine-room ventilator and stopped the engines."—Magazine story.

The subject of *Statics* is concerned with sets of forces which maintain systems of particles at rest.

The systems to be considered in this chapter include mainly single particles, single rigid bodies, and systems consisting of several rigid bodies.

The main theory is based on consideration of the case of a single rigid body. The treatment to be given will be two-dimensional.

A system of particles is said to be in *equilibrium* when the particles are all at rest or moving with a constant velocity. The set of forces maintaining this state is called a set of forces in equilibrium. (This accords with the earlier use of the term equilibrium (see e.g. § 12·221).)

15·1. Basic theory of statics of a single rigid body. For a rigid body to be maintained in equilibrium, it is necessary and sufficient that the acceleration of the centre of mass G and the angular acceleration of the body should be zero.

Thus by the principles of linear and angular momentum (extended form), it is necessary and sufficient for equilibrium that $\Sigma X = 0$, $\Sigma Y = 0$, $\Sigma N = 0$, where X, Y are the components of a typical external force and N is its moment about the centre of mass G.

Now, by § 12·17, these conditions hold if and only if the resultant (as defined in §§ 12·14, 12·22) of the external forces is zero; i.e., again by § 12·17, if and only if

$$\Sigma X = 0, \quad \Sigma Y = 0, \quad \Sigma N = 0, \tag{65}$$

where X, Y may here be interpreted as the resolved parts of a typical external force in *any* two directions (not necessarily at right angles), and N as its moment about *any* point (not necessarily G).

Thus a necessary and sufficient set of conditions for the equilibrium of a rigid body is that:

(i) the sum of the resolved parts, in any two different directions, of the external forces is zero;

(ii) the sum of the moments of the external forces about any point is zero.

15·1A. For readers who have omitted the main dynamical theory of Chapters XII, XIII, an alternative approach to that in § 15·1 is required. These readers should follow the directions given in § 10·4, and then regard the last paragraph of § 15·1 as a set of *postulates* enabling the theory of Statics to be developed; (65) is the analytical expression of these postulates.

15·11. Remarks.

(i) Elaborate techniques of solving statical problems have been evolved by engineers and others who have special interests in the domain of statics. In this book we shall largely be concerned with solving standard types of elementary problems by simple basic principles. We shall develop relatively little theory beyond the statements laid down in § 15·1.

(ii) In numerical problems in elementary statics, it is not usually necessary, as in dynamics, to change from practical to theoretical units, for there is the specially simple feature that forces and moments of forces are the only items entering the main statical equations.

15·12. Example.
Four coplanar forces act on a rigid body, the lines of action being the sides of a square ABCD. The forces are 10, 5, 15, 20 lb.wt. along and in the direction of AB, CB, CD, AD, respectively. Find the magnitudes and directions of two forces, one in the line AC and the other in a line through B, which together with the four given forces will produce equilibrium.

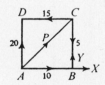

Fig. 163

The four given forces act as shown in Fig. 163. Let the force along AC be P lb.wt. in the sense shown, and let X, Y lb.wt. be the components along AB, BC of the remaining force through B.

Resolving along and at right angles to AB, we have

$$10 + X - 15 + P/\sqrt{2} = 0,$$
$$Y - 5 + 20 + P/\sqrt{2} = 0.$$

Taking moments about C, we have (cancelling a factor corresponding to the length of a side of the square)

$$10 + X - 20 = 0.$$

From these equations, we derive

$$P = -5\sqrt{2}, \quad X = 10, \quad Y = -10.$$

Hence the required forces are $5\sqrt{2}$ lb.wt. along CA, and $10\sqrt{2}$ lb.wt. along DB.

S.P.I.C.E. (α) The line of action of the required force through B was not given in the data. It is often convenient to represent such a force by two components (X, Y) as in Fig. 163.

(β) Moments could have been taken about any point. C was selected as preferable to some other points because four of the six acting forces are in lines through C and so have zero moments about C.

15·2. Statics of a particle. Since all the forces acting on a particle are concurrent, it is necessary and sufficient for the equilibrium of a particle that the sum of the resolved parts in any two directions should be zero. (In this particular case it may be checked that (ii) of § 15·1 automatically holds when (i) holds.)

At the same time, the moment relation (ii) of § 15·1 may be useful in some problems—see e.g. Ex. (3) of § 15·23.

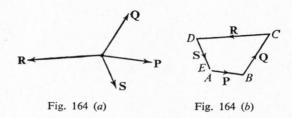

Fig. 164 (a) Fig. 164 (b)

15·21.
Instead of setting down the first two equations of (65), we may use an equivalent single free vector equation. Thus, e.g., for the forces **P, Q, R, S** of Fig. 164 (a) to be in equilibrium, it is necessary and sufficient that the polygon $ABCDE$, in which AB, BC, ... represent **P, Q**, ... in magnitude and direction, should be closed, i.e. that E and A of Fig. 164 (b) should coincide.

The auxiliary diagram (b) of course does not represent the actual *lines of action* of the forces. If it did, the forces would not be in equilibrium (see § 15·33).

15·22.
Among special theorems relating to concurrent forces in equilibrium may be mentioned *Lami's theorem*, according to which if there are just three forces, the magnitude of each is proportional to the sine of the angle between the other two.

The proof of this theorem is easily seen on drawing the closed triangle whose sides represent the forces in magnitude and direction.

15·23. Examples. (1) *Masses of 1, 3 lb are attached to the ends of a light string which passes over two fixed smooth pegs A and C. A third mass m lb is suspended from a point B of the string between the pegs, and the system is in equilibrium with AB inclined at $30°$ to the vertical. Find m and the inclination of BC to the vertical.*

The forces on the three masses are as shown in Fig. 165, T_1, T_2 pdl being the tensions of the strings.

Considering each of the two end masses, we have
$$T_1 - g = 0, \qquad T_2 - 3g = 0. \tag{i}$$
Let θ be the required inclination of BC to the vertical. Then resolving horizontally and vertically for the equilibrium of m, we have
$$T_2 \sin \theta - T_1 \sin 30° = 0, \tag{ii}$$
$$T_1 \cos 30° + T_2 \cos \theta - mg = 0. \tag{iii}$$
By (i) and (ii), we have $\sin \theta = \tfrac{1}{3} \sin 30° = \tfrac{1}{6}$. Hence $\theta = 9°\!\cdot\!6$.
By (iii), we then obtain $m = 3\cdot 82$.

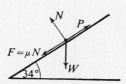

Fig. 165   Fig. 166

(2) *A mass is to be conveyed up a rough fixed inclined plane of inclination $34°$. Prove that the least force, acting up the plane, needed to drag the mass up the plane will be less than the least force needed to lift it if the angle of friction is less than $28°$.*

The least force needed to lift the mass is equal in magnitude to its weight, W say.

Let P be the least force (acting as in Fig. 166) needed to move the mass, and let N and F be the normal and frictional components of the force then exerted by the plane on the mass.

Since motion is on the point of occurring, $F = \mu N$, where μ is the coefficient of friction.

Taking resolved parts of the forces in Fig. 166, we have
$$P - \mu N - W \sin 34° = 0,$$
$$N - W \cos 34° = 0.$$
Eliminating N from these equations, we have
$$P = W(\mu \cos 34° + \sin 34°).$$
In order that $P < W$, we require, putting $\mu = \tan \epsilon$, so that ϵ is the angle of friction,
$$\tan \epsilon \cos 34° + \sin 34° < 1;$$
i.e. $\qquad \sin \epsilon \cos 34° + \cos \epsilon \sin 34° < \cos \epsilon;$
i.e. $\qquad \sin(\epsilon + 34°) < \sin(90° - \epsilon);$
i.e. $\qquad \epsilon + 34° < 90° - \epsilon$, (since $0 \leqslant \epsilon \leqslant 90°$);
i.e. $\qquad \epsilon < 28°.$

(3) *The poles N and S of a magnet respectively repel and attract a magnetic pole at any point P with forces inversely proportional to the squares of PN, PS, respectively. If the line of action of the force needed to keep the pole in equilibrium cuts NS produced in Q, prove that $NQ : SQ = PN^3 : PS^3$.*

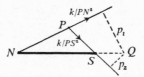

Fig. 167

The required force is equal and opposite to the resultant of the forces k/PN^2, k/PS^2, where k is constant, shown in Fig. 167. This resultant therefore acts along PQ.

The problem could be solved by resolving parallel and at right angles to NS, but some awkward trigonometry would be involved.

A better method is to take moments about Q. This gives

$$p_1 k/PN^2 - p_2 k/PS^2 = 0,$$

where p_1, p_2 are the perpendiculars from Q to PN, PS (produced). Then

$$PN^2 : PS^2 = p_1 : p_2$$
$$= (NQ \sin PNS) : (SQ \sin PSN)$$
$$= (NQ.PS) : (SQ.PN);$$

i.e., $NQ : SQ = PN^3 : PS^3$.

An alternative method which avoids the use of trigonometry altogether is to use equation (ii) of § 12·121, regarding NSQ as a transversal to the forces along NP and PS and their resultant along PQ.

We have (with the sign conventions of § 12·121)

$$\frac{(-k/PN^2).NQ}{PN} = \frac{(k/PS^2).QS}{PS},$$

which gives the required result forthwith.

S.P.I.C.E. (α) In considering the equilibrium of m in Ex. (1), Lami's theorem could have been readily used.

(β) In Ex. (2), it was permissible to put $F = \mu N$ because motion was on the point of occurring. Frequently in statical problems this is not the case.

(γ) In treating the trigonometry at the end of Ex. (2), use was made of the restriction $0 \leqslant \epsilon \leqslant 90°$ on values of ϵ. Restrictions on values of angles are often present in the solving of statical problems.

(δ) Ex. (3) is a problem on the statics of a particle which is better solved by taking moments than by taking resolved parts.

(ϵ) The alternative method used in solving Ex. (3) is interesting. But the student is advised to use the standard methods in general rather than waste a lot of time seeking 'clever' methods.

15·3. Problems involving a single rigid body. The essential theory of the statics of a rigid body has already been given in § 15·1, and is summed up in equations (65). The following subsections of § 15·3 contain some remarks and theorems which are useful in practice.

15·31.
The equations $\Sigma X = 0$, $\Sigma Y = 0$ may (as indicated in § 15·21) be replaced by the use of an equivalent single free vector equation. Only the equation $\Sigma N = 0$ is concerned with lines of action.

15·32. Theorem of three forces. If a rigid body is in equilibrium under just three external forces, these forces must be concurrent or parallel.

The proof is as follows. If two of the forces are parallel, the third force must be parallel to them; otherwise $\Sigma X \neq 0$ or $\Sigma Y \neq 0$. If no two are parallel, let the lines of action of any two intersect at A, say; then the line of action of the third force must also pass through A; otherwise $\Sigma N \neq 0$ when moments are taken about A.

This theorem is often the means of simplifying drastically the solution of a problem. It will be illustrated in later examples.

(The above proof is restricted to the two-dimensional case. The theorem is valid in three dimensions, but the proof is much more difficult.)

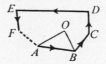

Fig. 168

15·33.
If a set of forces is represented *completely* by the sides AB, BC, CD, ... of a closed polygon $ABCDEF$ ... A, the set is statically equivalent to a couple whose moment is twice the area of the polygon.

The proof is as follows. Evidently, $\Sigma X = 0$ and $\Sigma Y = 0$. Hence the set of forces is equivalent to a couple or is in equilibrium. The moment of the couple (if any) can be determined—see § 12·132—by taking moments about any point O. The moment about O of the force in AB is $AB \times$ (perpendicular from O to AB); i.e., $2 \times$ Area of $\triangle OAB$. The sum of the moments of all the forces about $O = 2(\triangle OAB + \triangle OBC + \ldots)$ = $2 \times$ Area of polygon.

15·34.
Various sets of necessary and sufficient conditions for the equilibrium of forces acting on a rigid body may be set down as alternatives to the set in § 15·1.

One such set of conditions is that the sums of the moments of the external forces about any three non-collinear points should be zero. This is obviously a necessary set of conditions. It is also a sufficient set since the resultant, if a single force, could not have zero moment about all of three non-collinear points; and, if a couple, could not have zero moment about any point.

The brighter student will discern other such sets of conditions in the course of his working of examples; another set is stated in Exs. XV, No. 1. It is suggested that the average student should, however, in the main, rely upon the conditions stated in § 15·1, or upon the theorem of three forces when relevant.

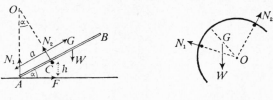

Fig. 169   Fig. 170

15·35. Examples. (1) *A beam AB of weight W rests in a vertical plane against and at right angles to a smooth fixed horizontal rail C, its lower end A being free to move on a smooth fixed horizontal plane distant h below C. The centre of mass of the beam is at a distance a from A. Find what horizontal force should be applied at A to keep the beam in equilibrium at a given inclination α to the horizontal.*

The external forces on the beam are: the weight W acting through the C.M. G; the reactions N_1 at A, N_2 at C; and the required force F; — as shown in Fig. 169.

Taking moments about A, we have

$$N_2 h \operatorname{cosec} \alpha - Wa \cos \alpha = 0. \tag{i}$$

Resolving horizontally, we have

$$F - N_2 \sin \alpha = 0. \tag{ii}$$

(Resolving vertically would introduce N_1, which is neither given nor sought.)

From (i), (ii), we derive $\quad F = (Wa/h) \sin^2 \alpha \cos \alpha.$ (iii)

Alternative method. By taking moments about the point O of intersection of the lines of action of N_1, N_2, we avoid introducing either N_1 or N_2. We have

$$F.OA - Wa \cos \alpha = 0,$$

where $OA = AC \operatorname{cosec} \alpha = h \operatorname{cosec}^2 \alpha$, from which (iii) immediately follows.

(2) *A piece of uniform wire, bent so as to form a semicircular arc, rests in a vertical plane on two smooth fixed horizontal pegs. Prove that the ends of the wire must be at the same level.*

The external forces on the wire are: its weight W acting through the C.M. G; and the reactions N_1, N_2, due to the pegs, passing through the centre O of the semicircle, as in Fig. 170.

These forces are three in number, and must therefore be concurrent.

Hence W must pass through O; i.e. GO must be vertical. Hence the result.

(3) *Two uniform rods, AB, BC of the same material and cross section are rigidly jointed at B with the angle ABC equal to* 120°. *The bent lever so formed is smoothly pivoted at B so that it can be turned in a vertical plane about B. In the equilibrium position, BC is horizontal. A load of weight W is now attached to C, and in the new equilibrium position, AB is horizontal. Find the ratio of W to the weight of BC.*

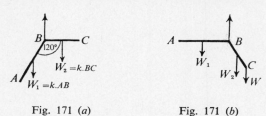

Fig. 171 (a)   Fig. 171 (b)

In the first set of circumstances, the forces on the lever are: the weights W_1, W_2 of AB, BC, viz. $k.AB$, $k.BC$, respectively, where k is constant; and the support at B (incidentally, by § 15·32, this support must act vertically).

Taking moments about B, we have
$$k.AB.\tfrac{1}{2}AB \cos 60° - k.BC.\tfrac{1}{2}BC = 0;$$
i.e. $\qquad AB = BC\sqrt{2}.$ \hfill (i)

In the second set of circumstances, there is the additional vertical force W at C.

Again taking moments about B, we have
$$W_1.\tfrac{1}{2}AB - W_2.\tfrac{1}{2}BC \cos 60° - W.BC \cos 60° = 0.$$ \hfill (ii)

From (i) and (ii), we derive $W/W_2 = 1\cdot 5$.

(4) *A uniform triangular lamina ABC is in equilibrium in a vertical plane with its base BC horizontal when two forces, equal in magnitude, act along the bisectors of the angles ABC, ACB. Prove that ABC is an isosceles triangle.*

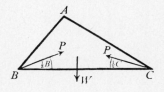

Fig. 172

The lamina is in equilibrium under its weight W and the two given forces, of magnitude P say (Fig. 172).

Resolving horizontally, we have $P \cos (\tfrac{1}{2}B) - P \cos (\tfrac{1}{2}C) = 0$; i.e. $\cos (\tfrac{1}{2}B) = \cos (\tfrac{1}{2}C)$, of which $B = C$ is the only relevant solution. Hence △ ABC is isosceles.

S.P.I.C.E. (α) Ex. (1) is an example of the fairly direct use of the basic equations (65). But it should be noticed that by exercising a wise choice of the point about which moments are taken, it is possible to avoid introducing certain unwanted forces into the equations. The working of many problems in statics can be shortened in this way.

(β) Ex. (2) shows the potency of the three-force theorem when it is relevant.

(γ) Ex. (3) is a typical elementary problem involving parallel forces. For these problems it is sufficient to use $\Sigma N = 0$, and just one of the equations $\Sigma X = 0$, $\Sigma Y = 0$.

(δ) In Ex. (4), a single resolution was sufficient to give the answer.

(ϵ) It will be noticed from the above examples that it is not always necessary to use all three of the equations (65), especially if there are forces present which are not given and not sought. On the other hand, whenever all three of (65) are used for one particular rigid body, no further information about the external forces on that body is derivable by resolving in a third direction, or by taking moments about a further point. (Students who e.g. resolve in three directions are sometimes surprised to find the (valid but not helpful) result '$0 = 0$' emerging (on those occasions when their trigonometrical manipulations are accurate).)

(ζ) In Ex. (2), Fig. 170 shows the wire convex upwards. Compatibly with the data, the wire could be concave upwards. The student might check that the required result follows in that case as well. Similarly, in Ex. (4), A could have been taken to be below BC.

15·4. Problems involving friction.
None of the problems in subsections of § 15·3 involves consideration of friction. The theory of friction already given in § 6·6 includes the case of statics, and in § 15·41 we merely add some notes that are helpful in solving statical problems involving friction.

15·41. Remarks.

(i) We remind the reader that (in usual notation—§ 6·6) $F = \mu N$ in statics only if relative motion at the point of contact involved is on the point of occurring. In practice, $|F| < \mu N$ in many cases.

(ii) In problems where limiting friction is involved, it is sometimes helpful to treat the force at a contact as a single force inclined at the angle of friction to the normal, rather than to work in terms of the components F and N. This is especially so when there are just two other external forces on the rigid body considered; for the three-force theorem of § 15·32 can then often be used to great advantage. (See Ex. (1) below.)

(iii) In statical problems, standard trigonometrical formulae are frequently needed. We add to these a trigonometrical theorem which is useful in a wide range of statical problems involving friction. We remark that, in mechanics, it is far better in general for the student to learn to rely on the valid use of principles rather than on 'ad hoc' formulae; but we think it worth while to make an exception in the case of this theorem.

The theorem is:
$$(m+n)\cot\theta = m\cot\alpha - n\cot\beta, \qquad (a)$$
$$= n\cot A - m\cot B, \qquad (b)$$
where the symbols are as indicated in Fig. 173, with $AP:PB = m:n$.

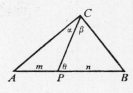

Fig. 173

To prove (a), we have
$$\frac{m}{n} = \frac{AP.PC}{PC.PB}$$
$$= \frac{\sin\alpha}{\sin(\theta-\alpha)} \cdot \frac{\sin(\theta+\beta)}{\sin\beta}$$
$$= \frac{\sin\theta\sin\alpha\cos\beta + \cos\theta\sin\alpha\sin\beta}{\sin\theta\cos\alpha\sin\beta - \cos\theta\sin\alpha\sin\beta}$$
$$= \frac{\cot\beta + \cot\theta}{\cot\alpha - \cot\theta},$$

which gives (a); (b) is similarly proved.

15·42. Examples. (1) *A uniform rod AB of length 2a rests in a vertical plane at right angles to, and with one end A against, a fixed rough vertical wall (coefficient of friction = 2·0), the other end B being fastened to a point C in the wall vertically above A by a light string BC of length 2a. Find the least angle the string can make with the wall.*

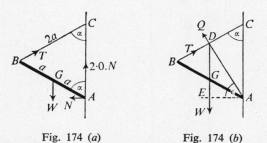

Fig. 174 (a)   Fig. 174 (b)

First Method.

The external forces on the rod AB are its weight W, the tension T of the string and the contact force at A. In Fig. 174(a), the contact force on AB is represented by the normal component N, and (the extreme case being taken in which α is least, so that the friction is limiting) by the frictional component $2·0.N$.

Resolving along and right angles to AC, we have
$$T\cos\alpha + 2N - W = 0, \qquad \text{(i)}$$
$$T\sin\alpha - N = 0. \qquad \text{(ii)}$$
Taking moments about A, we have, since $AC = 4a\cos\alpha$,
$$Wa\sin\alpha - T\cdot 4a\cos\alpha\sin\alpha = 0. \qquad \text{(iii)}$$
On eliminating T, N, W from (i), (ii), (iii), we obtain
$$\tan\alpha = 1\cdot 5; \quad \text{i.e. } \alpha \approx 56°.$$

Second Method.

In Fig. 174(*b*), the contact force at A (in the limiting equilibrium case) is represented by the force Q inclined at ϵ to the normal AE to the wall, where $\tan\epsilon = 2\cdot 0$.

The rod is thus in equilibrium under the three forces W, T, Q, whose lines of action must therefore be concurrent; and the point of concurrency is clearly the midpoint D of BC.

Hence we have
$$\begin{aligned}
2\cdot 0 &= \tan\epsilon \\
&= (DG + GE)/EA \\
&= (2a\cos\alpha + a\cos\alpha)/a\sin\alpha \\
&= 3/\tan\alpha,
\end{aligned}$$
giving $\tan\alpha = 1\cdot 5$, as before.

(2) *A carriage wheel of weight W and radius a, the axis being horizontal, is to be dragged over an obstacle of height h above the horizontal ground. Find the horizontal force, applied at the axis, under which the wheel will be on the point of moving over the obstacle.*

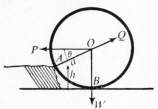

Fig. 175

The external forces on the wheel are its weight W, the required force P, and the contact forces at A and B. The force at B may be disregarded since it disappears at the instant when the wheel starts to move. By the three-force theorem, the contact force Q at A must then go through the centre O of the wheel.

Taking moments about A (this procedure avoids involving Q), we have
$$Pa\sin\theta - Wa\cos\theta = 0,$$
where θ is as in Fig. 175. Hence
$$\begin{aligned}
P &= W\cot\theta \\
&= \frac{W\sqrt{(2ah - h^2)}}{a - h}
\end{aligned}$$

(3) *The mass per unit length of a ladder increases uniformly from the top A to the bottom B, and is twice as great at B as at A. The ladder is stood on a horizontal pavement with the top resting against a vertical brick wall, the coefficient of friction at both ends being 0.2. Find the greatest angle the ladder can make with the wall without starting to slip.*

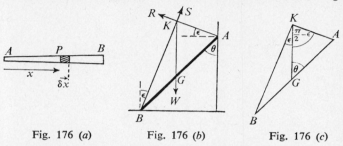

Fig. 176 (a)   Fig. 176 (b)   Fig. 176 (c)

It is first necessary to find the C.M. G of the ladder.

Let l be the length of the ladder, λ the mass per unit length at any point P distant x from A (see Fig. 176(a)). Then the data give
$$\lambda = \lambda_0(1 + x/l),$$
where λ_0 is the value of λ at A.

Thus the distance of G from A
$$= \int_0^l \lambda x\,dx \Big/ \int_0^l \lambda\,dx$$
$$= \int_0^l \left(x + \frac{x^2}{l}\right)dx \Big/ \int_0^l \left(1 + \frac{x}{l}\right)dx$$
$$= (\tfrac{1}{2}l^2 + \tfrac{1}{3}l^2)/(l + \tfrac{1}{2}l)$$
$$= 5l/9.$$

When the ladder is placed against the wall, the forces on it are its weight W acting vertically through G, and the contact forces R, S, say, at A, B. Fig. 176(b) is drawn for the case of limiting equilibrium, where $\tan \epsilon = 0.2$.

The forces W, R, S must be concurrent, say at K.

Let θ be the angle between the ladder and the wall.

In Fig. 176(c), the triangle ABK is drawn separately, showing the values of certain angles.

By the theorem (a) of § 15·41, we have, since $BG : GA = 4 : 5$,
$$(4 + 5) \cot \theta = 4 \cot \epsilon - 5 \cot(\tfrac{1}{2}\pi - \epsilon);$$
i.e., $9 \cot \theta = 4 \times 5 - 5 \times 0.2$ (since $\tan \epsilon = 0.2$);
i.e., $\tan \theta = 9/19$;   i.e., $\theta = 25°\!\cdot\!4$.

S.P.I.C.E. (α) The use of the angle of friction and the three-force theorem enabled Ex. (1) to be solved by very simple geometry and trigonometry in the second method, whereas more algebraic detail was needed in the first method. Nevertheless, the first method, even if more cumbrous, may indeed be the faster in practice when the geometry involved in the second method is not quickly discerned.

(β) In Ex. (2), moments were taken about A to avoid introducing the unknown force Q. It would have been equally expedient to resolve at right angles to AO; the reader might solve the problem this way as well.

(γ) There are many problems in which the trigonometrical theorem of § 15·41 (iii) is as effective as in Ex. (3).

(δ) In an example such as Ex. (1), students are sometimes puzzled as to why N in Fig. 174 (a) is taken along the normal to the wall AC and (allegedly) 'not along the normal to the rod AB'. The circumstances are that, at the end of an actual rod, the curvature of the surface of the rod changes rapidly, so that a suitable representation is:—

With this representation there is a common normal which of course is at right angles to the wall.

15·5. Problems involving several rigid bodies. The equations (65) may be applied not merely to a single body, but to any part, or to the whole, of a system of bodies in equilibrium.

An essential point to be watched is that forces which are *internal* for one part of the system, and which may hence be ignored in using (65) for that part, may be *external* for other parts of the system and hence must be taken into account in using (65) for these parts. The student is strongly advised to *draw separately* each part of a system to which (65) are applied, to show *all the external forces on that part*, and incidentally to *state which part* he is referring to when using (65).

Details of technique are indicated in the examples of §§ 15·51, 15·62.

15·51. Example. *Two uniform spheres of weights W, W', rest, both in limiting equilibrium, on different and differently inclined fixed rough planes which have a horizontal line l as their common upper boundary, the coefficients of friction being μ, μ'; and a horizontal light string perpendicular to l joins the highest points of the spheres. Prove that $\mu W = \mu' W'$.*

First method.
Consider the external forces on the sphere of weight W.

These forces are: the weight W passing through the highest point B; the tension T also passing through B; and the contact force at A (Fig. 177 (a)), which, by the three-force theorem, must also pass through B. Further, the angles OAB, OBA are each equal to ϵ, where $\tan \epsilon = \mu$.

Resolving horizontally and vertically, we then have
$$T - Q\sin\epsilon = 0, \qquad W - Q\cos\epsilon = 0;$$
and hence $\qquad T = W\tan\epsilon = \mu W.$

By similarly considering the second sphere, we can show that $T = \mu' W'$. Hence $\mu W = \mu' W'$.

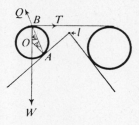
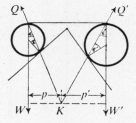

Fig. 177 (a) Fig. 177 (b)

Second method.

We now solve the problem by considering the forces on the whole system consisting of the two spheres and the string.

The *external* forces *on this system* are the weights W, W' and the contact forces Q, Q', acting as shown in Fig. 177(b).

By taking moments about the point K in which the lines of action of Q, Q' intersect, we have
$$Wp - W'p' = 0,$$
where p, p', the perpendiculars from K to the lines of action of W, W', are seen to be in the ratio $\tan\epsilon : \tan\epsilon'$.
Hence $\mu W = \mu' W'$.

S.P.I.C.E. (α) The tension T is an external force if the separate equilibrium of either sphere is being considered, but is an internal force, and hence ignored, if the whole system is being considered.

(β) The utility of the three-force theorem in relation to some friction problems should again be noted.

15·6. Problems on frameworks. An important class of problems involving several rigid bodies is concerned with plane frameworks consisting of rigid bodies (e.g. rigid rods) jointed together by pin-joints (as described in § 13·3), each body or member being pierced by two pins.

15·61. The following notes give a guide to the technique of solving elementary framework problems.

(i) Consider the case in which three members A, B, C of a framework have a common pin-joint at K, as in Fig. 178 (a), the pin being subjected to no forces except the reactions due to A, B and C.

We shall consider the forces on A, B, C, K, respectively. For this purpose we draw separate diagrams for each, as in Fig. 178 (b). We can represent the effect on A due to the pin by two force components (P, Q) and a friction couple L, on B by (R, S) and M, and on C by (X, Y) and N, as shown in Fig. 178 (b). By the action, reaction law, there will be equal and opposite effects on the pin, as also shown in Fig. 178 (b).

Since the pin is in equilibrium under the forces acting on it, we have, by (65), $P + R + X = 0$, $Q + S + Y = 0$, $L + M + N = 0$.

The forces on A, B, C at the pin may therefore be represented as shown in Fig. 178 (c), without the need for depicting the forces on the pin itself.

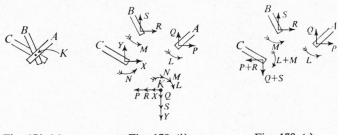

Fig. 178 (a)　　　　Fig. 178 (b)　　　　Fig. 178 (c)

(ii) Consider next the case as in (i), except that additional forces (e.g. an attached load or an external support) act on the pin K. Fig. 178 (b) would then be relevant, except that the additional forces acting on K would have to be shown. In this case, Fig. 178(c) does not apply, and, to begin with, the student should rely on Fig. 178(b), modified as stated.

(iii) If the number of members with a common pin-joint is other than three it is easy to adapt the above remarks.

If the number is two, and there are no additional forces on the pin, representation as in Fig. 179 is adequate.

Fig. 179　　　　　　　　　　Fig. 180

(iv) If the pin-joint is smooth, friction couples of the type L, M, N above are all zero and are of course omitted from the diagrams.

(v) If a member is 'light', and the two pin-joints attached to it are smooth, and if there are no forces acting on the member except those due to the pins, K_1, K_2, say, then these forces act in the line $K_1 K_2$ and may be represented as in Fig. 180. This follows since the member is, in this case, in equilibrium under just two external forces.

If X in Fig. 180 is positive (where X is exerted by a pin *on* the member), the member is a *strut*; if negative, a *tie* (see § 4·6).

Representation as in Fig. 180, when relevant, is important in solving problems. Representation as in Fig. 178 or 179 would of course be valid, but would be cumbrous in this particular case.

(vi) If there is symmetry in a framework (usually about a vertical line), this often enables the number of unknown symbols to be substantially reduced. (See Exs. (2), (3) below.)

(vii) If the equilibrium is being considered of a group of several members of a framework as a *whole*, say a group including all of A, B, C, K of Fig. 178, it is not necessary to take cognisance of any of the forces shown in Fig. 178 (b) or (c), since these forces would be internal for such a group.

15.62. Examples. (1) *ABC is a smoothly pin-jointed triangular framework of light rods; $AB = 3$ ft, $BC = 2$ ft, and the angle ABC is a right angle. The frame is at rest with the pin at B fastened to a vertical wall, and the pin at C pressing against the wall at a point vertically below B, the contact being smooth. A mass of 30 lb hangs from the pin at A. Find the forces in the rods and the forces exerted by the wall on the pins at B and C.*

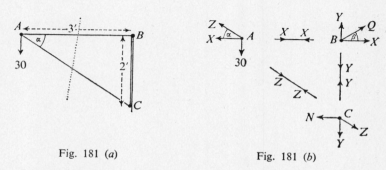

Fig. 181 (a) Fig. 181 (b)

Fig. 181(a) shows the framework. If the angle CAB is α, then
$$\cos \alpha = 3/\sqrt{13}, \quad \sin \alpha = 2/\sqrt{13}. \tag{i}$$
Since the three rods are light and the joints smooth, we can represent the forces on the rods as shown in Fig. 181(b).

Since there are additional forces at all the pins, it is desirable to show the forces on each pin, as in Fig. 181(b). These forces consist of: reactions equal and opposite to the forces X, Y, Z lb.wt., say, on the rods; the force 30 lb.wt. on the pin A; the force Q lb.wt. exerted by the wall on the pin B, inclined say at β to the horizontal; and the horizontal force N lb.wt. exerted by the wall on the pin at C.

The equilibrium conditions will be covered if we resolve in two directions for each pin; (the equilibrium conditions for each rod have already been in effect attended to).

Considering the pin A, we have, resolving vertically and horizontally,
$$Z \sin \alpha - 30 = 0, \quad Z \cos \alpha + X = 0;$$
i.e. (using (i)), $\quad Z = 15\sqrt{13}, \quad X = -45.$ \hfill (ii)

Similarly for C, we have
$$Y + Z \sin \alpha = 0, \quad N - Z \cos \alpha = 0;$$
i.e., $\quad Y = -30, \quad N = 45.$ (iii)

Hence the forces in the rods are:—
AB : tension of 45 lb.wt.
BC : tension of 30 lb.wt.
CA : thrust of $15\sqrt{13}$ lb.wt.

(AB, BC are ties; CA is a strut.)

Also, the force exerted by the wall on the pin C is 45 lb.wt.

For the equilibrium of the pin B, we have
$$Q \cos \beta + X = 0, \quad Q \sin \beta + Y = 0,$$
yielding $Q = 15\sqrt{13}$, $\tan \beta = 2/3$.

Hence the force exerted by the wall on the pin B is 54·1 lb.wt. in a direction $\tan^{-1}(2/3)$ above the horizontal, and directed into the wall.

(2) *Four heavy uniform rods each of weight w are smoothly pin-jointed to form a square framework ABCD. The pin at A is horizontal and smoothly attached to a fixed point, and the framework is kept in shape by a light rod smoothly jointed to the pins at B and D. A mass of weight W hangs in equilibrium from C. Find the force in BD.*

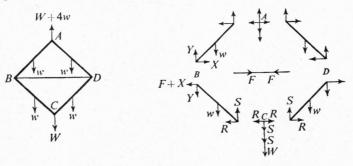

Fig. 182 (a) Fig. 182 (b)

Fig. 182(a) shows the forces that are external to the whole system; (it is immediately seen on resolving vertically that the external support on the pin at A is $W + 4w$).

In Fig. 182(b), the forces on various members of the system are indicated by arrows; (the pins at B and D are not included since there are no additional forces on these pins—see § 15·61 (i)).

Since BD is light, we may represent the forces on it by F, F as shown.

Symbols have been attached to only some of the arrowheads, the problem being solvable without considering all the forces present.

The force at B on AB is represented by (X, Y) as shown. The force at B on BC is then seen to be represented by $(F + X, Y)$ acting as shown, in accordance with § 15·61 (i).

The force at C on BC is represented by (R, S) as shown. Since there is complete symmetry about AC, the force at C on CD must then be representable by (R, S) as shown. The forces on the pin at C are then seen to be as shown.

We shall now start from the pin at C and work round the framework in the direction CBA. Let $AB = 2a$.

From the equilibrium of the pin at C, we have
$$2S + W = 0. \qquad \text{(i)}$$
Resolving vertically for the rod BC, we have
$$S - Y - w = 0; \quad \therefore Y = -(\tfrac{1}{2}W + w). \qquad \text{(ii)}$$
(Resolving horizontally for BC would introduce R which is not needed.)

Taking moments about C for BC, we have
$$wa/\sqrt{2} + Ya\sqrt{2} + (F + X)a\sqrt{2} = 0;$$
i.e.
$$F + X = -Y - \tfrac{1}{2}w = \tfrac{1}{2}(W + w). \qquad \text{(iii)}$$
Taking moments about A for AB, we have
$$wa/\sqrt{2} + Xa\sqrt{2} - Ya\sqrt{2} = 0;$$
i.e.
$$X = Y - \tfrac{1}{2}w = -\tfrac{1}{2}(W + 3w). \qquad \text{(iv)}$$
Eliminating X between (iii) and (iv), we have $F = W + 2w$.
Hence the force in BD is a thrust of $W + 2w$.

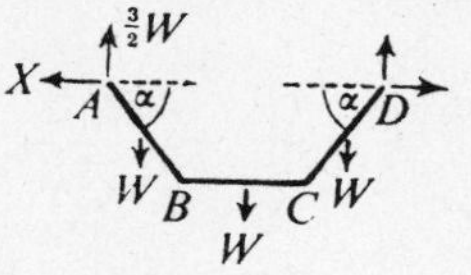

Fig. 183 (a) 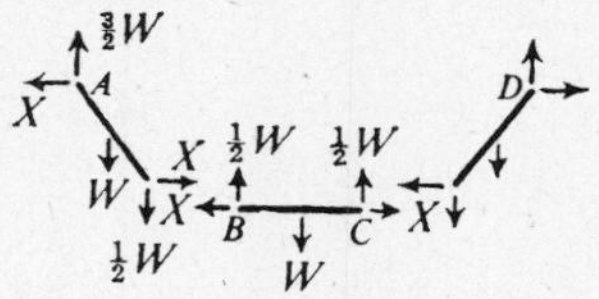   Fig. 183 (b)

(3) AB, BC, CD are three uniform rods of equal weights and lengths, smoothly pin-jointed at B and C. A and D are attached to fixed smooth horizontal pins at the same level, and the system hangs in equilibrium. If α is the inclination of AB (or CD), and β, γ the inclinations of the reactions at A, B, to the horizontal, prove that $3 \tan \alpha = 2 \tan \beta = 6 \tan \gamma$.

Fig. 183(a) represents the external forces on the whole system; the vertical component of the force at A is seen to be $3W/2$ on resolving vertically and noting the symmetry present.

Fig. 183(b) shows the forces on AB, BC, CD, respectively. The force on AB at B is seen to be representable by $(X, \tfrac{1}{2}W)$, as shown on resolving horizontally and vertically for AB. The force on BC at B is then representable by $(X, \tfrac{1}{2}W)$, and at C (by symmetry again) by $(X, \tfrac{1}{2}W)$, in the directions shown.

Taking moments about A for AB, we then have (letting $AB = 2a$)
$$X.2a \sin \alpha - \tfrac{1}{2}W.2a \cos \alpha - Wa \cos \alpha = 0;$$
i.e.,
$$X = W \cot \alpha. \qquad \text{(i)}$$
From Fig. 183(b), we derive
$$\tan \beta = 3W/2X, \qquad \tan \gamma = W/2X.$$
Hence
$$3 \tan \alpha (= 3W/X, \text{ by (i)}) = 2 \tan \beta = 6 \tan \gamma.$$

(4) *Two equal uniform rods AB, BC are pin-jointed at B. If AB is held fixed, BC can be just sustained (in the vertical plane containing AB) at $30°$ beyond the downward vertical. The system is now placed in a vertical plane with A and C resting on a rough horizontal table (coefficient of friction $\frac{3}{8}$ at both A and C) and with B above the table. If the system rests in equilibrium, find the greatest possible value of the angle ABC.*

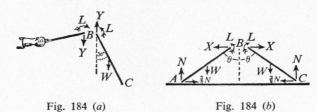

Fig. 184 (a)   Fig. 184 (b)

In this problem, there is a friction couple at B. Let L be its magnitude when slipping at B is about to occur.

The first set of data enables us to determine L.

Fig. 184(a) shows the forces on AB, BC in these circumstances, excluding the external forces applied by the hand on the rod AB. (By resolving horizontally for BC, it is seen that the force at B on BC (and hence also on AB) has no horizontal component.)

Taking moments about B for BC, we have
$$L - Wa/2 = 0, \tag{i}$$
where $BC = 2a$.

Fig. 184(b) represents the second set of circumstances, 2θ being the required angle. There is no vertical component of force at B on AB or BC; for if there were a component Q at B, say vertically upwards on AB, there would then be a component Q at B vertically downwards on BC; and this would constitute a breach of symmetry unless $Q = 0$.

Resolving vertically and horizontally, and taking moments about C for BC, we then obtain

$$N - W = 0; \quad X - \tfrac{3}{8}N = 0; \quad Wa\sin\theta - X.2a\cos\theta - L = 0. \tag{ii}$$

From (i) and (ii), we derive
$$4\sin\theta - 3\cos\theta = 2;$$
i.e., $\sin(\theta - \alpha) = 0.4$, where $\alpha = \sin^{-1} 0.6$;

Since θ is restricted to the range $0 \leqslant \theta \leqslant \tfrac{1}{2}\pi$, we thus have
$$\theta = \sin^{-1} 0.4 + \sin^{-1} 0.6$$
$$\approx 60°\cdot 5.$$

Hence the required greatest value of the angle ABC is about $121°$.

S.P.I.C.E. (α) The different modes of representing forces on light and heavy rods—see e.g. Figs. 181(b), 182(b)—should be noted.

(β) In Fig. 181(b), the arrowheads for X and Y happen to point in the opposite senses to those in which the actual forces act (and similarly in some other cases). This does not matter, as the algebraic equations simply reveal that X, Y are then negative.

(γ) In Ex. (1), Y could be equally well determined by considering the external forces on the system consisting of the three rods and the pins at A and C. There are, in fact, usually many routes by which a particular framework problem can be solved. By experience, the student will learn that some routes are much shorter than others; by adroit selection, one can often avoid introducing into the equations some of the forces which are not known and not sought.

(δ) Another technique often found useful is the 'method of sections'. Across the diagram picturing the system of bodies, a 'section line' is drawn with a view to 'isolating' one part of the system from the remainder. Equations are then formed for the equilibrium of the forces *external* to the isolated system. Several 'sections' may be needed to solve a problem.

Consider e.g. the section made by the dotted line in Fig. 181(a). The isolated system on the left of this line is in equilibrium under the external forces: 30 lb.wt. vertically at A, X along BA and Z along CA. This would yield the equations (ii) found in the previous solution. It is important to recognise that forces such as X and Z, acting at places where the section line cuts members of the original system, are *external* for the isolated system (though internal for the original whole system).

(There are complications in using the method of sections if a section line cuts members of the system that are not light. These complications will not be considered in this book.)

(ϵ) Use may often, as in Exs. (2), (3), be made of symmetry (see § 7·6, *S.P.I.C.E.* (ζ), p. 133); the treatment of the forces at C in Ex. (2) should be carefully noted. If symmetry were not made use of, it would be necessary to include more unknown symbols, and to form more equations; the same results would of course emerge finally. The student should, for this satisfaction, solve at least one example like Ex. (2) without recourse to symmetry considerations.

(ζ) Ex. (3) shows how the number of unknown symbols may be kept very low in solving some problems.

(η) Ex. (4) is included as a case in which a friction couple is involved.

(θ) In the first part of Ex. (4) (corresponding to Fig. 184(a)), it was not stated that AB was held at any particular angle; the solution did not need this datum.

(ι) For practical purposes, it would have to be presupposed in the second part of Ex. (4) that smooth constraints are present in order to keep the system in a vertical plane and prevent it from toppling over; otherwise the equilibrium would be unstable (see § 15·8). The presence of such constraints is often tacitly understood in exercises of this character.

($\varkappa$) The reason for omitting vertical reaction components at B in Fig. 184 (b) should be noted.

(λ) In some books, the word 'stress' is used where we have used 'tension' or 'thrust'. But 'stress' has an important meaning as force per unit area (see § 16·11), and is dimensionally distinct from force.

15·7. Graphical statics.
Engineers and technologists make varied and elegant use of graphical methods in solving many types of problems in statics. We shall here give one simple example.

15·71. Example. *The light framework shown in Fig. 185(a) is symmetrical about the vertical through A and rests on smooth horizontal platforms at B and C. The angles DBA, DAB, ADE are 20°, 25°, 60°, respectively. If a load of 10 cwt hangs from A, find by a graphical method the forces in the rods of the frame.*

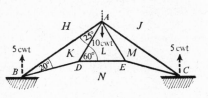
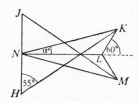

Fig. 185 (a)   Fig. 185 (b)

Fig. 185(a) shows the framework; it is obvious that the magnitudes of the supports at *B*, *C* are each 5 cwt.

The equilibrium of the set of forces acting on any one of the pins at *A*, *B*, *C*, *D*, *E* can be represented by a 'force diagram' in the form of a closed polygon whose sides are parallel to the relevant forces.

Since, however, the force acting along any one of the seven rods would be represented (in opposite senses) in two of these polygons, it is possible to economise by fitting the five polygons together in a single diagram. This is done in Fig. 185 (*b*); the five polygons are: *HKLMJH* (for the pin at *A*); *HNKH* (for *B*); *JMNJ* (for *C*); *KNLK* (for *D*); and *MLNM* (for *E*); each of *HN*, *NJ* corresponds to a force of 5 cwt.

With the use of *Bow's notation*, we can associate the points *H*, *J*, *K*, *L*, *M*, *N* of Fig. 185(*b*) with regions indicated by the same letters in Fig. 185(*a*).

The lines bounding the region *K* in (*a*) are parallel to lines through the point *K* in (*b*); and similarly for all points in (*b*). (The region *N* of (*a*) is regarded as bounded by *BD*, *DE*, *EC* and the dotted lines through *B*, *C*, imagined continued upwards indefinitely; and similarly in other cases.)

Further, the line between the regions *H* and *K* of (*a*) is parallel to the line *HK* of (*b*); and similarly for all lines in (*a*), (*b*).

It is now evident that the magnitudes of the forces in *AB*, *AC*, *AD*, *AE*, *BD*, *CE*, *DE* are represented by *HK*, *JM*, *KL*, *ML*, *KN*, *MN*, *LN*, respectively, in (*b*). By measurement on a carefully drawn diagram, these magnitudes are found to be 14·3, 14·3, 3·6, 3·6, 12·2, 12·2, 10·0 cwt, respectively.

By considering the separate polygons in (b), it is easy to infer that AB, AC are struts; and that AD, AE, BD, CE, DE are ties.

Hence the required forces are:—

AB, AC: thrusts of 14·3 cwt.
AD, AE: tensions of 3·6 cwt.
BD, CE: tensions of 12·2 cwt.
DE: tension of 10·0 cwt.

S.P.I.C.E. (α) The above example illustrates the use of Bow's notation. This notation can be applied to problems on light frameworks where all the external forces act at the pin-joints; (modifications can be made enabling Bow's notation to be used in some other cases).

The figures (a), (b) of Fig. 185 are 'reciprocal' in that not only do the points $J, K, \ldots$, of (b) correspond to regions of (a), but also the points $A, B, \ldots$ of (a) correspond to certain regions of (b), e.g. A to $HKLMJH$, B to $HNKH$, etc.

(β) The discrimination between struts and ties is made by considering individual force polygons. E.g. in the triangle HNK of Fig. 185(b), which represents the equilibrium of the pin at B, HN represents the vertically upward force of 5 cwt; hence NK, KH give the directions of the forces on the pin B due to the rods BD, BA, respectively; hence BD is a tie, and BA a strut.

(γ) In solving the above problem, recourse was had to graphical measurement. The answers could also have been obtained by trigonometrical calculation. In books (such as this one) on the theory of mechanics, it is normally understood that graphical measurements are to be used in solving problems only when specific reference to graphical methods is made.

(δ) Polygons in diagrams such as Fig. 185(b) are not necessarily convex; e.g., $HKLMJH$. Sometimes they enclose zero area; e.g. $JHNJ$ represents the equilibrium of the whole framework under the external forces of 10, 5, 5 cwt, these forces being parallel.

15·8. Stability of equilibrium.

Consider the equilibrium of a uniform solid cone supported by a fixed horizontal table: (i) with the base on the table; (ii) with the apex on the table (and base uppermost); (iii) with a generating line l of the curved surface on the table.

If in case (i) the cone is slightly tilted about a point on the rim of the base and released from rest, the cone moves back towards its original position. If in case (ii) the cone is slightly displaced about its apex and released from rest, it moves further away from its original position. If in case (iii) the cone is slightly displaced about l and released from rest, it stays in its new position.

We say that the equilibrium positions (i), (ii) and (iii) are *stable*, *unstable* and *neutral*, respectively. The stability of any other mechanical system in equilibrium is similarly classified according to the behaviour after applying appropriate small displacements and then releasing from rest.

When a system has more than one degree of freedom (see § 13·21), a single displacement may not be sufficient to test the stability of an equilibrium configuration. E.g., the equilibrium position of a particle on a saddle on a horse is stable for displacements along the line of the horse's back, but unstable for sideways displacements; in such a case the equilibrium is said to be on the whole unstable. In the sequel, we shall assume that there is just one degree of freedom unless the context indicates otherwise.

It is evidently a matter of some practical importance to know whether the equilibrium configuration of a system is stable or not.

In simple problems, stability can be tested by directly examining the forces present after the slight displacement is carried out. E.g., the equilibrium of a rigid body with one point A fixed is stable if, after a small rotation about A is made, the resultant of the moments about A of the forces now acting is in the sense to restore the equilibrium.

A more advanced method of investigating stability will be indicated in § 15·(10)5.

15·81. The above elementary method of investigating stability resembles the method used in investigating whether a body B placed in a given way on a fixed rough surface S will be in equilibrium.

E.g., an omnibus would not be in equilibrium on a sloping roadway if the inclination of the axles to the horizontal exceeded a certain value; and would topple.

Let C be a contour determined by drawing a flexible string tightly round the outer points of contact of B with S. Then a criterion for equilibrium is that the vertical line L through the centre of mass of the body should be encircled by C. When L is not encircled by C, the weight of B and the supports by S have moments about some point of C which are all in the same sense; the sum of these moments cannot then vanish, so that there cannot then be equilibrium.

15·82. Examples. (1) *A rigid body is in equilibrium under its weight and a smooth fixed support at O. Prove that the equilibrium is stable or unstable according as the centre of mass G is below or above O.*

Let G be below O. The corresponding equilibrium position is shown in Fig. 186(*a*).

Let the body be given a slight angular displacement about O. Then the body is subject to forces as shown in Fig. 186(*b*).

The resultant moment of these forces about O is the moment of the weight W, and the sense is that which will restore equilibrium.

Hence the equilibrium is stable.

If G is above O, it may similarly be shown that the equilibrium is unstable.

(If G is at O, the equilibrium is neutral.)

Fig. 186 (a)

Fig. 186 (b)

(2) *A uniform solid is made by joining the bases of a solid circular cylinder of height h and radius a and a hemisphere of radius a. It is placed in equilibrium on a horizontal table with the cylinder uppermost. Find the condition for stability.*

Fig. 187 shows the forces on the solid when displaced through a small angle, θ say, from the equilibrium position. W_1, W_2 are the weights and G_1, G_2 the centres of mass of the cylinder and the hemisphere, respectively.

The equilibrium is stable if the resultant of the moments about the point of contact P is positive in the anticlockwise sense; i.e. if $W_2 p_2 > W_1 p_1$, where p_1, p_2 are the perpendiculars from G_1, G_2 to the vertical PA through P.

Now $p_1 = AG_1 \sin \theta = \tfrac{1}{2} h \sin \theta$, and $p_2 = G_2 A \sin \theta = \tfrac{3}{8} a \sin \theta$.
Also $W_1 : W_2 = \pi a^2 h : \tfrac{2}{3} \pi a^3 = 3h : 2a$.

Hence the required condition is
$$2a . \tfrac{3}{8} a \sin \theta > 3h . \tfrac{1}{2} h \sin \theta \quad \text{(where } \theta \to 0\text{)};$$
i.e. $\quad a > h\sqrt{2}.$

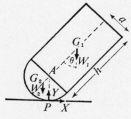
Fig. 187

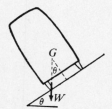
Fig. 188

(3) *A double-decker omnibus has a wheel span of 7 ft, and when the upper deck is fully loaded and the lower deck empty, the centre of mass (assumed symmetrically placed) is 4 ft 9 in from the ground. Neglecting compression of the springs, find the greatest inclination of a roadway on which the omnibus, loaded as described, can rest with its axles parallel to lines of greatest slope of the roadway.*

Let θ be the required greatest inclination.

The vertical through G will then cut the line which passes through the lowest points of contact of the wheels with the roadway.

We then have (Fig. 188) $\tan \theta = 3 \cdot 5 / 4 \cdot 75$, and hence $\theta \approx 36°$.

S.P.I.C.E. (α) In Ex. (1), it will be noticed that in the stable equilibrium position, the height of G above a fixed level (and hence also the P.E.) is a minimum; in the unstable position, a maximum. This will be elaborated in § 15·(10)5.

(β) In problems like Ex. (2), it is necessary to take the limit of an inequality where $\theta \to 0$. In Ex. (2), the presence of the same factor $\sin \theta$ on both sides of the equation made this operation specially easy. In harder problems, more effort may be needed in carrying out a limit operation.

(γ) Ex. (3) illustrates § 15·81. In practice, the angle at which toppling would start would be less than 36° due to unequal compression of the springs on the two sides of the omnibus; (in Ex. (3), we treated the omnibus as a rigid body). With London and Sydney omnibuses the angle is about 27°.

15·9. Machines. A mechanical machine is a device which receives mechanical energy (the *input*) from some external source, and delivers mechanical energy (the *output*) to some external receiving mechanism.

Let dE, dE' be the input, output, respectively, over the (differential) time-interval from t to $t + dt$. The ratio dE'/dE is not in general equal to unity, since (*a*) a proportion of the input energy will in practice be dissipated, (*b*) parts of the machine itself may receive or give out some energy (e.g., some massive parts may be raised or lowered during the time-interval dt). We shall ignore (*b*) except where we state to the contrary. The ratio dE'/dE, called the *efficiency* at time t, is then less than unity in practice.

We shall limit consideration to circumstances in which kinetic energies may be ignored; i.e., all velocities and angular velocities will be assumed steady at every instant, or, if a finite movement is considered, the initial values of the velocities and angular velocities will be taken to be equal to the final values (usually zero). In these circumstances, it is sufficient to use the equations of statical equilibrium in the problems discussed below; (the student should convince himself of this).

In simple machines, the input and output during time dt are respectively equal to the works done by and against forces F, F', say, acting at points which move through distances ds, ds', say, in the lines of action of F, F', respectively. Thus

$$\frac{dE'}{dE} = \frac{F'ds'}{Fds}.$$

The ratio F'/F is called the *mechanical advantage*, and ds/ds' the *velocity ratio*. Thus, at any instant,

$$\text{Efficiency} = \frac{\text{Mechanical advantage}}{\text{Velocity ratio}}.$$

Whereas the efficiency cannot exceed unity, it is evident that the mechanical advantage will exceed unity if the machine is constructed so that the velocity ratio exceeds unity by a sufficient margin.

In the ideal case in which the efficiency is equal to unity, the mechanical advantage is equal to the velocity ratio.

15.91. Some simple machines.

15.911. Levers.

A lever is a rigid body which can turn about a fixed axis (the *fulcrum*) and to which a force, of magnitude F say, is externally applied with a view to doing work against another external force, of magnitude F' say. The lines of action of these forces are in many cases, but not necessarily, parallel.

Simple devices using the idea of the lever are the see-saw, crowbar, wheel-barrow, pair of pliers, tongs and nutcrackers (the last three are 'double' levers), hinged safety-valve of a steam-chamber, etc.

Let a, a' be the distances of the fulcrum from the lines of action of F, F'. The mechanical advantage F'/F is by the theorem of moments equal to the velocity ratio a/a' if the fulcrum is smooth (the weight of the lever being neglected).

If there is a friction couple L at the fulcrum, then $Fa = L + F'a'$; in this case $F'/F < a/a'$, and the efficiency is less than unity. The equation is of course further complicated if the weight of the lever itself has to be taken into account.

Steelyards are levers used as weighing devices, the weight of the lever not being neglected in this case. In the common or Roman steelyard, the fulcrum is fixed. An object to be weighed is attached at a fixed place on the steelyard, and a rider (of relatively small but not negligible weight) is moved along a long straight arm of the steelyard until the whole system is in equilibrium with the arm horizontal. On the basis of the theorem of moments, the weight of the object is a simple function of the position of the rider, and is in practice read off a scale on the arm of the steelyard. (In practice the graduations would be determined empirically.) The range of utility of a given steelyard can be increased by providing for the attachment of known weights at an appropriate point of the arm.

In the Danish steelyard, there is no rider; instead, the fulcrum in the form of a pivotal support can be moved along an arm between the centre of mass of the steelyard and the (fixed) place where the object is attached. The weight of the object is indicated by the position of the fulcrum when the system is supported in equilibrium with the arm horizontal.

15.912. Pulleys.

A single fixed pulley in which the forces F, F' act on the string on opposite sides can be used as a machine. When friction is neglected, the efficiency is unity; and the mechanical advantage and velocity ratio are each obviously equal to unity. The essential practical advantage is that the pulley enables the effect of the force F to be applied in any desired direction.

When systems of pulleys are used, mechanical advantages exceeding unity may be realised.

For example, in the system depicted in Fig. 189, we have (see the method of Ex. (1) of § 4·72, if necessary) that $2x_1 - x_2$, $2x_2 - x_3$, $2x_3 + x$ are all constant. Hence

$$2dx_1 - dx_2 = 0, \qquad 2dx_2 - dx_3 = 0, \qquad 2dx_3 + dx = 0,$$

yielding $dx/dx_1 = -8$; thus the velocity ratio is 8. If we neglect friction and weights

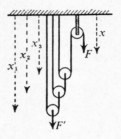

Fig. 189

of pulleys and strings, the mechanical advantage is thus 8; this may be independently deduced by considering the tensions in the strings and the forces on the three movable pulleys. In practice of course the mechanical advantage would be less than 8.

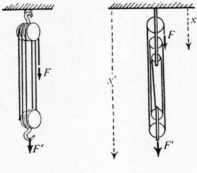

Fig. 190 (a)   Fig. 190 (b)

As a second example consider the 'block and tackle' as shown in Fig. 190(a); the theory of this pulley system is essentially the same as in the case of Fig. 190(b). From the latter figure, we derive $x + 5x' = $ constant; hence the velocity ratio $|dx/dx'| = 5$. If friction and the weight of the lower block of pulleys is neglected, the mechanical advantage F'/F is also 5. If we take into account the weight, w say, of the lower block, but not friction, we have $(F' + w)/F = 5$, so that the mechanical advantage is then $5 - w/F$.

Other pulley systems can be similarly investigated.

15·913. Wheel and axle.

In this machine, a wheel of radius a, say, and an axle of radius b, say, move as a single rigid body about a common fixed axis, and separate ropes are wound round them; forces F and F' act at the ends of the ropes as shown in Fig. 191. The velocity ratio is a/b; this is the value of the mechanical advantage if there is no friction.

Fig. 191

15·914. Differential wheel and axle.

The essence of this machine is indicated in Fig. 192. The machine is an extension of the wheel-and-axle idea, the axle here consisting of two parts of different radii b, c; the axle rope passes round a lower pulley D and when this rope winds up round the outer part of the axle, it simultaneously unwinds round the inner part. The velocity ratio is found to be $2a/(b-c)$; this is the mechanical advantage if we neglect the weight of the lower pulley and friction.

An important practical feature of this and other 'differential' machines is that the mechanical advantage may be made quite high; the denominator $b-c$ is the difference between two terms which may be made nearly equal.

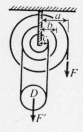
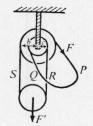

Fig. 192   Fig. 193

15·915. Weston differential pulley.

The differential pulley is closely related to the differential wheel and axle. Instead of the force F being applied as in Fig. 192, it is applied as indicated in Fig. 193, $PQRS$ being an endless chain. The velocity ratio is seen to be $2b/(b-c)$, by putting $a = b$ in the formula of § 15·914; b, c may be taken in practice as the numbers of teeth in the corresponding rims (see § 15·918).

15·916. Screw-jack.

A cylindrical screw is a cylindrical shaft whose outer surface is cut so that there is a uniform projecting ridge or 'thread' running round the shaft in the form of a helix; (the angle between the tangent at any point of a helix and the axis of the cylinder round which it runs is constant). In this context, the term 'pitch' is commonly used to denote the distance, measured in the direction of the axis, between two consecutive parts of the thread; (in some treatments, the pitch is $1/2\pi$ of that just defined).

In a simple screw-jack, a screw is moved inside an appropriately cut fixed helical groove by a force F applied to a lever rigidly attached to the screw at right angles to its axis; and the forward-moving end of the screw does work against an external force F'. If p is the pitch of the screw and a is the distance of the point of application of F from the axis of the screw, the velocity ratio is found to be $2\pi a/p$.

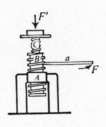

Fig. 194

15·917. Differential screw.

In this machine, the force F is applied to a lever attached to the screw B which moves inside a groove in A which is fixed. A second screw C moves relatively to B inside B, but there is a device at the top which prevents C from twisting relatively to A. Work is done against the force F' shown. If p, q are the outer pitches of B, C, respectively, the velocity ratio is found to be $2\pi a/(p - q)$, where a is as in Fig. 194.

15·918. Toothed wheels.

Suppose that a machine contains two toothed wheels which are geared together either by direct contact or by such means as the chain of a bicycle; and suppose that the axes of the wheels have no relative motion. Then in estimating the velocity ratio for the machine, it is necessary to include a factor equal to the ratio of the distances covered (when the machine is in action) by two points attached to the two wheels, respectively, at unit distances from their axes. This factor is equal to the ratio of the angular displacements of the two wheels, and hence equal to the reciprocal of the ratio of the numbers of teeth in the two wheels.

E.g., if the numbers of teeth in the larger and smaller toothed wheels of a bicycle are 48, 16, then on this account the velocity ratio must include a factor of $\frac{1}{3}$.

It is evident that the theory of toothed wheels is similar to that of wheels which have their rims pressed into contact, the rims being so rough that there is no slipping at the contact when the wheels rotate. In estimations of velocity ratio, the ratio of the radii of the wheels in this case corresponds to the ratio of the numbers of teeth in the toothed-wheels case.

It is easy to extend the results to cases of more than two wheels. Suppose for example three wheels A, B and C are geared together, energy being transmitted through A to C, via B. The velocity ratio factor associated with this system is equal to the reciprocal of the ratio of the numbers of teeth of (or radii of) A and C, for it is the product of the factors for A, B and B, C.

15·(10). The method of virtual work.*

It is possible to solve problems in statics by imagining the given system to be in motion and examining the rate of work of the forces present at an instant when the system is passing through the equilibrium configuration considered.

In the case of a single rigid body, the sum of the rates of work of the external forces present at such an instant, is zero, by (62a), p. 241, since ΣX, ΣY, ΣN are all zero in the equilibrium configuration.

Now if $d\phi/dt$ is zero when $t = t_0$ (where ϕ is a differentiable function of t), the increment $\delta\phi$ in ϕ when t changes from t_0 to $t_0 + \delta t$ is of higher order than δt. Hence the sum of the elements of work done by the external forces on a rigid body during a small displacement from equilibrium is zero, if we neglect small quantities of higher order than the first.

In applying the last statement, we may imagine any displacement from the equilibrium position to be made; and there is no need to introduce δt explicitly into the equation obtained. The displacement is called a 'virtual' displacement, and the equation a *virtual work* equation. We shall demonstrate *elementary* features of the method of virtual work in the following example.

(The statement in the second paragraph of this section could also be used; the corresponding equation is called a *virtual velocities* equation).

Since a system may in general be imagined to move through an equilibrium configuration in more than one way, it is possible in general to form more than one virtual work equation in a given problem. (In harder problems, more than one virtual work equation may indeed have to be formed to get the answer.) Much of the art in using the method of virtual work lies in choosing that particular virtual displacement (or displacements) from equilibrium which will lead most readily to the answer.

* § 15·(10) and its subsections depend on theory given in Chapter XIII, and may be omitted by readers to whom § 15·1A is relevant.

15·(10)1. Example. *Solve Ex. (2) of § 15·42 by the method of virtual work.*

In Fig. 175 (p. 268), take axes AX horizontally to the right, and AY vertically upwards; and let (x, y) be the coordinates of O referred to these axes.

Consider a virtual displacement in which the wheel turns slightly about A from the equilibrium position in which B is on the point of leaving the ground. In standard notation, the components of the displacement of O are: δx horizontally to the right, δy vertically upwards; (that δx is negative does not matter).

The virtual work equation is then
$$-P\delta x - W\delta y = 0. \qquad \text{(i)}$$

(Note.—(a) The words 'neglecting higher order quantities' are understood in this and similar equations. (b) The work done by the force Q is zero since the wheel turns about A as about a fixed axis.)

Since O moves in a circle of centre A and radius a,
$$x^2 + y^2 = a^2.$$
Hence
$$x\delta x + y\delta y = 0, \qquad \text{(ii)}$$
where (x, y) correspond to the initial configuration.

By (i) and (ii),
$$P = Wx/y \qquad \text{(iii)}$$
$$= \frac{Wa\cos\theta}{a-h}$$
$$= \frac{W\sqrt{(2ah-h^2)}}{a-h}.$$

S.P.I.C.E. (α) The solution just given may be a little harder to follow than the solution in § 15·42. This is because the example is particularly simple. The method of virtual work, however, often gives solutions to *complicated* problems much more readily than other methods.

(β) A feature of the virtual work method is that it often results in the exclusion from the equations of more unknown unsought forces than is the case with other analytical methods.

(γ) The question of signs in forming (i) above should be carefully studied.

(δ) Although (i) and (ii) are correct only to the first order, the result (iii) is precise; for, in effect, a limit process is entailed in getting (iii), and small errors in (i), (ii) disappear in this process. The student should show this in detail in the case of the above example.

15·(10)2. The virtual work equation for systems of rigid bodies. An equation of virtual work for a system of rigid bodies may be formed by the addition of equations for the separate bodies. There will frequently be cancelling out of the works of action, reaction forces at contacts between the bodies; but this is by no means always the case—see the Ex. of § 15·(10)3.

The method of virtual work can be very fruitfully applied to framework problems. In these problems, an *internal* thrust, such as F in Ex. (2) of § 15·62, is often sought. Now this F cannot give an uncancelled contribution to a virtual work equation for any displacement in which BD is treated as a rigid body. We circumvent this apparent impasse by solving an *equivalent problem*: the effect of the rod BD is replaced by two equal and opposite forces of magnitude F applied externally, as shown in Fig. 182(c), at B and D on the remaining system. We then invoke the result (i) proved in § 13·61, by which the net work done by these forces when BD increases by $\delta(BD)$ is equal to $F\delta(BD)$. The course of the solution as shown below is typical of such cases.

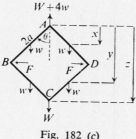

Fig. 182 (c)

15·(10)3. Example. *Solve Ex.* (2) *of* § 15·62 *by the method of virtual work*.

The set of rods AB, BC, CD, DA may be regarded as being in equilibrium under the forces shown in Fig. 182(c).

Give this system a displacement from the equilibrium configuration such that: AB, BC, CD, DA keep fixed lengths; A is kept fixed; symmetry about AC is preserved; and the angle θ of the figure ($\theta = \tfrac{1}{4}\pi$ in the equilibrium configuration) is increased by the small quantity $\delta\theta$.

Let x, y, z be as in Fig. 182(c). The virtual work equation is then
$$2w\delta x + 2w\delta y + W\delta z + F\delta(BD) = 0; \qquad (i)$$
(it is seen that the net work of all other forces present is zero).

In order to use (i), we must connect $\delta x, \delta y, \delta z, \delta(BD)$. We have
$$x = a\cos\theta,\ y = 3a\cos\theta,\ z = 4a\cos\theta,\ BD = 4a\sin\theta;$$
and hence
$$\left.\begin{array}{l} \delta x = -a\sin\theta.\delta\theta,\ \delta y = -3a\sin\theta.\delta\theta,\ \delta z = -4a\sin\theta.\delta\theta, \\ \delta(BD) = 4a\cos\theta.\delta\theta. \end{array}\right\} \qquad (ii)$$

By (i) and (ii),
$$-(2w + 6w + 4W)\sin\theta.\delta\theta + F\times 4\cos\theta.\delta\theta = 0.$$
Now replacing θ by $\tfrac{1}{4}\pi$, we obtain $F = W + 2w$, as previously.

S.P.I.C.E. ⁻(α) It will be noticed that the virtual displacement given to the system was selected so that F, w, W were the only forces entering the virtual work equation. The student should satisfy himself that there is no net contribution to the equation from other forces, e.g. reactions between rods other than BD.

(β) As pointed out in § 15·(10)2, we managed to introduce F into the equations by solving an 'equivalent problem' to the original. Alternatively, we may say that we gave the original system of five rods a displacement not compatible with the constraints present. This procedure is common in the method of virtual work.

(γ) In expressing the virtual work of the forces F in BD, the high importance of the result (i) of § 13·61 should be appreciated.

(δ) Other virtual displacements than the one selected could be made, e.g. one in which the rods BC, CD were moved apart at C. But the reaction forces at C would enter the corresponding virtual work equation and add unnecessary complications.

(ϵ) In using the method of virtual work, pains should be taken as in the above solution to describe the virtual displacement fully so that there can be no residual ambiguity.

(ζ) The angle θ in Fig. 182(c) is equal to $\frac{1}{4}\pi$ in the equilibrium configuration. But θ must not be put equal to $\frac{1}{4}\pi$ until *after* the differential operations have been carried out. Otherwise, quantities that vary during the displacement would have been falsely treated as constants.

15·(10)4. The method of stationary potential energy.

The method of stationary potential energy, to be now introduced, is related to the method of virtual work, but is more special in its application. Small displacements from an equilibrium position, C_0 say, are again considered, but it is now stipulated that the displacements are compatible with the constraints present. The method is limited to cases in which first-order work during displacements is done only by forces which are conservative.

By the method of virtual work, the net work done during any small displacement from C_0 is zero, correct to the first order. In the present case, this work is equal to the accompanying loss of potential energy. Hence the potential energy of the system is stationary in the configuration C_0.

This gives a simple means of solving some statical problems and is illustrated in § 15·(10)6. Further, the method is a first step towards investigating the stability of equilibrium where conservative forces are involved. In § 15·(10)5, the question of stability is considered for the case of one degree of freedom.

15·(10)5. Testing stability of equilibrium. In the circumstances of § 15·(10)4, let a system be given a small displacement from the configuration C_0 to C' say, and then released from rest in the configuration C'. Let C be the configuration shortly afterwards (before the system has returned to C_0, if it goes that way).

Let V_0, V', V be the potential energies in C_0, C', C, respectively. Let T be the kinetic energy of the system in the configuration C.

Since $V + T = V'$, and since the kinetic energy cannot be negative, it follows that $V \leqslant V'$. Thus (excluding the case $V = V'$, which applies to neutral equilibrium) the system moves away from C' in a direction which reduces its potential energy.

The criterion for C_0 to be a stable equilibrium configuration is, by § 15·8, that the system when released from rest in the configuration C' should start to return towards C_0; i.e., that V should lie between V_0 and V'; i.e., by the last paragraph, that $V_0 < V'$; i.e., that V_0 (already known to be a stationary value) should be a *minimum* value of the potential energy.

If on the other hand C_0 is unstable, it is seen by a similar argument that it is sufficient that V_0 should be a *maximum* value of the potential energy.

This leads us to the following rule for the case of a system with one degree of freedom, i.e. a system for which the potential energy, V say, is expressible as a function of one variable, z say: First put

$$dV/dz = 0,$$

finding values of z, say z_1, z_2, ..., which correspond to equilibrium configurations. Then form

$$d^2V/dz^2;$$

if the value of this expression is positive when z is put equal to z_1, then $z = z_1$ gives a stable equilibrium configuration; if negative, unstable; and similarly for z_2, z_3, ... (If $d^2V/dz^2 = 0$ when $z = z_0$, higher derivatives need to be considered.)

In following out the argument in § 15·(10)5, the student may find it helpful to consider, by way of illustration, the stable and unstable positions occupied by a heavy uniform sphere on a fixed rough corrugated surface.

If the system has more than one degree of freedom, the potential energy V will be a function of two or more variables, and partial derivatives will be needed in place of the ordinary derivatives above.

It can be shown that the equilibrium is on the whole unstable unless the value of V is an *absolute* minimum.

15·(10)6. Example. *Particles of weights W, w are joined by a light inelastic string of length $\tfrac{1}{2}\pi a$, and the string is slung over the upper part of the surface of a fixed smooth cylinder of horizontal axis and radius a. Find the equilibrium configuration, and investigate its stability.*

Fig. 196 shows a general position of the system specified by the angle θ.

(The cylinder being smooth, it is sufficient to restrict consideration to a plane at right angles to the axis of the cylinder.)

Since, by the data, the string subtends a right angle at O, the potential energy V in the general position is given, taking the level of O as standard level, by
$$V = Wa \cos \theta + wa \sin \theta. \tag{i}$$
Hence
$$\frac{dV}{d\theta} = -Wa \sin \theta + wa \cos \theta, \tag{ii}$$
and
$$\frac{d^2V}{d\theta^2} = -Wa \cos \theta - wa \sin \theta. \tag{iii}$$

By (ii), there is equilibrium when $Wa \sin \theta = wa \cos \theta$, i.e. when θ is the acute angle given by $\tan \theta = w/W$.

By (iii), $d^2V/d\theta^2$ is negative in this configuration, which is therefore unstable.

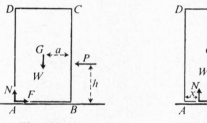

Fig. 195 (a)   Fig. 195 (b)

15·(11). Miscellaneous examples. (1) *A uniform heavy rectangular prism, initially in equilibrium on a rough horizontal table (coefficient of friction μ), is subjected to a horizontal force P acting, as shown in Fig. 195, at a height h above the table. If P is steadily increased from zero, prove that the equilibrium will be broken by the prism slipping on the table or turning about an edge, according as $a/h > \mu$ or $a/h < \mu$, where $2a$ is the width of the prism.*

Denote by (a), (b) the cases in which the prism starts to turn about the edge through A, and starts to slip, respectively.

Suppose that (a) is the case. Then the forces on the prism are as shown in Fig. 195(a).

Resolving and taking moments about A, we have
$$F - P = 0, \quad N - W = 0, \quad Ph - Wa = 0, \tag{i}$$
yielding $F/N = a/h$.

Now F/N cannot exceed μ. Hence (a) cannot be the case if $a/h > \mu$.

Suppose that (b) is the case. The forces on the prism are as indicated in Fig. 195(b); N is the resultant of the normal components of the forces exerted by the table and acts at a distance, x say, from A, while F is the resultant of the frictional components.

Resolving and taking moments about A, we have
$$F - P = 0, \qquad N - W = 0, \qquad Ph + Nx - Wa = 0. \qquad \text{(ii)}$$
Since slipping is about to occur in case (b), we have $F = \mu N$. From this and (ii), we derive $a - \mu h = x$.

Since x cannot be negative, (b) cannot happen if $a/h < \mu$.

(a) and (b) are the only possible cases (excluding the case in which turning and slipping start together; it is quickly deducible from above that this case could not arise unless $a/h = \mu$).

Hence $a/h > \mu$ entails the case (b), and $a/h > \mu$ entails the case (a).

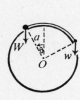

Fig. 196   Fig. 197

(2) *Investigate circumstances in which a horizontal drawer, with two handles, running loosely in its guides, jams when pulled by one handle.*

Let a force P be applied as in Fig. 197. We assume that there is then contact with the guides at H and C, and nowhere else, but ignore any inclination of the drawer to HK, LM.

Suppose the drawer does move outwards. The forces on the drawer are as shown in Fig. 197, and
$$F = \mu N, \qquad F' = \mu N', \qquad \text{(i)}$$
taking the same coefficient of friction μ at both contacts.

Resolving parallel to AD, we have
$$N - N' = 0; \qquad \therefore F = F'. \qquad \text{(ii)}$$
Taking moments about the C.M. G, we have
$$Pc - Nx = 0, \qquad \text{(iii)}$$
since the two forces N, N' constitute a couple of moment Nx; (the angular acceleration of the drawer is assumed negligible).

Resolving parallel to DA, we have
$$P - F - F' > 0. \qquad \text{(iv)}$$
From (i), (ii), (iii), (iv), we obtain
$$Nx/c - 2\mu N > 0; \quad \text{i.e.} \quad x > 2\mu c.$$
It follows that, on the stated assumptions, the drawer cannot be pulled out by any force, however large, applied at one handle, if $x < 2\mu c$.

(3) *A uniform girder of length 2a and weight W rests on rough horizontal ground (coefficient of friction μ), and a force P just sufficient to move the girder acts horizontally at right angles to the girder at one end. Assuming the girder to be uniformly supported by the ground throughout its length, find P and the point about which the girder starts to turn.*

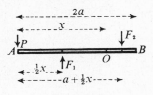

Fig. 198

Let AB be the girder, and let O be the point about which it starts to turn when P is applied at A. Let $AO = x$; other lengths are then as indicated in Fig. 198.

There is limiting friction at all points except O, but all the friction forces on elements of the girder to the left of O act oppositely to P; and, on elements to the right of O, in the same sense as P.

Since the support exerted by the ground is uniformly distributed over the length AB, the friction force per unit length of the girder is constant from A to O and from O to B.

Thus the resultant friction force is equivalent to F_1, F_2, acting in the senses shown, at the midpoints of AO, OB, respectively, where

$$F_1 = (x/2a)\mu W, \qquad F_2 = \{(2a - x)/2a\}\mu W. \qquad \text{(i)}$$

Resolving parallel to P, we have $P - F_1 + F_2 = 0$,
and hence, by (i),

$$P = \mu W(x - a)/a. \qquad \text{(ii)}$$

Taking moments about A, we have

$$\tfrac{1}{2}xF_1 - (a + \tfrac{1}{2}x)F_2 = 0,$$

and hence, by (i),

$$x^2 - 2a^2 = 0. \qquad \text{(iii)}$$

In (iii), only the positive root is relevant. Hence O is distant $a\sqrt{2}$ from A.

Putting $x = a\sqrt{2}$ in (ii), we obtain $P = \mu W(\sqrt{2} - 1) = 0{\cdot}41\mu W$.

(4) *A light string is wrapped round a rough circular cylinder (coefficient of friction μ) and is in a plane at right angles to the axis of the cylinder. The portion AB of the string in contact with the cylinder subtends an angle ψ at the centre. If the string is on the point of slipping in the direction AB, prove that the ratio of the tensions T_1, T_2 at A, B is given by $T_2/T_1 = e^{\mu\psi}$.*

Fig. 199(a) shows the given conditions.

Fig. 199(b) shows the forces on an element PQ of the string between A and B, where $PQ = \delta s$ and $\angle POQ = \delta\theta$.

T and $T + \delta T$ are the tensions at P and Q.

$N\delta s$ is the normal component of the force exerted by the cylinder on PQ; (thus N here denotes force per unit length).

Since slipping is on the point of occurring, the frictional component of the force on PQ is $\mu N\delta s$, acting as shown.

Resolving tangentially and normally for the equilibrium of PQ, we have, to sufficient accuracy,
$$(T + \delta T)\cos(\tfrac{1}{2}\delta\theta) - T\cos(\tfrac{1}{2}\delta\theta) - \mu N\delta s = 0,$$
$$(T + \delta T)\sin(\tfrac{1}{2}\delta\theta) + T\sin(\tfrac{1}{2}\delta\theta) - N\delta s = 0;$$
i.e., neglecting $(\delta\theta)^2$ and the product $\delta T.\delta\theta$,
$$\delta T - \mu N\delta s = 0,$$
$$T\delta\theta - N\delta s = 0.$$
Eliminating N, and letting $\delta\theta \to 0$, we obtain
$$\frac{1}{T}\frac{dT}{d\theta} = \mu.$$
Integrating from $\theta = 0$ to $\theta = \psi$ then gives
$$\left[\log_e T\right]_{T_1}^{T_2} = \mu\psi;$$
i.e., $\qquad \log_e(T_2/T_1) = \mu\psi;$

i.e., $\qquad T_2 = T_1 e^{\mu\psi}.$

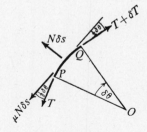

Fig. 199 (a)   Fig. 199 (b)

S.P.I.C.E. (α) In Ex. (1) it should be noted that consideration of the case $a/h > \mu$ is not alone sufficient to establish the answer. The logical point is that if a proposition p implies the contradictory of a proposition q (i.e., implies that q is false), then it does not follow that the contradictory of p implies q. (In the symbolism of mathematical logic, '$p \supset \bar{q} . \supset . \bar{p} \supset q$' is not a valid proposition; the symbol '$\supset$' denotes 'implies', and $\bar{p}$ denotes the contradictory of p.) This point commonly arises in problems where equilibrium may be broken in various ways.

(β) It may be noted, however, that in the particular circumstances of Ex. (1), the second section of the argument may be adapted as follows to give the full result: The equations (ii), which include (i) as a particular case, yield $x = a - (F/N)h$. If $a/h > \mu$, then x must be positive since F/N cannot exceed μ; in this case, (b) must hold. If $a/h < \mu$, then $F/N < \mu$, since x cannot be negative; in this case, (a) must hold.

(γ) It may be noted in Ex. (2) that when the drawer jams, increasing P merely increases F, F', and does not (on our model assumptions) result in the drawer being pulled out. In practice of course a sufficiently great force P would pull the drawer out—but only by doing structural damage not contemplated in our model treatment.

(δ) It may be shown that on similar assumptions to those in Ex. (2), it is not possible to close the drawer by pushing on one handle if the length of drawer inside the guides is less than $2\mu c$.

(ϵ) The result in Ex. (3) on 'continuous friction' rests very much on the assumption that the girder is uniformly supported along its length. A small unevenness in the ground might cause a marked deviation from the result found.

(ζ) In the answer to Ex. (4), it will be noticed that the ratio T_2/T_1 is very large if ψ is appreciable and μ not very small. E.g., if $\mu = 0.5$ and $\psi = 2\pi$, then $T_2 \approx 23T_1$; if $\mu = 0.5$ and $\psi = 4\pi$, then $T_2 \approx 534T_1$. This indicates why knots can hold so tightly, and also how large vessels can be securely moored.

EXAMPLES XV

1. Prove that it is necessary and sufficient for the equilibrium of a rigid body (treated two-dimensionally) that the sum of the moments of the external forces about each of two points A, B should be zero and the sum of the resolved parts in some direction not perpendicular to AB should be zero.

2. Prove that if the sums of the moments of a coplanar set of forces about three non-collinear points are equal, the set is equivalent to a couple or is in equilibrium. What further condition is necessary for equilibrium?

3. A, B, C are the points $(-4, 3)$, $(-5, -12)$, $(3, 4)$, respectively, referred to rectangular axes OX, OY. Forces 8, 11, 13, 8 lb.wt. act along OY, OA, OB, OC, respectively. Find their resultant.
[Answer: $9\sqrt{2}$ lb.wt. along a line through O inclined at $\tfrac{3}{4}\pi$ to OX.]

4. $ABCDE$ is a pentagon in which the length of each side is 1 metre, and A, B, C, D are the corners of a square. Forces of magnitude 1, 2, 3, X, Y Kg act along AB, BC, CD, DE, EA, respectively, and are equivalent to a couple. Find X, Y and the moment of the couple.
[Answers: $2(1 - \sqrt{3}/3)$, $2(1 + \sqrt{3}/3)$ Kg; $(4 + \sqrt{3})$ Kg-metres.]

*5. Two sets of forces, of magnitudes P, Q, R and P', Q', R', respectively, act along the sides, taken in order, of lengths a, b, c, respectively, of a triangle. Prove that their resultants are parallel if
$$a(QR' - Q'R) + b(RP' - R'P) + c(PQ' - P'Q) = 0.$$

6. A body of mass M can just be held without slipping on a fixed rough inclined plane by a force P, and just dragged up by a force Q, where P, Q act up a line of greatest slope. Find the coefficient of friction.
[Answer: $(Q - P)/\sqrt{\{4M^2g^2 - (P + Q)^2\}}$.]

7. A particle of weight w is supported by two light inelastic strings, each of length l, which are attached respectively to two small rings, each of weight $\tfrac{1}{4}w$, which can slide along a fixed rough horizontal rod, the coefficient of friction being μ in each case. Find the greatest distance apart of the rings, consistent with equilibrium, and find the angle which the strings then make with the vertical.
[Answers: $6\mu l/\sqrt{(9\mu^2 + 4)}$; $\tan^{-1}(3\mu/2)$.]

8. A light inelastic string ABC of length l has two light rings, A, C fastened at its ends, and the rings A, C can slide on a fixed rough straight wire inclined at β to the horizontal. A heavy smooth ring B is threaded on the string and hangs in equilibrium. Prove that the greatest distance between the rings A, C consistent with equilibrium is $l \sin(\lambda - \beta) \sec \beta$, where λ is the angle of friction between the rings and the wire, given that $\lambda > \beta$.

9. Four equal lamps are suspended across a road by four light vertical strings whose upper ends B, C, D, E are attached to a light string $ABCDEF$, the ends A, F being fixed at the same level on opposite sides of the road. The distances between consecutive members of the verticals through A, B, C, D, E, F are equal. Prove that the distance between the levels of AF, BE is twice that between the levels of BE, CD.

10. A straight uniform horizontal rod ACB of weight W can turn smoothly about the point C which is fixed; and bodies of weights P, Q are attached at A, B, respectively. Prove that if the rod is in equilibrium, then $AC : CB = (2Q + W) : (2P + W)$.

11. A light rod AB of length 0·5 m is suspended from a fixed point O by two light inelastic strings OA, OB of lengths 0·4, 0·3 m, respectively. A particle of mass 3 kg is attached to the rod at A, and one of mass 2 kg at B. In the position of equilibrium, find the tensions in the strings OA, OB and the force in the rod AB.
[Answers: $2\sqrt{5}$, $\sqrt{5}$, $\sqrt{5}$ Kg.]

12. A uniform horizontal diving board AB of length 10 ft and mass 200 lb is clamped at the end A, and has a man of 25 stone standing at the end B. What is the support at the end A ?
[Answer: A vertical force of 550 lb.wt., together with a couple of 4500 ft-lb.wt.]

13. A uniform bar AB, of length 12 ft and line-density 1 lb/ft, rests on supports P and Q which are at the same level; and $AP = 4$ ft. The bar carries loads of 4, 11 lb at R, S, respectively, where $AR = 5$ ft, $AS = 10$ ft. (a) Find the smallest value of AQ compatible with equilibrium. (b) If $AQ = 8$ ft, find the greatest length which can be cut off from the end at A without disturbing the equilibrium.
[Answers: (a) 7·48 ft; (b) 2 ft.]

14. A gate of weight W, which may be treated as a rectangular lamina of breadth b, hangs on two hinges in a vertical line, distant h apart, in such a way that there is no vertical support at the upper hinge. Find the magnitudes of the forces at the hinges.
[Answer: $bW/2h$; $W\sqrt{(b^2 + 4h^2)}/2h$.]

15. A uniform straight rod ABC of length 3 ft and mass 6 lb is smoothly pivoted to a fixed point at B, distant 9 inches from A. A mass of 12 lb is suspended from C, and the rod is held horizontally by means of a light string AD, 15 inches long, where D is fixed vertically below B. Find the tension in the string and the reaction at B.
[Answers: $52\frac{1}{2}$ lb.wt.; 67·8 lb.wt., inclined at $62°·3$ to the horizontal.]

16. Two spheres, of equal weight and radius 2 in, rest in equilibrium inside a smooth spherical basin of radius 6 in. Prove that the force between the basin and either sphere is double the force between the spheres.

17. Representing the human arm by two rods AB, BC, each of length 15 in and smoothly jointed at B, and the biceps muscle and tendon by a light string connecting C (the shoulder) to a point in AB distant 2 in from the elbow-joint B, find the tension in the tendon and the force at the elbow-joint when a mass of 10 lb is held at A, AB being horizontal and BC vertical, the mass of the forearm being 3 lb, and its centre of mass distant 6 in from B.

[Answers: 85 lb.wt.; 72 lb.wt. inclined at about 9° to the vertical.]

18. At a vertical pole AB, carrying at its top A four horizontal telegraph wires, each at a tension of 50 lb.wt., the direction of the wires changes by 15°. A light stay is attached to the pole half-way between A and B (which is at ground level). The stay is inclined at 30° to the downward vertical and is in the vertical plane which bisects the angle between the wires. Prove that if the effect of the ground on the pole is a single force passing through B, the tension in the stay is about 209 lb.wt.

Is it necessary for equilibrium that the tension in the stay should be about 209 lb.wt.?

[Answer: No, because the effect of the ground on the pole would in general be equivalent to a force at B, together with a couple.]

19. A motor-cycle which is too long to go on a weighing machine is weighed by placing first one and then the other wheel on the scales. If the platform of the scales is level with the surrounding ground, prove that the weight of the motor-cycle is given as the sum of the two weighings.

If the platform is above the level of the surrounding ground, prove that the sum of the two weighings is less than the actual weight W by $(2Wh/c) \tan \theta$, where c is the distance between the centres A, B of the wheels (which are of equal diameter), h is the distance of the centre of mass of the motor-cycle from AB, and θ is the inclination of AB when either wheel is on the platform.

20. The handle of a garden roller balances about the (assumed smooth) axle of the cylinder of the roller. The effect of the (horizontal) ground on the roller when moving may be assumed to be equivalent to a force at the lowest point, of which the horizontal component is equal to k times the vertical (where k is constant), together with a couple. Find the ratio of the vertical components of force between the roller and the ground when the roller is pushed and when it is pulled (with steady speeds), the handle being inclined at a given angle α to the ground in both cases.

[Answer: $(1 + k \tan \alpha) : (1 - k \tan \alpha)$.]

21. A rod AB of weight W is in equilibrium with its ends on the inside surface of a smooth fixed spherical bowl. The centre of mass of the rod divides AB in the ratio $m : n$. If θ is the inclination of the rod to the horizontal, and 2α is the angle subtended at the centre of the sphere, prove that the reactions at A, B are $W \sin (\alpha \pm \theta)/\sin 2\alpha$, and that $\tan \theta = |(m - n)/(m + n)| \tan \alpha$.

22. A uniform wire is bent into the form of a square, and is hung on a rough fixed horizontal peg. If the square rests in equilibrium whatever the distance of the peg from the centre of the side it touches, prove that the coefficient of friction is not less than unity.

23. A uniform beam AB of weight W is standing with the end B on a horizontal floor and the end A leaning against a vertical wall, and is in a plane perpendicular to the wall. The coefficient of friction μ is the same at both contacts, and the beam is on the point of slipping when its inclination is $45°$. Find μ and the horizontal component of the force exerted by the wall. [Answers: $\sqrt{2}-1$; $W/2\sqrt{2}$.]

24. A board, movable about a fixed horizontal line in its plane, is supported by resting against a rough uniform sphere which lies on a fixed horizontal table on which it can roll but not slip. If ϵ is the angle of friction between the board and the sphere, find the greatest inclination of the board to the horizontal. [Answer: 2ϵ.]

25. A uniform solid hemisphere rests in equilibrium with its spherical surface supported on a rough inclined plane and with its plane face vertical. If slipping is on the point of occurring, find the coefficient of friction. [Answer: $3/\sqrt{55}$.]

26. The cross-section of a piece of gutter, made of thin uniform sheet-metal, is a semicircular arc. The gutter rests with its straight edges horizontal and its curved surface in contact with a perfectly rough fixed plane inclined at $30°$ to the horizontal. Show that the inclination to the horizontal of the plane containing the edges of the gutter is about $52°$.

27. A uniform rod AB, of mass $4M$ and length l is smoothly jointed at the end A to a fixed straight horizontal wire AC. The end B is attached by means of a light inelastic thread of length l to a bead of mass M which can slide on the wire, the coefficient of friction between the bead and the wire being μ. Find the inclination of the thread to the horizontal in the configuration of limiting equilibrium.
[Answer: $\tan^{-1}(2\mu)^{-1}$.]

28. A particle of weight w rests on the inside of a rough thin uniform hemispherical bowl of weight W which rests on a fixed horizontal plane. If μ is the coefficient of friction between the particle and the bowl, prove that the greatest angular distance θ of the particle from the axis of the bowl is given by $\mu^{-1}\sin\theta - \cos\theta = 2w/W$.

29. A uniform solid hemisphere of weight W rests with its curved surface on a fixed horizontal plane, and a particle is placed on its base close to the rim. The coefficient of friction between the particle and the base of the hemisphere is μ. If the system rests in equilibrium, what is the greatest possible weight of the particle?
[Answer: $\tfrac{3}{8}\mu W$.]
Why, in this and Ex. 28, is it not necessary to state whether the plane is smooth or rough?

30. A ladder of weight W, whose centre of mass is distant b from the foot, stands on rough horizontal ground and leans in equilibrium against a rough cylindrical pipe of radius a, lying on the ground and wedged to prevent it rolling. The ladder projects beyond the point of contact and is at right angles to the axis of the pipe. If μ is the coefficient of friction at both points where friction acts, and 2α is the inclination of the ladder to the horizontal, prove that a man of weight W' can climb a distance x up the ladder before the ladder starts to slip, where x is given by
$$2(Wb + W'x)\sin^2\alpha \cos 2\alpha = (W + W')\mu a/(1 + \mu^2).$$

*31. Two uniform beams AB, AC of equal length are smoothly hinged together at A, and placed standing in a vertical plane with the ends B, C on a rough horizontal plane and the angle BAC equal to 2θ. The weight of AC is double that of AB. Prove that the least value of the coefficient of friction necessary for equilibrium is $0\cdot6\tan\theta$. If the coefficient of friction has this value, at which end is the friction limiting?

[Answer: B.]

32. A light horizontal straight rod ABC has one end A in contact with a rough vertical wall, the coefficient of friction being μ. A point B of the rod is supported by a light string which is kept at an inclination α to the horizontal. Prove that, if the rod is sufficiently long, a load may be suspended at any point of the rod which lies within distances $AB\tan\alpha/(\tan\alpha\pm\mu)$ from A, without causing the rod to slip at A.

*33. A uniform rod AB, of weight W and length 2α, is placed over one fixed horizontal rough peg and under another, the mid-point of AB being half-way between the pegs which are distant $2c$ apart. The inclination of the rod to the horizontal is α, and the rod is kept in equilibrium by a weight w at the lower end of the rod. Prove that if μ is the coefficient of friction at the contact with each peg, the conditions $aw > c(W+w)$, $\mu aw > c(W+w)\tan\alpha$ must both be satisfied.

*34. The edge of a uniformly rough table is rounded off in a continuous curve. A mass m on the table is connected with a hanging mass M by a light inelastic string passing over the edge, the plane of the string being at right angles to the edge. A vertical force F applied to M will just pull M down, and a horizontal force F' applied to m will just pull M up. Prove that the coefficients of friction between the table and m, and between the table and the string are $M\alpha/m$ and $(2\log_e\beta/\pi)$, where α, β are the roots of the equation $Mgx^2 - F'x + F + Mg = 0$.

35.† In a light smoothly-jointed plane framework $ABCD$, with a light vertical diagonal bar AC, the lengths AD and CD are equal, the lengths AB, BC and CA are equal, and $AD = 2AB$. The framework carries a load of 200 lb at C and is supported by vertical forces at B, D. Draw Bow's diagram for the system, and find the forces in the five bars.

[Answer: AB, 138·2 lb.wt.; BC, $-138\cdot2$ lb.wt ; CA, -100 lb.wt.; AD, 123·6 lb.wt.; CD, $-123\cdot6$ lb.wt.]

36.† A framework consists of seven smoothly-jointed light bars forming a Warren girder: the bars AB, BE, EC, CD are of equal length a and all inclined at $45°$ to the horizontal ; the bars BC, AE, ED are of equal length $a\sqrt{2}$ and horizontal; A and D are on opposite sides of E in the same straight line, and BC is vertically above AED. The frame rests on supports at A and D, and carries a load of 1 ton at E. Find the forces in the seven bars.

[Answer: AB, CD, $10\sqrt{2}$ cwt; BC, 1 ton wt.; AE, ED, -10 cwt; BE, EC, $-10\sqrt{2}$ cwt.]

† Plus and minus signs in the answer correspond to struts and ties, respectively.

37. AC, BC, respectively, represent the jib and tie-rod of a crane, and are each 4 cwt; A, B are fixed at the same level, and C is vertically above a point on AB produced; $AB = AC$, and $\angle ABC = 30°$. A load of 30 cwt is suspended from C. Treating AC, BC as uniform bars smoothly pin-jointed at A, B, C, find the forces at C on the jib AC.

[Answer: 57 cwt, acting at 59° below the horizontal.]

38. A framework consisting of seven smoothly-jointed light rods is in the form of a regular pentagon $ABCDE$ with two of its diagonals AC, AD. The framework is in a vertical plane with the lowest rod CD horizontal, and is simply supported at C, D. Loads of 10, 20, 10 lb are suspended from B, A, E, respectively. Find the tension in CD.
[Answer: 1·24 lb.wt.]

39.† A smoothly-jointed framework consists of four light rods in the shape of a square $ABCD$, with a light diagonal rod BD. The framework is simply suspended at A, and loads of 10, 20, 30 kg are suspended from B, C, D, respectively. Prove that in the equilibrium position, AD is inclined at about 31° to the vertical. Then determine graphically or otherwise the forces in the five rods AB, BC, CD, DA, BD.

[Answer: -31, -17, -10, -51, 36 Kg, approximately.]

40. AB, BC, CD are three equal uniform rods, each of weight W, smoothly jointed together at B, C. The ends A, D are fitted with small light rings; the ring at A can slide on a smooth fixed horizontal wire AO, and the ring at D can slide on a smooth fixed vertical wire OD. A horizontal force of magnitude F and direction OA is applied to the ring at A, and the system rests in equilibrium with AB, BC, CD inclined at angles α, β, γ, respectively, to the horizontal. Prove that $2 \tan \beta = \tan \alpha + \tan \gamma$.

If $F = W$, find the reactions at B, C, D.

[Answer: $W\sqrt{5}$, $W\sqrt{2}$, W, inclined to the horizontal at $\tan^{-1} 2$, $\frac{1}{4}\pi$, zero, respectively.]

41. A bracket ABC consists of two uniform rods, AB of mass 4 lb and BC of mass 6 lb, smoothly jointed at B. The bracket is smoothly pin-jointed at A and C to points of a vertical wall, A being above C; AB is horizontal and at right angles to the wall, and $\angle ABC$ is 30°. A mass of 10 lb is attached to the pin at B. Prove that the force due to the wall on AB is about 26·1 lb.wt. acting at 4°·4 above the horizontal, and find the magnitude of: (a) the force at B on AB; (b) the force due to the wall on BC.

[Answers: 26·1, 31·6 lb.wt.]

42. A triangular framework consists of three equal uniform rods AB, BC, CA, each of weight w, smoothly jointed together at their ends. The framework is placed so that BC rests symmetrically on two smooth fixed pegs at the same level, with A vertically below BC. A load of weight W hangs from A. Find: (a) the force at A on AB; (b) the force at B on AB.

[Answers: (a) the resultant of $(w + W)/2\sqrt{3}$ parallel to BC, and $\frac{1}{2}W$ vertically down; (b) the resultant of $(w + W)/2\sqrt{3}$ in the direction CB, and $w + \frac{1}{2}W$ vertically up.]

43. A smoothly-jointed square framework, formed of four uniform rods each of weight W, is held up by one corner, and a weight W is suspended from each of the other three corners. The square shape is preserved by a light rod along the horizontal diagonal. Find the thrust in this rod: (a) using the method of virtual work; (b) by a more elementary method. [Answer: $4W$.]

44. Four rods are smoothly jointed to form a rhombus $ABCD$. The rods BC, CD are light, while AB, AD are uniform rods each of weight W. A light rod BD is smoothly jointed to the rhombus at B, D, so that the angle BAD is 2α. The framework is suspended freely from a fixed point at A, and supports in equilibrium at C a load of weight P. Find the force in the rod BD: (a) using the method of virtual work; (b) by a more elementary method. [Answer: $\frac{1}{2}(W+ 2P)\tan \alpha$.]

*45. A uniform flexible wire hangs in equilibrium in a vertical plane with its end points fixed (not necessarily in a vertical line). Prove that the tension at any point P is wy, where w is the weight-per-unit-length of the wire, and y is the height of P above a fixed level.

(Hint.—Consider separately a portion PQ of the wire, where P, Q are at different levels, and apply the method of virtual work, giving every element of PQ a small tangential displacement.)

46. A light lever in the form of the letter L, with arms AB, BC of lengths a, b, is smoothly pivoted at B so that it can turn in a vertical plane. Masses m, m' are attached at A, C, respectively. Use the method of stationary energy: (a) to show that, in an equilibrium position, the inclination θ of AB to the vertical satisfies the equation $\tan \theta = m'b/ma$; (b) to show that there are just two possible equilibrium positions, one stable and one unstable.

47. A rhombus framework $ABCD$ of four uniform smoothly-jointed rods, each of length a and weight w, is hung up at A, and inside it is placed a smooth uniform circular vertical disc of radius b and weight W, resting symmetrically on the two lower rods BC, DC. If θ is the inclination of the rods to the vertical in the equilibrium configuration, prove that
$$2a(W+ 2w)\sin^3 \theta = Wb \cos \theta.$$

*48. A particle P of weight w newtons is supported by two light perfectly elastic strings AP, BP, each of natural length a m and modulus $15w/16$ newtons, the other ends A, B of the strings being fixed to points at the same level, distant $2a$ m apart. By the method of stationary energy, or otherwise, prove that in the equilibrium position the inclination θ of either string to the vertical is given by
$$15(\operatorname{cosec} \theta - 1) \cos \theta = 8,$$
and verify that $\theta = \cos^{-1}(4/5) \approx 37°$. Show also that the work done in stretching either string is $5aw/24$ joules.

*49. A uniform rod AB rests with its ends A, B on two smooth planes inclined at $30°$, $45°$, respectively, to the horizontal. Using the method of stationary energy, or otherwise, prove that the inclination of the rod to the horizontal is $\cot^{-1}(1 + \sqrt{3})$.

*50. A uniform bar of length $2a$ and weight W can turn smoothly about the end A which is fixed. To the end B is tied a light inelastic cord BCD of fixed length which passes over a small smooth pulley C fixed vertically above and distant b from A. The portion CD of the cord hangs vertically and carries a load of $\frac{1}{2}W$ at D. Prove that when $BC = x$, the potential energy V of the system differs from a constant by $\frac{1}{2}Wx - \frac{1}{4}Wx^2/b$. Hence show that, provided $a \leqslant b$, an unstable equilibrium position exists in which $x = b$.

Explain why the determination of dV/dx fails to reveal the equilibrium positions in which AB is vertical. (Hint.—If θ is the inclination of AB to the vertical, show that $dx/d\theta$ is zero when $\theta = 0$.)

Prove that vertical equilibrium positions are stable if $a < b$.

51. A frustum is cut from a right circular uniform solid cone by a plane parallel to the base through the midpoint of the axis of the cone. Prove that the frustum will rest with its slant side on the surface of a table if the ratio of the height of the cone to the diameter of the base exceeds $\sqrt{7} : \sqrt{17}$.

52. The cross-section of a uniform solid prism of weight W is a triangle ABC obtuse-angled at B. The prism is placed with the rectangular face containing AB on a horizontal table. Find the least weight which when suspended from C will cause the prism to overturn. [Answer: $\frac{1}{3}W(a^2 - b^2 + 3c^2)/(b^2 - c^2 - a^2)$.]

53. A closed vessel of uniform thickness consists of a hollow cone and hollow hemisphere joined together at their circular rims. The densities of the materials of the cone and hemisphere are in the ratio of 2 : 1. Find the least semi-vertical angle of the cone if the vessel can stand upright on a horizontal table in stable equilibrium with the vertex of the cone uppermost. [Answer: $43°\cdot 9$.]

54. A uniform circular disc of radius a rests in equilibrium in a vertical plane on two fixed similar rough horizontal pegs, distant $a\sqrt{2}$ apart and at the same level. A gradually increasing horizontal force in the plane of the disc is applied to the highest point of the disc. Prove that when the equilibrium is first broken the disc will start to rotate with its centre fixed, or will start to rotate about one of the pegs, according as the coefficient of friction between the disc and a peg is less or greater than $\sqrt{2} - 1$.

55. The diameters in a wheel-and-axle machine are 1 ft and 3 in, and a load of 250 lb is raised by applying a force of 75 lb.wt. What is the efficiency ? What additional load could have been raised if the machine had been frictionless ?
[Answers: $83\cdot 3\%$; 50 lb.]

56. In the weighing of objects of weights nW, where $n = 1, 2, 3, \ldots$, prove that in an ideal Roman steelyard, the positions of the rider are equally spaced; and that in an ideal Danish steelyard, the distances of the fulcrum from a certain fixed point of the arm are proportional to $\frac{1}{2}, \frac{1}{3}, \frac{1}{4}, \ldots$, if W is the weight of the steelyard resting on the fulcrum.

57. A differential pulley, the two rims of which have 24, 25 teeth, is used to raise a load of 12 cwt. Find the velocity ratio, and find the necessary force to be applied if the efficiency is 60%. [Answers: 50; $44\cdot 8$ lb.wt.]

58. If the two screws in a differential screw have 2, 3, threads to the inch, respectively, and a couple of 20 lb.wt.-ft when applied to the larger screw produces a thrust of $\frac{1}{2}$ ton wt., find the efficiency. [Answer: 0·124.]

59†. A balance consists of a beam of weight W', which can turn freely about a fulcrum O, with scale-pans each of weight S suspended from points A and B of the beam. The centre of mass G of the beam, the midpoint C of AB, and O are in a straight line perpendicular to AB; and $OC = h$, $OG = k$. If loads of weights P, W are placed in the respective scale-pans, prove that for equilibrium, the inclination θ of AB is given by

$$\tan \theta = |P - W|a/\{W'k + (P + W + 2S)h\},$$

(the beam being treated as a non-yielding rigid body).

Hence observe that the *sensitivity*, taken as $\theta/|P - W|$, is increased by increasing a/k, a/h, or decreasing W'.

Show also that decreasing a/h, a/k increases the *stability* of the balance, taken as proportional to the frequency of small oscillations about the equilibrium position when $P \approx W$.

60.† Masses m, m', where $m > m'$, are suspended vertically from the ends of a light inelastic string which passes over a fixed rough pulley which cannot rotate about its bearings. Prove that the system will stay in equilibrium unless $m > m'e^{\mu\pi}$, where μ is the coefficient of friction between the string and the pulley. If this inequality holds, prove that the system will move with acceleration $g(m - m'e^{\mu\pi})/(m + m'e^{\mu\pi})$.

† The last parts of Exs. 59, 60 involve some dynamical theory.

CHAPTER XVI

HYDROSTATICS

"One may not doubt that, somehow, good
Shall come of water and of mud;
And, sure, the reverent eye must see
A purpose in liquidity."

—Rupert Brooke.

In Chapter XV, the bodies considered were rigid. The next step forward is to consider *deformable bodies*, which include fluids and elastic solids. We shall treat all bodies in Chapter XVI as continuously distributed mass-systems (see § 11·2), and after § 16·12 shall consider only fluids in equilibrium, i.e., the subject of *Hydrostatics*.

When we speak of a fluid 'at rest', we shall treat the fluid as if all its particles are at rest. This again is a model representation; experimental evidence indicates that the 'molecules' of 'real' gases are always in rapid motion.

16·1. Stress and pressure.

16·11. Stress. Let δS be a small plane element of area containing a point P in the interior of a given body. Let A, B be two portions of the body with a common boundary which includes δS. Let $\mathbf{v}$ be a unit vector normal to δS in the sense away from A.

Fig. 200

The vector sum of the forces on A due to its contact with B across δS is called the *traction*, $\delta \mathbf{F}$ say, across δS from A to B.

The limit, as $\delta S \to 0$, of $\delta \mathbf{F}/\delta S$ is called the *stress* at P corresponding to the direction $\mathbf{v}$. In general, the stress at P is inclined to the direction $\mathbf{v}$.

The dimensions of stress are those of force per area, i.e. $[ML^{-1}T^{-2}]$. The m.k.s., c.g.s. and f.p.s. units of stress are the N/m^2, the dyn/cm^2 and the pdl/ft^2.

16·12. Solids and fluids. In the particular case in which the stress at P is parallel to $\mathbf{v}$ for all points P of a given body and all directions $\mathbf{v}$, whenever the body is at rest, the body is called a *fluid*. Otherwise, it is a *solid*.

Experiments indicate that 'real' fluids conform very closely to the property just stated (provided the words 'at rest' are interpreted as applying to large-scale, as opposed to molecular, behaviour).

16·13. Forces on a portion of fluid. The external forces acting on a portion of fluid (or other substance) may be classified into: (*a*) 'body forces' which act on the contained particles independently of boundary conditions; and (*b*) forces acting at the boundary.

In this book, the only forces of type (*a*) will be the weights of fluid particles. Forces of the type (*b*) will consist of tractions across elements of the boundary surface area; additional forces such as those due to 'surface tension' will here be ignored.

16·14. Theorem on stress in fluid at rest. We now prove that at any point P in the interior of a fluid at rest, the magnitude of the stress is independent of the particular direction $\mathbf{v}$ taken.

Fig. 201

Consider the forces acting on the fluid contained in a small tetrahedron $PABC$ (Fig. 201), of volume V say, where P is at one corner and PA, PB, PC are mutually perpendicular. These forces consist of : (*a*) the weight W of the contained fluid; (*b*) the tractions F_1, F_2, F_3, F', say, acting across the faces PBC, PCA, PAB, ABC, respectively, in the directions of the outward normals.

Resolving parallel to PA for the equilibrium of the fluid $PABC$, we have $W_1 - F_1 + F' \cos \theta = 0$, where W_1 is the resolved part of the weight W, and θ is the angle between PA and the outward normal to the face ABC.

Since $\cos \theta =$ (area PBC)/(area ABC), we thus have

$$\frac{F'}{\text{area } ABC} = \frac{F_1}{\text{area } PBC} - \frac{W_1}{\text{area } PBC}. \qquad \text{(i)}$$

Now let the side-lengths of the tetrahedron approach zero, the orientations of the four faces remaining unchanged. Let ρ be the (finite) fluid-density at P. Then $\rho V/(\text{area } PBC) = \tfrac{1}{3}\rho.PA \to 0$. Since $W_1 \leqslant W$ and $W \approx \rho V$, it then follows that $W_1/(\text{area } PBC) \to 0$.

Hence, by (i), the stresses at P in directions corresponding to the outward normals to the faces ABC, PBC are equal.

The theorem follows since the faces ABC, PBC may have any relative orientation.

16·15. Pressure. The term 'stress' is used regularly in the theory of elasticity. In hydrostatics it is more usual to use the term *pressure*, which is the stress (in hydrostatic conditions) reversed in sign. Thus the pressure on $PABC$ across the small face ABC (Fig. 201) has the direction of the *inward* normal to that face.

The units and dimensions of pressure are as for stress.

The magnitude of the pressure at a point (sometimes called the 'intensity of pressure') will be denoted by the symbol p.

The theorem of § 16·14 may now be restated in the form that *the intensity of pressure at any point P in a fluid at rest is independent of the orientation of the small area taken through P.*

16·16. Compressibility. The *incompressibility* (or *bulk modulus*) k at any point P of a substance is defined by

$$k = \rho \frac{dp}{d\rho}, \qquad (66)$$

where ρ is the density and $dp/d\rho$ is the limit of the ratio of a small pressure increase to the accompanying density increase of a small element of mass of the fluid containing P.

The dimensions of k are those of stress and pressure.

The *compressibility* is defined as $1/k$.

It will be seen in §§ 16·64, 16·65 that the functional relation between p and ρ, and hence the value of $dp/d\rho$, can be different in different thermodynamical conditions. Thus k, as defined by (66), depends not only on the particular substance, but also on the thermodynamical conditions under which p and ρ change. In subsections of § 16·6, particular thermodynamical conditions will be specified.

16·17. Liquids and gases. Fluids are classified into *gases* and *liquids* according to the value of k, gases being much more compressible than liquids and therefore having much smaller values of k. (In ordinary conditions, k is of the order of 10^6, 10^{10} dyn/cm^2 for air, water, respectively.)

We shall artificially simplify by considering model liquids for which $k = \infty$ and therefore, by (66), $d\rho/dp = 0$; i.e., the density is unchanged on application of pressure.

With gases, the density will vary with the pressure.

We shall further simplify by taking particular liquids (e.g. water, mercury, alcohol) to be *homogeneous*, i.e. of uniform density throughout the volume considered.

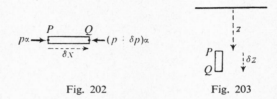

Fig. 202   Fig. 203

16·2. Variation of pressure in fluid under gravity.

16·21. Zero variation of pressure in horizontal direction. Let P, Q be two neighbouring points, δx apart, and at the same level, inside a given fluid; and let p, $p + \delta p$ be the pressures at P, Q.

Consider the forces on the fluid contained in a cylinder (Fig. 202) of axis PQ, small cross-section α, and plane ends at right angles to PQ. Resolving horizontally for the equilibrium of this fluid, we have (to sufficient accuracy) $p\alpha - (p + \delta p)\alpha = 0$; (the other forces present have no horizontal component). Dividing by δx and letting $\delta x \to 0$, we have $dp/dx = 0$.

Thus the rate of change of pressure horizontally is zero.

16·22. Variation of pressure with depth. Now take circumstances as in § 16·21, except that P, Q are in a vertical line, at depths z, $z + \delta z$ below some fixed level. Let p, $p + \delta p$ again denote the pressures at P, Q, and let ρ be the fluid-density at P (We exclude here the case of a discontinuous jump in ρ in the neighbourhood of P.)

We consider the equilibrium of the fluid in a cylinder (Fig. 203) with PQ as axis, and otherwise taken as in § 16·21. The resultant vertical force includes the weight, $(\alpha \delta z)\rho g$ vertically downwards, in addition to the tractions across the plane ends, and we obtain (to sufficient accuracy)

$$p\alpha - (p + \delta p)\alpha + g\rho\alpha\delta z = 0.$$

Hence $$\frac{dp}{dz} = g\rho. \tag{67}$$

16·23. Remarks.

(i) The intensity of pressure is the same at all points of a given body of homogeneous liquid which are at the same level. This theorem follows immediately from § 16·21 if it is possible to join any two points at the same level by a horizontal curve lying entirely in the liquid; (in this case the conclusion would also hold for a non-homogeneous fluid). In other cases, e.g. when the two points are in different arms of a U-tube, the theorem follows on bringing the result (67) to bear, as well as § 16·21.

(ii) Across a surface, S, say, of separation between two fluids that do not mix, there is in general an abrupt change in density. By (67), it follows that the rate of increase of pressure with depth is abruptly different on the two sides of S. Hence S must be horizontal; otherwise there would exist points on opposite sides of S which were at the same level and at which the pressures were unequal.

(iii) If p_1, p_2 are the pressures at two points P_1, P_2 at depths z_1, z_2 below a fixed level, we have, by (67),

$$p_2 - p_1 = \int_{z_1}^{z_2} g\rho \, dz. \tag{68a}$$

In the particular case in which P_1, P_2 are in a homogeneous liquid, so that $g\rho$ is constant, (68a) becomes

$$p_2 - p_1 = g\rho(z_2 - z_1). \tag{68b}$$

(iv) Since, by (68b), the difference in pressure at any two points in a given body of liquid depends only on the difference of their levels, it follows that if by any means the pressure at any point in the liquid is increased, the body forces being unchanged, then the pressures at all points in the liquid are equally increased. This is *Pascal's principle*, and is involved in the *hydraulic press*.

(v) The *specific weight* w of a substance is defined as the weight per unit volume; thus $w = g\rho$. The dimensions of w are $[ML^{-2}T^{-2}]$. In some treatments, w is used regularly in place of $g\rho$.

The *specific gravity* (sp.gr.) or *relative density* of a substance is the ratio of its density to that of some standard substance, commonly water. Specific gravity is dimensionless.

16·24. Atmospheric pressure.

The Earth's atmosphere will in this book be treated as a fluid at rest. In ordinary conditions the atmospheric pressure at sea-level, which we shall denote by p_0, is about 14·7 lb.wt./in² ($\approx$ 2120 lb.wt./ft²), or $1 \cdot 01 \times 10^6$ dyn/cm² ($\approx$ 1030 g.wt./cm²).

This pressure is sometimes taken as a practical unit of pressure and referred to as 'one atmosphere' (atm). One atm $\approx$ one *bar*, a 'bar' being defined as 10^6 dyn/cm² or 10^5 N/m².

Sometimes reference is made to a 'homogeneous atmosphere'. This is a hypothetical atmosphere (very different from reality— see § 16·66), of *constant* density equal to that of air at ground-level and such that the pressure at ground-level would be p_0. The density of the air at ground-level being about 0·00129 g/cm³, it follows, using (68*b*), that the height of the 'homogeneous atmosphere' is about 8 km.

The pressure at a depth of about 30 in or 76 cm below the upper surface (above which there is a vacuum) of a column of mercury (under 'normal' thermodynamical conditions) is approximately one atm. This follows from (68*b*), using the value of 13·6 g/cm³ for the density of mercury. In the *mercurial barometer*, such a column of mercury is supported by atmospheric pressure. Changes in the height of the top of the column above a lower level which is exposed to atmospheric pressure, indicate changes in the atmospheric pressure.

The height of the 'water barometer', i.e. a column of water (for which $\rho \approx 1$ g/cm³ $\approx$ 62·5 lb/ft³) that can be supported by atmospheric pressure, is about 34 ft or 1030 cm.

By (68*b*), it follows that if the surface of a homogeneous liquid is exposed to atmospheric pressure p_0, then the pressure p at depth z below the surface is given by

$$p = p_0 + g\rho z. \tag{68c}$$

Now suppose that a homogeneous liquid of density ρ and vertical thickness h has its upper (horizontal) surface exposed to the atmosphere, and its lower (horizontal) boundary in contact with a second homogeneous liquid, below, of density ρ'. Then the pressure p at depth z' below the surface of separation of the two liquids is given by

$$p = p_0 + g\rho h + g\rho' z'. \tag{68d}$$

16·25. Examples. (1) *An object rests on a frictionless lid which fits exactly into one of the tubes shown in Fig. 204, and is supported in equilibrium by liquid pressure as shown. The combined weight of the lid and the object is W, the density of the liquid is ρ, and the cross-sectional area of the tube containing W is A. Find the difference h in the levels of the liquid in the two tubes.*

The vertical forces on the object and lid consist of: (i) the weight W; (ii) the downward force $p_0 A$ due to the atmospheric pressure; (iii) the upward force pA due to the liquid pressure, where p is the pressure in the liquid at a point R just below the lid.

Hence $\qquad\qquad W + p_0 A - pA = 0.$ \hfill (i)

By § 16·23 (i), p is also the pressure at Q (Fig. 204), where Q is at the same level as R.

Thus $\qquad\qquad p = p_0 + g\rho h.$ \hfill (ii)

By (i), (ii), we then have

$\qquad\qquad W = g\rho A h;\qquad$ i.e. $\quad h = W/g\rho A.$

(2) *A U-tube whose arms are vertical and of equal length has mercury poured in until the level is 60 cm from the top in each tube. Water and alcohol (neither of which mixes with mercury) are poured each into one branch of the tube until the tube is filled. The level of the surface of separation between the alcohol and mercury is then 0·71 cm above that for the water and mercury. Find the specific gravity of the alcohol, assuming the specific gravity of the mercury to be 13·6.*

By the data, the lengths of the columns of alcohol and water are seen to be 59·6, 60·4 cm, respectively.

The pressures at P, Q (Fig. 205) are equal since P and Q are at the same level, and can be joined by a curve lying entirely in one body of homogeneous liquid (mercury).

Hence, by (68d),
$$p_0 + 59 \cdot 6gs\rho + 0 \cdot 71g \times 13 \cdot 6\rho = p_0 + 60 \cdot 4g\rho, \qquad \text{(i)}$$
where s is the specific gravity of the alcohol, and ρ g/cm^3 is the density of water.

From (i), we find that $s \approx 0 \cdot 85$.

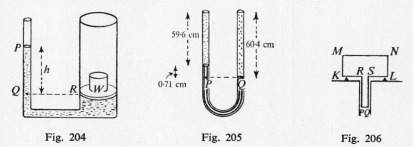

Fig. 204　　　　　Fig. 205　　　　　Fig. 206

(3) *The safety-valve of a steam boiler consists of a mass of 20 lb which has a horizontal lower face of cross-section 0·2 sq. in. in contact with the steam. The valve fits smoothly into the boiler as shown in Fig. 206. Find in atmospheres the pressure of the steam in the boiler when the valve just opens.*

When the valve is about to open, the forces on it are: its weight W pdl; the net force, F_0 pdl say, due to the atmospheric pressure p_0 pdl/ft^2; and the force F pdl due to the steam pressure, p pdl/ft^2, say.

We have $\qquad\qquad F = F_0 + W.$ \hfill (i)

Now $F = Ap$, where A ft^2 is the area corresponding to PQ in Fig. 206. Also $F_0 = Ap_0$ (taking into account the downward atmospheric thrust on MN and the upward thrust on KR, SL).

Hence, by (i), $\quad p = p_0 + W/A = p_0 + 100 \times 144g,$
since $W = 20g$, and $A = 0 \cdot 2/144$.

Hence the required pressure = 1 atm + $(100 \times 144g)$ pdl/ft^2
　　　　　　　　　　　= 1 atm + 100 lb.wt./in^2
　　　　　　　　　　　= $(1 + 6 \cdot 9)$ atm
　　　　　　　　　　　= 7·9 atm.

S.P.I.C.E. (α) In Ex. (1), the height h in Fig. 204 is independent of the cross section of the smaller tube, which may (we are neglecting surface tension) therefore be exceedingly small. It is sometimes thought surprising that such a small column of liquid should 'balance' a heavy object, and the result is sometimes referred to as the 'hydrostatic paradox'.

(β) In Exs. (1) and (2), the atmospheric pressure p_0 cancels out from the equations. In practice in working such problems, p_0 is commonly ignored.

(γ) In Ex. (3), a practical man would sometimes give the answer as 6·9 atm, or 100 lb.wt./in², tacitly understanding that he was concerned with the excess pressure over and above that of the atmosphere.

(δ) In Ex. (3), the significance of the upward atmospheric thrust on KR, SL, should be noted.

(ϵ) An important practical point in Ex. (2) is that only two-figure accuracy is obtainable from the data; and the method would not be used to determine the sp.gr. in practice. Better methods will be indicated in § 16·71.

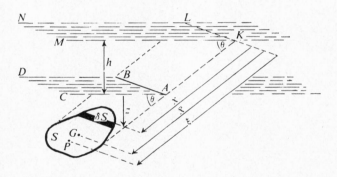

Fig. 207

16·3. Thrusts on immersed surfaces. The resultant force due to the pressure of a fluid on an immersed surface is called the *fluid thrust* on the surface.

A simple case is that where a plane *horizontal* surface, of area S say, is exposed to fluid pressure. In this case the pressure p is, by § 16·23 (i), constant over the surface, and so the resultant thrust is of magnitude pS, the line of action being at right angles to the area and passing through its centroid; this case has already been illustrated in Exs. (1), (3) of § 16·25.

Equally simple in practice is the case where a limited plane surface (not necessarily horizontal) is exposed to *gaseous* pressure. In this case, the pressure can in many ordinary problems be treated as constant over the area (see § 16·6), since the density ρ is very small and hence, by (67), dp/dz is very small. Hence the formula pS is again closely relevant.

In the following subsections, we consider less simple cases.

16·31. Thrust on a plane surface immersed in a homogeneous liquid.

16·311. Magnitude of thrust. Let S be the area of a plane surface immersed in a homogeneous liquid, and $\bar{p}$ the pressure at the centroid G of S. We shall show that the magnitude of the liquid thrust on S is $\bar{p}S$.

Let $ABCD$ be any fixed horizontal plane which is at or below the highest level reached by the liquid, but not below the highest point of S; and let p_1 be the liquid pressure at the level of $ABCD$. Let δS be an element of S, shown in black in Fig. 207, the upper and lower edges of δS being horizontal and at depths z, $z + \delta z$ below $ABCD$.

The magnitude of the thrust on δS is (to sufficient accuracy) $p\delta S$, where p is the pressure at the upper edge of δS. Hence the magnitude of the thrust on S (the integrations below range over the area of S)

$$\begin{aligned} &= \int p dS \\ &= \int (p_1 + g\rho z) dS \\ &= p_1 \int dS + g\rho \int z dS \\ &= p_1 S + g\rho \bar{z} S \\ &= \bar{p} S, \end{aligned} \qquad (69)$$

since $\bar{p} = p_1 + g\rho\bar{z}$, where $\bar{z}$ is the depth of G below $ABCD$.

16·312. Line of action of thrust. We next investigate the line of action of the resultant thrust on S. The point P of S through which this thrust passes is called the *centre of pressure* (C.P.).

Let $KLMN$ (Fig. 207) be a second horizontal plane distant h above $ABCD$, where $p_1 = g\rho h$, and let the plane containing S cut this plane in KL at an angle θ. Let x, $\bar{x}$, ξ be the distances from KL of the upper edge of δS, G and P, respectively.

Now $p = p_1 + g\rho z = g\rho(h + z) = g\rho x \sin\theta$. Similarly, $\bar{p} = g\rho\bar{x}\sin\theta$.

Taking moments about KL and using (69), we have

$$\bar{p}.S\xi = \int p x dS;$$

i.e., $$g\rho\bar{x}\sin\theta.S\xi = \int g\rho x^2 \sin\theta\, dS;$$

i.e., $$\xi = \frac{\int x^2 dS}{\bar{x} S};$$

i.e., $$\xi = k^2/\bar{x}, \qquad (70a)$$

where k is the radius of gyration of S about KL.

The equation (70a) determines the distance of the centre of pressure P from KL. This is sufficient to determine P when the area S is symmetrical about a vertical plane through G; this will be the case in all the applications we shall make in this book.

(In more complicated cases, it is further necessary to take moments about a line in the plane of S at right angles to KL.)

16·313. Remarks.

(i) As an alternative to (70a), we may write

$$GP = k_G^2/\bar{x}, \tag{70b}$$

where k_G is the radius of gyration of S about the horizontal line of S which passes through G. This follows by (70a) and the theorem of parallel axes (§ 11·71), which give $\bar{x}\xi = k_G^2 + \bar{x}^2$, and hence $\bar{x}(\xi - \bar{x}) = k_G^2$, i.e. $\bar{x}.GP = k_G^2$.

(ii) The above theory and results are relevant whether we are considering the liquid thrust on the upper or the lower side of S; (the only difference in the two cases is in the sense of the resultant thrust). The lower side is involved e.g. with the liquid thrust on a side of a liquid container which makes an acute angle with the base.

(iii) The construction in Fig. 207 is equivalent to supposing the whole space up to the level $KLMN$ to be filled with the same liquid, and the space above $KLMN$ to be a vacuum. The plane $KLMN$ is commonly called the *effective surface* of the liquid.

(iv) In simple problems, $ABCD$ will frequently correspond to the surface of a liquid exposed to the atmosphere. The effective surface would then be about 34 ft or 1030 cm above the actual liquid surface in the case of water, and about 30 in or 76 cm in the case of mercury.

(v) In problems where one liquid L_1 overlies a second liquid L_2, then if S is entirely immersed in L_2, $ABCD$ would ordinarily be taken as the surface of separation of L_1 and L_2; the effective surface would be determined from knowledge of the pressure at the level of $ABCD$.

(vi) Next suppose that S is immersed partly in L_1 and partly in L_2, the areas of S in the two liquids being S_1, S_2, respectively. Then the resultant thrust on S is determined by finding separately the resultant thrusts on S_1 and S_2, and combining by ordinary statical methods.

(vii) When a surface is subject to fluid pressure on both sides, it is permissible in many problems to ignore the influence of atmospheric pressure. An example is the case of the resultant pressure on the wall of a tank. By Pascal's principle, the contribution of the atmospheric pressure to the liquid thrust on the inside surface would be precisely cancelled by the thrust due to the atmospheric pressure from outside. The liquid surface inside the tank could for many purposes be treated as the effective surface, the atmosphere being ignored altogether.

(viii) Certain aspects of the resultant thrust on S are alternatively investigated by considering the liquid, U say, *superincumbent* to S, i.e. the liquid occupying the region bounded by S, vertical lines through the rim of S, and the effective surface. (This consideration can of course be applied directly only to cases where the plane of S is not vertical.)

The vertical component of the resultant thrust on S is equal to the weight of U. This is seen on resolving vertically for the equilibrium of U.

Further, the centre of pressure P of S is vertically below (and at twice the depth of) the centre of mass, K say, of U. The proof is as follows: Consider the weights of columns of the liquid superincumbent to elements of the area S. The resultant of this set of forces acts vertically through K, cutting S in P' say. Now consider a second set of forces, all perpendicular to S, derived from the first set by turning each member about the point in which it cuts S and dividing each magnitude by $\cos \theta$. The resultant of this second set must also pass through P'. But (on applying the statement in the preceding paragraph to elements of S) it is seen that this second set of forces is actually the set of liquid thrusts on elements of S, the resultant of which must intersect S in P. Hence P, P' coincide.

(It may be remarked that the above definition of superincumbent liquid and the property stated in the second (but not the third) paragraph of (viii) are also relevant if the given surface S is curved—see § 16·33.)

(ix) From the third paragraph of (viii), or by (70b), it is evident that unless S is horizontal the centre of pressure P is at a lower level than the centroid G of S.

By (70b), if S occupies different positions in the liquid, the same lines of S being always horizontal, then $GP.\bar{x}$ is constant. (Incidentally, the position of P relative to S, unlike that of G, depends on where S is immersed in the liquid.)

In particular, if the depth of S is increased, the inclination θ being kept unchanged, then $\bar{x}$ is increased and so GP is diminished. Further, if the depth of S is indefinitely increased, then $P \to G$. These results may also be deduced using (viii).

(x) It follows from (70a) that, for different positions of S obtained by rotating S about the line KL in which its plane cuts the effective surface, the distance ξ of P from KL would be the same.

(xi) Expressions giving the depth of P below the effective surface $KLMN$ in terms of the depths of specific points Q_1, Q_2, ..., say, of S (see the examples in § 16·314 below), are independent of the inclination θ. For, by (x), if S were rotated about KL, the distance ξ of P from KL would be unchanged, and hence the ratios of the depths of P, Q_1, Q_2, ..., below $KLMN$ would also be unchanged.

16·314. Centres of pressure in particular cases. We now find centres of pressure of immersed surfaces in some simple cases, taking the liquid surface to be identical with the effective surface.

When we use (70a), we shall, if no special mention is made, assume the immersed surface to be vertical; but, by virtue of § 16·313 (x), (xi), the stated results will fit the case of any inclination.

When we use the method of § 16·313 (viii), the immersed surface is treated as other than vertical; but the results obtained will, by § 16·133 (x), (xi), also apply to the vertical case.

Centres of pressure.

(i) *Parallelogram with edge in surface.* Let α be the depth of the lowest side. Then, using (70a) and § 11·82, we see that the depth of the C.P. is $\{\frac{1}{3}(\frac{1}{2}\alpha)^2 + (\frac{1}{2}\alpha)^2\}/(\frac{1}{2}\alpha) = \frac{2}{3}\alpha$.

This result may be independently obtained using the method of § 16·313 (viii).

(ii) *Parallelogram with horizontal edges at depths α, β.* The depth of the C.P. $= \{\frac{1}{3}[\frac{1}{2}(\alpha - \beta)]^2 + [\frac{1}{2}(\alpha + \beta)]^2\}/[\frac{1}{2}(\alpha + \beta)]$
$$= \frac{2(\alpha^2 + \alpha\beta + \beta^2)}{3(\alpha + \beta)}.$$

(iii) *Triangle ABC with base BC in surface.* It will be sufficient to consider the case in which the plane of the triangle is vertical. Consider two elements δS_1, δS_2 of the area of the triangle ABC (Fig. 208), each bounded by horizontal lines δx apart, δS_1 being at a distance x from BC, and δS_2 at the same distance x from A. Let D be the midpoint of BC. Then, to sufficient accuracy, the thrusts on these elements are $g\rho x \delta S_1$, $g\rho(d-x)\delta S_2$ respectively, where d is the distance of A from BC. But $\delta S_1 : \delta S_2 = (d-x) : x$. Hence the two elementary thrusts are equal. Hence the resultant thrust acts through the midpoint of AD; this point is therefore the C.P.

(iv) *Triangle ABC with A in surface and BC horizontal.* By the method of § 16·313 (viii), it is quickly seen that the C.P. is $\frac{3}{4}$ of the way down AD, where D is the midpoint of BC.

(The results in (iii) and (iv) can be derived using (70a) and the methods of Chapter XI, but in these cases the process would be a little longer.)

(v) *Triangle ABC, with A, B, C at depths α, β, γ.* We shall use (70a) and treat the triangle as in a vertical plane. By § 11·(11), the radius of gyration of a uniform triangular lamina of mass M about any line is equal to that of a system of three particles, each of mass $\frac{1}{3}M$, situated at the midpoints D, E, F of the sides; also, the centres of mass of these two systems coincide.

Now the depths of D, E, F in our problem are $\frac{1}{2}(\beta + \gamma)$, $\frac{1}{2}(\gamma + \alpha)$, $\frac{1}{2}(\alpha + \beta)$. Hence, by (70a), the depth of the C.P. is $k^2/\bar{x}$, where
$$Sk^2 = \tfrac{1}{3}S\{[\tfrac{1}{2}(\beta + \gamma)]^2 + \ldots + \ldots\},$$
$$S\bar{x} = \tfrac{1}{3}S\{\tfrac{1}{2}(\beta + \gamma) + \ldots + \ldots\};$$

i.e., the depth of the C.P. $= \dfrac{\alpha^2 + \beta^2 + \gamma^2 + \beta\gamma + \gamma\alpha + \alpha\beta}{2(\alpha + \beta + \gamma)}$

(vi) *Circle of radius a, with centre (O say) at distance $\bar{x}$ from KL.* By (70b) and § 11·85, the distance of the C.P. from O is $\tfrac{1}{4}a^2/\bar{x}$.

16·315. Examples. (1) *A conical vessel whose height is 10 cm and base-radius 5 cm is just full of mercury. Find the thrust, over and above that due to atmospheric pressure, exerted by the mercury on the base of the vessel.*

The 'superincumbent' mercury occupies the volume indicated by $DECB$ in Fig. 209. Its weight is $\pi 5^2 \times 10 \times g\rho$ dynes, where $\rho \approx 13\cdot 6$.
The required thrust is equal to this weight, i.e. 10·7 Kg, approx.

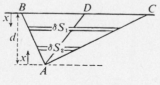

Fig. 208   Fig. 209

(2) *A vertical trapezoidal plate in the side of a ship has its parallel sides horizontal, the upper, of length a, being in the water surface, and the lower, of length b, being at a depth h. Find the depth of the centre of pressure associated with the excess of the liquid pressure over the contribution due to the atmosphere.*

Let ρ be the density of the water.
The resultant thrust on the plate $ABDC$ (Fig. 210) is the resultant of the thrust $ah.g\rho.\tfrac{1}{2}h$ acting at P_1 on the parallelogram $ABEC$, and $\tfrac{1}{2}(b-a)h.g\rho.\tfrac{2}{3}h$ acting at P_2 on the triangle BED, where the depths of P_1, P_2 are $\tfrac{2}{3}h$, $\tfrac{3}{4}h$, respectively.
Let x be the depth of the C.P. of the plate.
Then, by the theorem of moments,
$$\{\tfrac{1}{2}ah^2 g\rho + \tfrac{1}{3}(b-a)h^2 g\rho\}x = \tfrac{1}{2}ah^2 g\rho.\tfrac{2}{3}h + \tfrac{1}{3}(b-a)h^2 g\rho.\tfrac{3}{4}h;$$
hence
$$x = \frac{(a+3b)h}{2(a+2b)}.$$

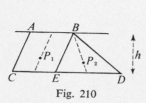

Fig. 210   Fig. 211

(3) *The heights of water on the two sides of a rectangular dock-gate 32 ft wide, are 20 ft, 10 ft, respectively. Find the magnitude and line of action of the resultant thrust on the gate.* (Fig. 211.)

The atmospheric pressure may be disregarded, since its contributions to the thrusts on the two sides cancel out.

On the deeper side a thrust F_1 pdl acts at P_1, and on the shallower side a thrust F_2 pdl at P_2, where
$$BP_1 = 20/3 \text{ ft}, \qquad BP_2 = 10/3 \text{ ft};$$
$$F_1 = 640g\rho \times 10, \qquad F_2 = 320g\rho \times 5; \qquad \rho = 62\cdot 5.$$
The magnitude of the resultant thrust on the gate
$$= (6400 - 1600)62\cdot 5g \text{ pdl}$$
$$= 300{,}000 \text{ lb.wt.} \approx 134 \text{ tons wt.}$$

Let x ft be the height above B of the line of action of the resultant thrust on the gate. Then, taking moments about B, we have
$$(6400 - 1600) \times 62\cdot 5gx = 6400 \times 62\cdot 5g \times 20/3 - 1600 \times 62\cdot 5g \times 10/3;$$
whence $\qquad\qquad\qquad x = 7\cdot 78.$

Thus the line of action of the resultant is about 7 ft 9 in above B.

S.P.I.C.E. (α) Ex. (1) illustrates the use of the idea of superincumbent liquid. It will be noticed that the thrust on the base of the vessel exceeds the weight W of the mercury in the vessel. In fact, the thrust is quite independent of the shape of the vessel between the levels BC and DE. The curved part of the vessel exerts a resultant downward thrust on the mercury; the sum of this and W is equal to the resultant thrust on the base BC.

(β) Ex. (2) is an exercise involving the use of formulae in § 16·314. The same result could be derived independently, using (70a) and moment of inertia theory.

(γ) In Ex. (3), it should be noted that the atmospheric pressure p_0 is safely ignorable. In practice in many problems, it is only the excess effects above those due to the atmosphere that need to be determined.

16·32. The principle of Archimedes. Consider a solid body B wholly immersed in fluid. The resultant thrust on B due to the fluid pressure is equal to $\int \mathbf{p}dS$ evaluated over the surface of B, the pressure $\mathbf{p}$ being treated as a vector acting normally into the solid.

This resultant thrust is no different from what would be the resultant thrust on the *displaced fluid*, i.e. a body of fluid which would fully occupy the space of B (imagined removed) and which would have the same density at all levels as the actual fluid surrounding B.

The displaced fluid would be in equilibrium under this resultant thrust and the weight, W' say, of the displaced fluid, acting through its centre of mass, H say.

Thus the resultant thrust on the solid is equivalent to a force of magnitude W' acting vertically upwards through H. This is the principle of Archimedes.

In the case of a solid body floating in a liquid (or otherwise only partly immersed), the displaced fluid will in general consist partly of displaced air and partly of displaced liquid. The displaced air is often ignored in practice when the principle of Archimedes is used, because its weight is so small compared with the weight of the displaced liquid.

16·321. Examples. (1) *A common hydrometer consists of a long graduated tube of cross-section A with a dense bulb at the bottom, and is used to determine specific gravities of liquids by being floated vertically in them. When the hydrometer floats in water, a volume V cm³ is immersed and a zero mark on the stem is at the water surface. When it floats in another liquid the stem is submerged to a point x cm above the zero mark. Find the specific gravity s of the liquid.*

Let W dyn be the weight of the hydrometer, and ρ g/cm³ the density of water.

Then, by Archimedes' principle (see Fig. 212),
$$W = g\rho V,$$
$$W = gs\rho(V + xA),$$
(the weight of the air displaced being neglected).

Hence $\qquad s = (1 + xA/V)^{-1}.$

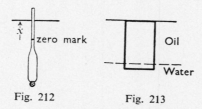

Fig. 212   Fig. 213

(2) *The apparent weight of a body, of sp.gr. s, surrounded by air of sp.gr. σ, is W_1. What is the 'true' weight of the body?*

Let W be the 'true' weight (as determined say in a vacuum), and V the volume of the body. Let ρ be the density of water.

When surrounded by air, the body will be in equilibrium under: W downwards; the supporting force W_1 upwards; and the resultant (upward) force, W' say, due to the pressure of the air.

Hence $\qquad W = W_1 + W'.$
Also $\qquad W = gs\rho V, \qquad W' = g\sigma\rho V.$
Hence $\qquad W = W_1(1 - \sigma/s)^{-1}.$

(3) *A uniform cylindrical piece of wood of sp.gr. 0·8 floats, with its axis vertical, partly immersed in water. Oil is poured upon the water until the wood is completely covered. It is observed that one-tenth of the wood is then in water and nine-tenths in oil. Find the specific gravity of the oil.* (Fig. 213.)

Let s be the sp.gr. of the oil, ρ the density of water, and V the volume of the wood.

By the principle of Archimedes, the weight of the wood is equal to the sum of the weights of the oil and water displaced.

Hence
$$0{\cdot}8\rho V g = 0{\cdot}9 V s \rho g + 0{\cdot}1 V \rho g.$$
Hence $\qquad s = 7/9.$

(4) *A thin uniform rod AB is in equilibrium in an inclined position with the portion AC immersed in a bowl of water, and one point D supported by the (rough) edge of the bowl, as indicated in Fig. 214. If $AB = l$, $AC = x$, $AD = y$, find the specific gravity s of the material of the rod. Find also, in terms of s, the greatest fraction of the length of the rod that could be immersed in the given circumstances.*

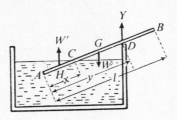

Fig. 214

Let ρ be the density of water, and α the cross-sectional area of the rod.

The rod is in equilibrium under: its weight W; the resultant liquid thrust W' acting vertically upward through H, the midpoint of AC; and the support Y at D (which by the three-force theorem of statics must incidentally be vertical).

Now $W = \alpha l s \rho g$, and $W' = \alpha x \rho g$, by the principle of Archimedes.

Since $GD = y - \tfrac{1}{2}l$, $HD = y - \tfrac{1}{2}x$, we have, taking moments about D,
$$ls(y - \tfrac{1}{2}l) - x(y - \tfrac{1}{2}x) = 0;$$

i.e.,
$$s = \frac{x(2y - x)}{l(2y - l)}. \tag{i}$$

Multiplying (i) through by $l(2y - l)$, and differentiating with respect to y, we have (l, s being assigned constants)
$$2sl = (dx/dy)(2y - x) + x(2 - dx/dy);$$

i.e.,
$$\frac{dx}{dy} = \frac{sl - x}{y - x}.$$

Now $sl/x = W/W'$, which is seen, by resolving vertically in Fig. 214, to exceed unity; hence $sl > x$. Also $y > x$, obviously. Hence dx/dy is always positive.

Hence x is greatest when y is greatest, i.e. when $y = l$.

The required greatest fraction, β say, is therefore the value of x/l when y is put equal to l. Thus by (i),
$$s = \beta(2 - \beta); \tag{ii}$$
i.e.,
$$\beta = 1 - \sqrt{(1 - s)}, \tag{iii}$$
the other root of (ii) being irrelevant since β cannot exceed unity.

S.P.I.C.E. (α) In Exs. (1), (3), (4), the weight of the displaced air has been neglected. The influence of atmospheric pressure on weights of objects has in fact been ignored in all the preceding chapters of this book. This is usual in practice except in very special work. In Ex. (2), it is shown how the necessary correction could be made.

(β) The formula obtained in Ex. (1) could be used in graduating a common hydrometer.

(γ) Ex. (3) is a case in which a solid is immersed in two liquids.

(δ) Ex. (4) is a statical problem in which a resultant liquid thrust constitutes one among several forces acting on the given rod.

(ϵ) In Ex. (4), it should be noted that the edge of the bowl has to be rough. Otherwise the resultant force on the rod would have a non-vanishing horizontal component which would make the rod slide entirely into the water.

(ζ) The second part of Ex. (4) is a case in which a maximum is not obtainable by equating a differential coefficient to zero. This is because x steadily increases as y increases until y reaches the terminal value l. The required maximum occurs at this terminal. The details of argument in the second part of Ex. (4) should be carefully followed.

16·33. Resultant thrust on an immersed curved surface. A direct method of finding the magnitude and direction of the resultant thrust due to a liquid on an immersed curved surface, S say, would be to form the integral $\int p\,dS$.

The vertical component of the resultant thrust on S may be found, independently, as the weight of the superincumbent liquid (see § 16·313 (viii)).

A horizontal component of the resultant thrust on S may be found by considering the equilibrium of a portion of the liquid bounded by S, by horizontal lines through all points of the rim of S, and by a plane vertical surface at right angles to these lines; and then resolving parallel to the horizontal lines.

If S is bounded by a plane curve, C say, the resultant thrust on S is often determined by considering the equilibrium of the portion of the liquid contained between S and the plane through C.

We shall illustrate below.

16·331. Example. *A thin hemispherical shell of radius a is totally immersed in a liquid of density ρ so that its axis of symmetry makes an angle θ with the vertical (as shown in Fig. 215) and its centre is at depth h below the effective surface. Find the resultant thrust on a side of the shell.*

First method.

Consider the equilibrium of the liquid, L say, contained between the hemispherical surface S and the diametral plane S' (Fig. 215(a)). Let O be the centre of S'.

The resultant thrust, **R** say, exerted by L on its whole boundary surface (i.e. S and S') is equivalent to the weight of L, i.e. $\tfrac{2}{3}\pi a^3 g\rho$ acting vertically downwards through G, where $OG = \tfrac{3}{8}a$.

The resultant thrust, $\mathbf{F}'$ say, exerted on S' by the liquid outside L is, by (69), $g\rho h.\pi a^2$ acting at right angles to S' at P', where, by § 16·314 (vi), $OP' = \frac{1}{4}a^2/(h \operatorname{cosec} \theta)$.

The resultant thrust, $\mathbf{F}$ say, on the inner side of S is given in magnitude and direction by $\mathbf{R} + \mathbf{F}'$.

Hence, the vertical and horizontal components of $\mathbf{F}$ are

$$\tfrac{2}{3}\pi a^3 g\rho + \pi a^2 hg\rho \cos \theta, \qquad \pi a^2 hg\rho \sin \theta;$$

hence the magnitude of $\mathbf{F}$ is

$$\pi a^2 g\rho \sqrt{\{(\tfrac{2}{3}a + h \cos \theta)^2 + (h \sin \theta)^2\}},$$

i.e. $\qquad \tfrac{1}{3}\pi a^2 g\rho \sqrt{(4a^2 + 9h^2 + 12ah \cos \theta)}.$

The inclination of $\mathbf{F}$ to the horizontal is

$$\tan^{-1}\{(2a + 3h \cos \theta)/3h \sin \theta\}.$$

The line of action of $\mathbf{F}$ can be found by noting that: (*a*) by the three-force theorem in statics, it must pass through K, the intersection of the lines of action of $\mathbf{R}, \mathbf{F}'$; (*b*) it must pass through O, since the thrust on every element of S being normal to S must pass through O.

Second method.

A second method would be to consider the vertical component $\mathbf{V}$ and the horizontal component $\mathbf{X}$ of the liquid thrust on the shell S, lines of action being taken into account.

$\mathbf{V}$ is equal to the weight of the superincumbent liquid: i.e. (see Fig. 215(*b*)) the liquid L inside the hemisphere DEF, together with the portion $ABFD$, of which the horizontal section is an ellipse of semi-axes a, $a \cos \theta$.

Thus $\mathbf{V}$ is the resultant of $\mathbf{R}, \mathbf{Y}$ acting vertically through G, P', respectively, where the magnitude of $\mathbf{Y}$ is $\pi a^2 \cos \theta.hg\rho$. (The area of an ellipse of semi-axes a, b is πab.)

$\mathbf{X}$ is given by considering the equilibrium of the liquid contained inside DEF, together with the portion $DFNM$ which has a vertical elliptic cross-section of semi-axes $a, a \sin \theta$.

On resolving horizontally, it is seen that $\mathbf{X}$ is equal to the thrust on the ellipse MN, i.e. $\pi a^2 \sin \theta.hg\rho$ acting through P'', where, by (70*b*) and Routh's rule, $CP'' = \tfrac{1}{4}(a \sin \theta)^2/h$. ($P''$ is thus at the same level as P', as is otherwise obvious.)

The required answer is then given as the resultant of $\mathbf{V}$ and $\mathbf{X}$.

S.P.I.C.E. (α) The above example was chosen to demonstrate two methods of finding resultant thrusts on curved surfaces.

In this example $\mathbf{X}$ and $\mathbf{Y}$ are together equivalent to $\mathbf{F}'$, and there is no special advantage in using the second method. In more complicated problems, steps such as those illustrated in the second method may, however, be of distinct help.

(β) In the steps used in the second method, it should be noted that it does not matter that the hemisphere DEF bulges to the left of AD and below the level of FN. The student should assure himself of this point.

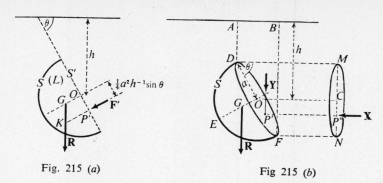

Fig. 215 (a) Fig 215 (b)

16·4. Work and energy considerations.

16·41. Work on fluid due to movement of a boundary over which the pressure is uniform. Let S be part or the whole of the area of the boundary of a body of fluid, and let S be subject to an applied pressure p which is always uniform over S. Suppose that S suffers a change of configuration between the times t_0 and t_1.

(This involves fluid motion, which, strictly speaking, belongs to *hydrodynamics*. We assume here that the fluid is 'perfect', i.e. that the stress at any point has the property stated in § 16·12, even while the fluid is in motion.)

We shall show that the work done on the fluid by the applied pressure forces across S is

$$-\int_{V_0}^{V_1} p\, dV, \qquad (71)$$

where dV is the volume traced out by S in time dt in the sense tending to increase the volume of the fluid; the integration is over the time-range $t_0 \leqslant t \leqslant t_1$, and V_0, V_1 are the volumes at the times t_0, t_1.

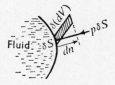

Fig. 216

To prove (71), let δS be an element of the area S. Then at time t, the force across δS is $p\delta S$. During the ensuing small time dt, the work done on the fluid by this force is $-p\delta S dn$ (where dn is the normal component of the displacement of δS); i.e., is $-p\delta(dV)$, where $\delta(dV)$ is the volume corresponding to the more darkly shaded region in Fig. 216. Since p is uniform over S, the total work done on the fluid in time dt is $-p\{\Sigma\delta(dV)\} = -pdV$; hence the result (71).

If the configuration of the fluid changes sufficiently slowly, i.e. if the conditions are *quasi-statical*, the pressure at any point just inside the fluid is virtually the same as the pressure p applied from outside. (This can be seen by forming the equation of motion for an elementary volume of the fluid and neglecting its acceleration.)

We can therefore re-state (71) in the following form: the work done *by* the fluid during a quasi-statical volume change is

$$\int_{V_0}^{V_1} p\,dV, \tag{71a}$$

where p now denotes the pressure at any point just inside the fluid.

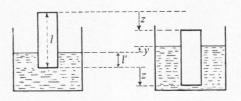

Fig. 217

16·42. Use of principle of energy. A body of fluid under gravity possesses potential energy just as any other mass-system.

Suppose that the fluid suffers a change of configuration and is at rest initially and finally. It then follows from (22b), § 12·53, that the sum of the works done by the applied forces across the boundary and by internal forces inside the fluid is equal to the gain in potential energy.

In the particular case of a (perfect) liquid, the net work of the internal forces is zero. This follows (we shall not give the proof) as a consequence of: (*a*) the absence of tangential stress, (*b*) the constancy of the volume of each element of mass of a liquid.

In the case of a gas, the net work of the internal forces is in general not zero, for (*b*) no longer holds.

16·43.

When a solid is floating in a liquid under gravity, the potential energy of the whole system is stationary for displacements about an equilibrium configuration. For, by § 16·42, the only forces contributing to the energy equation are the weights of the liquid and solid.

As in § 15·(10)5, the equilibrium is stable if the potential energy of the system is a minimum.

16·44. Example. *A cylindrical solid of height l is floating in a cylindrical vessel of liquid. The cross sections of the solid and vessel are α, β, and the densities of the liquid and solid are ρ, σ. Find the work done in slowly pressing the solid down through a distance z, where z is not sufficient for the solid to be totally immersed.*

The solid loses P.E. of amount $\alpha l \sigma g z$.

The liquid gains P.E. corresponding to the raising of the C.M. of a portion of volume αz through a height $\tfrac{1}{2}z + l' + \tfrac{1}{2}y$, where l', y are as in Fig. 217.

Now, by Archimedes' principle, $l'\rho = l\sigma$.

Also, $y(\beta - \alpha) = z\alpha$.

Hence, $\tfrac{1}{2}z + l' + \tfrac{1}{2}y = \tfrac{1}{2}z\beta/(\beta - \alpha) + l\sigma/\rho$.

The required work = the net gain in P.E.; i.e., (neglecting complications —very small in practice—due to the disturbance of the air above)
$$= -\alpha l \sigma g z + \alpha z \{\tfrac{1}{2} z \beta/(\beta - \alpha) + l\sigma/\rho\} g \rho$$
$$= \tfrac{1}{2} \alpha \beta z^2 g \rho / (\beta - \alpha).$$

16·5. Equilibrium of floating solids. We have already considered (§§ 16·32, 16·43) some aspects of the equilibrium of a solid floating in a liquid. We now proceed to some further detail. We shall, as on most previous occasions, ignore the air above.

In subsections of § 16·5, ρ, σ will denote the densities of the liquid, solid, respectively; G, the centre of mass of the solid; H, the centre of mass of the displaced liquid; and V, the volume of the displaced liquid in an equilibrium configuration. We shall assume the surface area of the liquid to be great and shall neglect any disturbances in its level. All displacements made will be small, and higher powers of displacements will be ignored.

16·51. Centre and force of buoyancy. H is called the *centre of buoyancy*; and the resultant liquid upthrust on the solid, equal to the weight W' of the displaced liquid, is called the *force of buoyancy*.

When the solid, of weight W say, is floating in equilibrium, H and G are in the same vertical line; and $W = W' = V\rho g$.

16·52. Stability of equilibrium for vertical displacements. Let α be the area of the section of the solid made by the *plane of flotation* (i.e. the plane through the horizontal liquid surface) when the solid is floating in equilibrium. Now release the solid from a position at a small distance z below its equilibrium position. The resultant upward force now acting on the solid is (to sufficient accuracy) the excess, $\alpha z g \rho$, of the force of buoyancy now acting, over that in the equilibrium position.

Since this resultant force is in the direction tending to restore equilibrium (this is so whether z is positive or negative), it follows that the equilibrium position is stable for vertical displacements. (The equilibrium may nevertheless be on the whole unstable when angular displacements (§ 16·53) are taken into account.)

16·521. The mass of the solid being $V\rho$, its equation of motion when it is released as in § 16·52 is

$$V\rho\ddot{z} \approx -\alpha g \rho z.$$

Hence (ignoring hydrodynamical complications) the solid will oscillate vertically in approximate simple harmonic motion of period $2\pi\sqrt{(V/\alpha g)}$.

16·522.
The results of §§ 16·52, 16·521 may be deduced from the Ex. of § 16·44. Taking the limit $\beta \to \infty$ in the result of this Ex., we see that, due to a small vertical displacement z, the P.E. is increased by $\frac{1}{2}\alpha g\rho z^2$. (The displacement being small, this result holds to sufficient accuracy whether the sides of the solid are vertical or inclined to the vertical.)

Because this increase of P.E. is positive (whether z is positive or negative), the P.E. in the equilibrium configuration is a minimum for vertical displacements, giving stability as already found in § 16·52.

Also, the energy equation for the motion is

$$\tfrac{1}{2}V\rho\dot{z}^2 + \tfrac{1}{2}\alpha g\rho z^2 = \text{constant},$$

which gives the results of § 16·521.

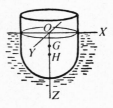

Fig. 218

16·53. Stability of equilibrium for angular displacements. We here limit consideration to a floating solid which is symmetrical with respect to a vertical plane, taken as OYZ, where O is in the plane of flotation, OY is horizontal, and OZ passes through G in the equilibrium configuration, as shown in Fig. 218; and we consider the stability for small angular displacements of the solid about OY.

Let θ be a small angular displacement about OY in the anti-clockwise sense in Fig. 218. The equilibrium will be stable for such a displacement if the horizontal component, x_G say, of the change in position of G

exceeds that, x_H say, of H. This follows because the forces on the solid would then have a resultant moment about G in the sense tending to restore equilibrium.

Now $x_G = OG.\theta$ (to sufficient accuracy).

Also, $x_H = OH.\theta + x'$, where x' arises from the shift in the centre of mass of the displaced liquid due to the removal, during the rotation, of a wedge-shaped portion of displaced liquid to the right of OY (Fig. 218), and the corresponding addition of a wedge-shaped portion to the left of OY. By the usual centre of mass theory, x' is given by
$$V\rho x' = \int(-x\theta d\alpha)\rho x,$$
the integral being taken over the area α, where x is the distance from OY of a typical element $d\alpha$ of α; in the integral, $-x\theta d\alpha$ is the added element of volume standing on $d\alpha$, x being positive to the right and negative to the left of OY. Hence
$$Vx' = -\theta \int x^2 d\alpha.$$
I.e.,
$$x' = -\alpha k^2 \theta/V,$$
where k is the radius of gyration of the area α *about OY*.

Hence, for stability, we require
$$OG.\theta > OH.\theta - \alpha k^2\theta/V;$$
i.e.,
$$\alpha k^2/V > HG. \tag{72}$$

16·531. The point M vertically above H, such that $HM = \alpha k^2/V$, is called the *metacentre*. The criterion for angular stability about OY is that M must be above G; i.e. that the *metacentric height*, i.e. the height of M above G, must be positive.

It is evident from § 16·53 that M may be regarded as the limiting position, as $\theta \to 0$, of the intersection of GO and the vertical through the centre of buoyancy in the displaced configuration.

If the metacentric heights for small rotations about OX and OY are both positive, then, taking § 16·52 into account, the equilibrium is stable.

16·532.

The formula (72) may also be established using the energy criterion. The brighter student may interest himself in showing that the increase in P.E. consequent upon the angular displacement θ of § 16·53 is
$$\tfrac{1}{2}g\rho\theta^2(\alpha k^2 - V.HG).$$
The condition for stability is that this expression should be positive, i.e. that $\alpha k^2/V > HG$.

16·533.

It should perhaps be pointed out that the above theory is only a first step towards investigating the stability of ships at sea. The reader who has been to sea will have learned that angular displacements are not always small.

16·534. Examples. (1) *The displacement of an object of weight w through a distance x amidships across the deck of a symmetrically-laden ship causes the ship to heel through a small angle θ. Prove that the metacentric height is wx/Wθ, where W is the total weight of the ship and cargo, and w/W is small.*

Suppose that initially the deck is horizontal, with w situated at A on the vertical through the C.M.; and that when w is moved to Q_1 (Fig. 219), where $AQ_1 = x_1$, the ship heels through a small angle θ_1. In Fig. 219, G is the position, relative to the ship, of the C.M. of the ship and cargo (including w), prior to the moving of w.

After the displacement, the ship and cargo may be regarded as being in equilibrium under: W at G, $-w$ at A, w at Q_1, all acting vertically downward; and the force of buoyancy, acting vertically upward through the metacentre M.

Taking moments about M, we then have (noticing that w, $-w$ constitute a couple)
$$W.GM \sin \theta_1 - wx_1 \cos \theta_1 = 0,$$
yielding $GM.\theta_1 = wx_1/W$, to sufficient accuracy.

Similarly if w had been moved to Q_2, where $AQ_2 = x_2$, we should have $GM.\theta_2 = wx_2/W$, where θ_2 is analogous to θ_1.

If Q_1, Q_2 are taken as the initial and final positions of w, we then have, by subtraction, $GM.(\theta_2 - \theta_1) = w(x_2 - x_1)/W$, which gives the required result.

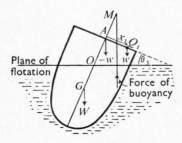
Fig. 219

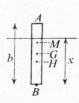
Fig. 220

(2) *A common hydrometer of mass M consists of a tube of radius a, length b and mass M/n, at the centre of the bottom of which is a dense bulb. Treating the bulb as of negligible volume, prove that the hydrometer can be used to determine the density of any liquid for which $\rho \geqslant M/\pi a^2 b$, provided $n > b/a\sqrt{2}$.*

If the hydrometer can be used for a liquid of density ρ, obviously it must not sink in that liquid. The condition for this is that Mg must not exceed the greatest possible weight of displaced liquid, i.e. $\pi a^2 b \rho g$. Hence we require $\rho \geqslant M/\pi a^2 b$.

Secondly, the hydrometer would be useless if it floated unstably in the liquid, and this we now investigate.

Let x be the depth immersed when the hydrometer (AB) floats vertically in a liquid of density ρ, and let G be the C.M. of the hydrometer, H the centre of buoyancy, and M the metacentre, as indicated in Fig. 220.

By ordinary C.M. theory, $BG = \frac{1}{2}b/n$. Also, $BH = \frac{1}{2}x$. Hence
$$HG = \frac{1}{2}(b/n - x).$$

By § 16·531, $HM = \pi a^2(\frac{1}{4}a^2)/\pi a^2 x = \frac{1}{4}a^2/x$.

(In these equations, HG and HM are understood to denote the heights of G and M above H.)

Hence for stability, we require
$$\tfrac{1}{4}a^2/x > \tfrac{1}{2}(b/n - x);$$
i.e.,
$$n > 2bx/(a^2 + 2x^2). \tag{i}$$

Now (i) is required to hold for all values of x over the range $0 \leqslant x \leqslant b$.

The maximum value, c say, of the right-hand side of (i) is found, by differentiation, to occur when $a^2 - 2x^2 = 0$, i.e. $x = a/\sqrt{2}$; hence $c = b/a\sqrt{2}$.

Hence we require $n > b/a\sqrt{2}$.

S.P.I.C.E. (α) In Ex. (1), the movement of w involves departure from the symmetrical conditions taken in the theory of § 16·53; but this does not matter, within the order of accuracy involved, since w/W is small.

(β) Ex. (2) demonstrates the use of the formula (72), and also gives a good indication of the practical importance of the stability theory.

16·6. Special consideration of gases.

The elementary theory of the hydrostatics of gases in ordinary conditions differs from that for liquids in two significant respects: (i) we cannot treat the density ρ of a gas as constant when the pressure p changes; (ii) at any instant, we may in practice treat p as constant throughout a volume of gas (unless the extent is very great as, for example, in the circumstances of § 16·66).

On account of (i), we have to consider the question of functional relations between p and ρ for gases. (For a *liquid* as artificially defined in § 16·17, the relation is simply that ρ is independent of p.)

Whereas (i) makes the elementary treatment of gases more complicated than that of liquids, (ii) is a simplifying feature; (ii) is a consequence of (67), coupled with the postulate that, for a gas, ρ is very small (this being an 'observed' property of 'real' gases).

16·61. Temperature.

For a 'real' gas, the form of the relation between p, ρ depends very significantly on the temperature.

(The density of a liquid also depends on the temperature; but—see § 16·16—we are ignoring thermal influences in the case of liquids.)

We do not here attempt to give a definition of temperature, but, as we did with time and space, shall rest on intuition. We assume that the reader has some understanding of the Centigrade and Fahrenheit scales: a temperature of $x°F$ is equal to a temperature of $[5(x - 32)/9]°C$.

In thermodynamical theory, it is customary to use an 'absolute' (or Kelvin) scale of temperature. We state here without giving detail that a temperature of $T°$ absolute (denoted as $T°K$) is approximately equal to a temperature of $(T- 273)°C$. Henceforth, except where we indicate otherwise, temperatures will be understood to be in $°K$.

16·62. Ideal gas. The relation between p, ρ, T for a gas is a 'characteristic gas equation'. We define an *ideal gas* (another mathematical model) as one whose characteristic equation is of the form

$$p/\rho T = K, \tag{73}$$

where K is constant for the particular gas.

For a number of 'real' gases, including air, (73) gives a good representation of 'observed' behaviour under pressures not too great and temperatures not too low. Henceforth, we shall treat all gases as ideal gases (cf. § 1·42); but it should be noted that other model forms are also used for various purposes.

For air, K is about $2\cdot88 \times 10^6$ cm^2-sec^{-2}-deg^{-1}; this corresponds to a density of $0\cdot00129$ g/cm^3 at a pressure of $76\cdot0$ cm of mercury and a temperature of $0°C$, i.e. at 'normal' temperature and pressure.

If M, V are the mass, volume of a given body of gas, then $M = V\rho$. Hence (73) may be written in the form

$$pV/T = MK. \tag{73a}$$

(In thermodynamical theory, $K = R/M_0$, where R is the same for all ideal gases, and M_0 g is the 'molecular weight' of the particular gas. The value of the 'universal gas constant' R is about $8\cdot314 \times 10^7$ g-cm^2-sec^{-2}-deg^{-1}. For hydrogen, $M_0 = 2\cdot016$; for oxygen, $M_0 = 32$; etc. For one 'mole' or 'gramme-molecule', i.e. M_0 g, of an ideal gas, we may write: $pV/T = R$.)

16·621. The relation (73) includes 'Boyle's law': for a given gas,

or
$$\left.\begin{array}{l} p/\rho = \text{constant, if } T \text{ is constant,} \\ pV = \text{constant, if } T, M \text{ are constant;} \end{array}\right\} \tag{73b}$$

and 'Charles's law': for a given gas,

or
$$\left.\begin{array}{l} \rho T = \text{constant, if } p \text{ is constant,} \\ V/T = \text{constant, if } p, M \text{ are constant.} \end{array}\right\} \tag{73c}$$

The second equation in (73c) may also be written in the form
$$V = V_0(1 + \alpha\theta),$$
where V_0 is the volume at 0°C, V is the volume at θ°C, and $\alpha \approx 1/273$; α is the 'coefficient of cubical expansion'.

16·63. Mixtures of gases.

Suppose that equal volumes V of two different gases (assumed not to interact chemically), initially at pressures p_1, p_2, respectively, are brought together to form a homogeneous gas mixture occupying the same volume V, the temperature being fixed throughout. We postulate, in accordance with experimental results embodied in 'Dalton's law', that the pressure of the mixture is $p_1 + p_2$.

Next suppose that given quantities of several gases (not interacting chemically), initially at pressures $p_1, p_2, \ldots$, volumes $V_1, V_2, \ldots$, and temperatures $T_1, T_2, \ldots$, respectively, are mixed so as to form a homogeneous gas of pressure p, volume V and temperature T. Then

$$\frac{pV}{T} = \frac{p_1 V_1}{T_1} + \frac{p_2 V_2}{T_2} + \cdots \qquad (74)$$

To prove (74) for the case of a mixture of just two gases, let the two gases be first brought separately to temperature T and volume V, their pressures being now p_1', p_2', where $p_1' V/T = p_1 V_1/T_1$ and $p_2' V/T = p_2 V_2/T_2$; and let the gases be now mixed as in the case of the preceding paragraph. Then, by Dalton's law, we have $p = p_1' + p_2'$, which leads by simple algebra to the result (74). It is easy to make the extension to the case of more than two gases.

16·64. Isothermal conditions.
When changes take place at a fixed temperature, the thermodynamical conditions are said to be *isothermal*. Thus for a given (ideal) gas in isothermal conditions, p/ρ is constant, by (73b).

As an exercise on isothermal conditions, suppose the volume of the gas changes quasi-statically from V_0 to V_1 under isothermal conditions. Then, by (71), the work done by the external forces acting across the boundary

$$= -\int_{V_0}^{V_1} p \, dV$$
$$= \int_{V_1}^{V_0} \frac{MKT}{V} \, dV,$$

in the notation of (73a). Since M, K, T are here constant, the work in question

$$= MKT \int_{V_1}^{V_0} dV/V$$
$$= MKT \log_e (V_0/V_1).$$

16·65. Adiabatic conditions.
Another important class of thermodynamical conditions is that in which changes occur *adiabatically*, i.e. without heat entering or leaving the system considered.

It has been pointed out (§ 4·54) that heat is a form of energy, distinct*
from kinetic and potential energy, but convertible into and from mechanical energy in appropriate circumstances. A commonly used unit of heat is the *calorie*, where 1 calorie $\approx$ 4·185 joules.

A study of the properties of heat belongs to thermodynamics. Here we merely state that, in adiabatic conditions, an ideal gas is found to satisfy the relation
$$p/\rho^\gamma = \text{constant}, \tag{73d}$$
where γ is a constant for the gas; ($\gamma \approx 1·4$ for hydrogen, nitrogen and oxygen). For a given mass of the gas, we may write $pV^\gamma = $ constant.

16·66. Use of barometer to determine altitude. We now proceed to find the variation of the atmospheric pressure p with the height z above sea-level, on the model assumptions that: (a) air is an ideal gas of uniform composition; (b) the temperature is uniform; (c) the value of g is independent of z.

By (67), we have (z being here measured *upwards*)
$$dp/dz = -g\rho,$$
where ρ is the density of air at height z. Also, by (73b),
$$p = C\rho, \tag{i}$$
where C is constant. Eliminating ρ, we have
$$\frac{1}{p}\frac{dp}{dz} = -\frac{g}{C}.$$
Integrating from sea-level, where $p = p_0$ say, to any height z, we have
$$\log_e p - \log_e p_0 = -gz/C;$$
i.e.,
$$p = p_0 e^{-gz/C} = p_0 e^{-z/H}, \tag{ii}$$
where H is the height of the 'homogeneous atmosphere' (§ 16·24). (By (i), $C = gH$ since $p_0 = g\rho_0 H$, where ρ_0 is the density at sea-level.)

From (ii), z can be determined if p_0 and C (or H) are known, and if p is given by a barometer reading at the height z.

In practice, because of variation of p_0 in changing meteorological conditions, the method is used to determine *differences* in altitude from barometer readings taken at times not too far apart. If p, p' are the pressure readings at heights z, z', then (ii) gives
$$z' - z = (C/g)\log_e(p/p') = H\log_e(p/p').$$
This formula is found to give fair results for moderate changes of altitude. Corrections may be made in practice for some of the departures from the stated assumptions (a), (b) and (c).

It should be noted that the density variation corresponding to (ii) is given by
$$\rho = \rho_0 e^{-z/H}.$$

* We are here disregarding the kinetic theory of gases.

16·67. Examples. (1) *A mountaineer finds that between two levels the barometric pressure changes from 30·0 to 26·1 inches of mercury. Taking the specific gravity of air at the lower level to be 0·00129, and that of mercury to be 13·6, and making assumptions as in § 16·66, find the change in altitude.*

Taking the pressure-density relation as $p = C\rho$, we have (on dividing through by the density of water)

$$(30 \cdot 0/12) \times 13 \cdot 6g = 0 \cdot 00129C;$$

i.e., $\qquad C/g = 2 \cdot 64 \times 10^4 \text{ (ft)}.$

Hence, by § 16·66, the required altitude change is an increase

$$= 2 \cdot 64 \times 10^4 \times \log_e(300/261) \text{ ft}$$
$$= 2 \cdot 64 \times 10^4 \times 2 \cdot 3026 \times \log_{10}(300/261) \text{ ft}$$
$$\approx 3700 \text{ ft}.$$

(2) *The envelope of a balloon of total mass M contains hydrogen and is gas-tight and of fixed volume. The balloon requires a force mg to prevent it from rising when its lowest point touches the ground. The atmospheric pressure and density at the ground-level are p_0, ρ_0. On the assumptions of § 16·66, find the height of the balloon above the ground when floating in equilibrium in the air.*

Considering the equilibrium of the sphere in the two positions, we have

$$Mg + mg = V\rho_0 g, \qquad Mg = V\rho g, \qquad \text{(i)}$$

where V is the volume of the sphere and ρ is the density of the air at the required height, h say.

By § 16·66, since ρ is proportional to the atmospheric pressure, we have

$$\rho = \rho_0 e^{-gh\rho_0/p_0}.$$

Hence
$$h = (p_0/g\rho_0) \log_e (\rho_0/\rho)$$
$$= (p_0/g\rho_0) \log_e (1 + m/M), \text{ by (i)}.$$

16·7. Applications. We conclude the chapter with a brief review of common elementary practical devices and methods which are based on hydrostatic principles.

16·71. Density determination.

We have already indicated in § 16·321, Ex. (1), how a *common hydrometer* may be used to determine the density of a liquid. In § 16·534, Ex. (2), we referred to the practical question of the stability of the hydrometer when floating.

Another type, Nicholson's hydrometer, is always immersed so that a fixed point A of the stem is in the surface of the liquid. There is a tray above A and a cup below A, either of which can carry a solid. During any operation, sufficient known weights are added to the tray to bring the mark A to the liquid surface. Nicholson's hydrometer can be used to determine the specific gravity of a liquid or a solid.

A *specific-gravity bottle* is a specially made glass bottle with a perforated stopper, enabling it to be entirely filled with a liquid. It can obviously be used directly for comparing the weights of equal volumes of two liquids and hence their densities. It can also be used to determine the specific gravity of a solid in powdered form which can be made to occupy part of the space in the bottle.

If two liquids which do not mix are poured into the two arms of a *U-tube*, it is evident by (68b) that their densities are inversely proportional to the heights of the columns above the horizontal plane through the surface of separation.

A *hydrostatic balance* is an ordinary balance with a hook attached to one scale-pan enabling a solid suspended therefrom to be 'weighed' when entirely immersed in a liquid.

The following results, involving the use of a hydrostatic balance, are easily checked using the principle of Archimedes:

(i) If a (uniform) solid of weight W appears to be of weight w when immersed in water, its sp. gr. is $W/(W-w)$. (If w is to be positive, the solid must of course be denser than water.)

(ii) Let a solid, of weight W and less dense than water, be attached to a 'sinker'; let w' be the apparent weight of the sinker and solid in water, and w'' that of the sinker alone in water. Then the sp.gr. of the solid is $W/(W+w''-w')$.

(iii) If a solid of weight W has apparent weights of W_1, W_2 in two different liquids (each less dense than the solid), the ratio of the densities of the liquids is

$$(W-W_1)/(W-W_2).$$

The use of some of the above devices is further illustrated in examples set at the end of the chapter.

16·72. Barometers.

We have already referred (§16·24) to the essential principle of the mercurial barometer. In practice, many refinements are introduced with a view to determining atmospheric pressure to good precision. In the Ex. of § 16·77 below, we shall show how the presence of a small quantity of air above the main mercury column can affect the barometer reading.

Other barometers measure atmospheric pressure by other means. In the aneroid barometer, for instance, a thin corrugated metal container, partially exhausted of air, is deformed when the atmospheric pressure changes, and the deformations correspond to divisions on a scale calibrated by comparison with a mercurial barometer. The aneroid principle is involved in *altimeters*, used to determine changes in altitude (see § 16·66).

16·721.

Pressure gauges measure pressures other than atmospheric pressures. Their construction differs in practical details from the construction of barometers in so far as they are required to measure different ranges of pressure and are used in different practical conditions.

16·73. Balloons and submarines.

In § 16·67, Ex. (2), we have already indicated how to determine (on simplified conditions) the equilibrium height of a balloon which is 'lighter than air' at ground-level. We add that if the balloon is below its equilibrium height, then, neglecting air friction, it is subject to a 'lift' force equal to the excess of the weight of the displaced air over that of the balloon (including its appendages).

A submarine below sea-level is subject to a positive or negative 'lift' according as the excess of the weight of the displaced water over that of the submarine is positive or negative. This excess is controllable by pumping sea-water out of or into ballast tanks inside the submarine, so that the submarine can be made to ascend or descend.

16·74. Pumps.

An important feature in many pumps is a *valve* which closes when the pressure on one side, A say, exceeds that on the other side B, but opens when the pressure on the side B exceeds that on A by a sufficient margin. In the case of pumps this margin is ordinarily small; but in other devices, e.g. the safety-valve (see Ex. (3) of § 16·25), the margin may be substantial.

It is possible to transmit fluid between two regions in contact with a common valve by contriving to change the pressure on one side of the valve, e.g. with the use of a piston moving in a cylindrical barrel.

16·741.

In the ordinary *lift-pump* (village pump), water is raised from a well by decreasing the pressure above a valve A in a vertical pipe extending down below the surface S of the well water. The decrease in pressure takes place during the up-stroke of a piston above the valve; the valve A is closed during the down-stroke. A second valve B, in the piston itself, opens during the down-stroke and enables water to be passed into the pipe above the piston and thence to be lifted upwards to the outlet. It is evident from § 16·24 that the height above S of the lowest position of B must not exceed about 34 ft (in practice, the height is usually some feet less).

In a *force-pump*, the valve B is not present; instead, there is a side-spout joined at C say to a pipe at a place between the valve A and the lowest position of the piston, and there is a valve at the junction C. Water can be raised through A during an up-stroke of the piston and then forced out through C during the down-stroke. Additional devices are present in practice, e.g. a compressed air chamber in the delivery pipe enabling the water to be delivered continuously instead of intermittently.

16·742.

Gas pumps are used in practice either to evacuate a confined region of gas (so far as possible), or to compress a gas into a small volume.

In an evacuating air pump, a piston moves in a cylindrical barrel through a volume, V' say, during each stroke. A valve at the bottom of the barrel separates the barrel from the region R, of volume V say, to be evacuated, and opens and closes during an up-stroke and down-stroke respectively. A valve in the piston itself separates the air below the piston from that above and closes and opens during an up-stroke and down-stroke, respectively. It is evident that the density in R is multiplied by the factor $V/(V+V')$ with each pair of up-and-down-strokes. (In *Smeaton's* air pump there is an additional valve, opening outwards, at the top of the barrel and separating the air above the piston from the atmosphere; this valve enables the piston to be raised with less effort than would otherwise be needed.)

A compressing air pump differs from an evacuating pump in that the two main valves open in opposite directions to those in the last paragraph. The density is evidently increased by the amount $\rho_0 V'/V$ with each pair of up- and down-strokes, where ρ_0 is the density of atmospheric air.

16·743.

Other types of pumps, not necessarily involving valves, are based on hydrodynamical principles.

16·75. The siphon.

The siphon consists of a bent pipe, AB say, which has both ends below the plane S containing the surface of a liquid, one end A being inside the liquid and the other end B in the outside atmosphere. When the pipe is full of liquid and is open at A and B, liquid flows out through B, provided: (i) B is below the level of S; (ii) the height of the highest point of the pipe above S does not exceed the height of a column of the liquid that could be supported by the atmosphere.

To demonstrate this, suppose that there is a light smooth stopper C of smooth sides held so as to close the end B, the pipe being full of the liquid. C is then exposed to atmospheric pressure p_0 on the outside, and to a liquid pressure on the inside which exceeds p_0 since B is below S; hence for equilibrium C must be held by a further force directed into the pipe. If this force be removed, C and the liquid in contact with C must move out through the pipe.

Fig. 221

16·76. The diving bell.

This is a hollow metal vessel closed at the top and open at the bottom, its weight, W say, being greater than the weight of water it would displace when its interior is full of air. The internal volume, V say, is large enough to contain a man who will in practice be engaged in some operation below sea-level.

As the bell is lowered into the sea, the volume of air in the bell is diminished under the increasing pressure. The reader may be interested to show that for an unoccupied cylindrical bell of length l:

(i) If the bell is initially full of air at atmospheric pressure, and no air is pumped in, then the length x of the bell occupied by air when the depth of the top of the bell is y, is the positive root of the equation

$$x^2 + x(y + h) - lh = 0,$$

where h is the height of the water barometer.

(ii) The volume of air at atmospheric pressure needing to be pumped into the bell in order to force the encroaching water down to the bottom of the bell is $V(l + y)/h$.

16·77. Example. *A mercury barometer has an imperfect vacuum above the mercury column. When the readings are h_1, h_2, the readings of an accurate barometer would be $h_1 + \epsilon_1$, $h_2 + \epsilon_2$, respectively. Find the correction to be applied to any reading h, assuming the tube to be cylindrical above the mercury column and neglecting changes in the lower level.*

When the reading is h, let the reading of an accurate barometer be $h + \epsilon$; let BC be the mercury column in these circumstances. Let A be the top of the tube, and let $AC = l$, taken constant. (Fig. 221.)

The pressure at B is equal to the pressure at C less that due to a column of mercury of height h; i.e., is proportional to $h + \epsilon - h$, i.e. to ϵ.

The volume of the gas in AB is proportional to AB, i.e. to $l - h$.

Hence, assuming the temperature always to be the same, we have

$$\epsilon(l - h) = \text{constant.}$$

Hence $\quad \epsilon(l - h) = \epsilon_1(l - h_1) = \epsilon_2(l - h_2).$

Eliminating l, we obtain

$$\epsilon = \frac{\epsilon_1 \epsilon_2 (h_1 - h_2)}{\epsilon_1(h_1 - h) - \epsilon_2(h_2 - h)},$$

the required correction to be added to h.

EXAMPLES XVI

1. A uniform fluid mixture is formed by mixing homogeneous liquids of densities $\rho_1, \rho_2, \ldots$, which interact chemically so as to cause a reduction of $x\%$ in the resulting volume. Find the density of the mixture given that: (*a*) the volumes of the original liquids are $V_1, V_2, \ldots$; (*b*) their weights are $W_1, W_2, \ldots$

[Answers: (*a*) $100(V_1\rho_1 + V_2\rho_2 + \ldots)/\{(100 - x)(V_1 + V_2 + \ldots)\}$;
(*b*) $100(W_1 + W_2 + \ldots)/\{(100 - x)(W_1/\rho_1 + W_2/\rho_2 + \ldots)\}$.]

2. A uniform mixture formed of equal masses of three non-interacting homogeneous liquids is divided into three portions. To each portion is then added its own mass of one of the three given liquids; and the densities of the resulting mixtures are as $3 : 4 : 5$. Find the ratio of the densities of the three given liquids.

[Answer: $73^{-1} : 43^{-1} : 25^{-1}$.]

3. If the pressure 44 ft below the surface of a lake is twice the pressure 5 ft below the surface, find the atmospheric pressure. [Answer: 14·8 lb.wt./in².]

4.† A cistern with vertical sides, and base in the form of an equilateral triangle of side-length 6 ft is filled with water to a depth of 4 ft, and a board of mass 90 lb is then floated in it. Find the pressure-intensity at the base. [Answer: 256 lb.wt./ft².]

5. The arms AB, BC of an open, thin uniform V-shaped tube meet at B at a fixed angle of $60°$, and contain two homogeneous liquids that do not mix. When AB is vertical, the common surface of the liquids is at B and the lengths occupied in AB, BC are 24, 30 cm, respectively. Prove that if the tube is tilted until BC is vertical, the occupied length of BC becomes 15 cm.

† Signifies (in Exs. 1–27) that the influence of the atmosphere is to be ignored.

*6. Treating the Earth as a homogeneous self-gravitating liquid sphere of radius $6\cdot4\times10^8$ cm and density $5\cdot5$ g/cm^3, prove that the pressure at the centre would be about $1\cdot7\times10^{12}$ dyn/cm^2.

(Notes.—(i) Use the result that at an internal point distant r from the centre of a homogeneous sphere the gravitational attraction is proportional to r. (ii) The student may be interested to know that in the actual Earth the pressure at the centre is about $3\cdot9\times10^{12}$ dyn/cm^2.)

7. Atmospheric air is above the surface of liquid in a vessel in which a cork is floating. If the air is pumped out, does the cork move up or down ? [Answer: Down.]

A vessel containing a piece of iron is floating in a bath of water. If the iron is removed from the vessel and placed in the water, does the level of the water in the bath change, and, if so, does it rise or fall ? [Answer: Falls.]

8. Hiero bade his goldsmith convert a mass M of gold into a gold crown. Archimedes found that the crown was of the correct mass M, but that the crown and masses M of gold and silver displaced different volumes V, V', V'' of water, respectively. Assuming the crown to contain no other metals than gold and silver, show that the goldsmith cheated Hiero; and find by how much, taking P, Q as the prices per unit mass of gold and silver, respectively. [Answer: $M(P-Q)(V-V')/(V''-V')$.]

9. A uniform cylinder of wood, of density σ, is fixed on the top of a block of ice, and the system is floating in sea-water of density ρ with a length l of the cylinder, whose axis always remains vertical, exposed above the water level. When half the block of ice has melted, a length l' of the cylinder is exposed. Find the total length of the cylinder. [Answer: $(2l'-l)\rho/(\rho-\sigma)$.]

10. A cylindrical piece of metal whose axis AB is 5 cm long floats in a large vessel of mercury with AB vertical and half immersed. Water is now poured on top of the mercury until the whole cylinder is covered. How far does the cylinder rise ? [Answer: 0·198 cm.]

11. A uniform solid cylinder floats with $\frac{2}{3}$ of its axis (which is vertical) immersed in one homogeneous liquid, and with 4/5 immersed when in another homogeneous liquid. What fraction of the axis would be immersed when the cylinder floats in a uniform mixture of equal volumes of the two liquids (assuming there is no interaction) ? [Answer: 8/11.]

12. A rectangular block is held completely immersed in water with four edges horizontal, and two faces inclined at θ to the horizontal. Prove that the resultant of the pressures on these two faces is of magnitude $W\cos\theta$, where W is the weight of water displaced by the block, and acts through the centre of the block.

13. A 2-cwt spherical mine of diameter 2 ft is submerged below sea-level and is kept in position by being attached by a chain of length 10 fathoms and mass 20 lb to a sinker of mass 50 lb which rests on the ocean floor at a depth of 12 fathoms. (Take 1 fathom = 6 ft.) The specific gravity of both chain and sinker is 7. Find the tensions at the top and bottom of the chain and the force between the sinker and the ocean floor. [Answers: 38, 21, 22 lb.wt., approx.]

14. A bucket containing water, the total weight being W, is attached at A to a light string AB which passes over a smooth fixed pulley small enough to let the other end B fall vertically into the bucket. To the end B is attached a uniform ball, and the system is in equilibrium with the ball entirely immersed and not touching the bucket. Prove that the specific gravity s of the ball exceeds 2, and that its weight is $Ws/(s-2)$.

*15. A circular cylinder, symmetrical about its axis, has its mass M assigned, but its radius and height are not assigned. It is required to float with its axis vertical in a homogeneous liquid of density ρ with the least possible surface area in contact with the liquid. Prove that its radius r must satisfy the equation $\pi r^3 \rho = M$.

16.† A dam of a reservoir is 200 yards wide, and its face towards the water is rectangular and inclined at 30° to the horizontal. Show that the thrust on the dam due to the water when 60 ft deep is a little more than 60,000 tons wt.

17. It is desired to test whether the vertical wall of a closed cubical tank of side 4 ft can withstand a fluid thrust of 10 tons wt. The test is made by attaching a vertical tube to an orifice at the top of the tank and pouring water in through the tube. How high should the tube be? [Answer: 20·4 ft.]

18.† Find the depths of the centres of pressure of: (a) a vertical isosceles triangular area ABC immersed in water with BC horizontal at twice the depth h of A; (b) a vertical square area $ABCD$, of side $2a$, immersed in water with A in the surface and BD horizontal. [Answers: (a) $1\cdot 7h$; (b) $7a\sqrt{2}/6$.]

19.† A rectangular lamina $ABCD$, where $AB = 2a$, $AD = 2b$, is immersed in a homogeneous liquid with AB in the surface. Find the distance between the centres of pressure of the triangular areas ABD, DBC. [Answer: $\tfrac{1}{4}\sqrt{(9a^2 + 4b^2)}$.]

20.† A square plate of side-length a is lowered slowly with two edges vertical, at uniform speed V into a homogeneous liquid. Prove that the centre of pressure moves with speed $2V/3$ for time a/V, and then approaches the centre O of the plate with a speed inversely proportional to the square of the depth of O.

21. A vertical rectangular dockgate $ABCD$, where $AB = 20$ ft, $BC = 15$ ft, can move smoothly about its upper horizontal edge AB, and is then fastened at the midpoint of the lower edge CD. Find the force on the fastener when the water levels on the two sides of the gate are 15, 3 ft, respectively, above CD. [Answer: 39·5 tons wt.]

22†. A vertical rectangle $ABCD$, where $AB = 2a$, $BC = 2b$, is attached to a semi-circular disc of diameter $2a$ so that AD, BC fit tangentially and rigidly at D, C on to the semicircular boundary of the disc. The whole area is vertically immersed in a homogeneous liquid with AB in the surface. Find the depth of the centre of pressure. [Answer: $\{128b^3 + \pi(3a^3 + 48ab^2)\}/\{8(2a^2 + 12b^2 + 3\pi ab)\}$.]

23.† A circular piece of centre B of radius b is removed from a circular disc of centre A and radius a, where $AB = c$ and $b + c \leqslant a$. The remaining piece is immersed in water with AB vertical. If the centre of pressure is at A, prove that the depth of A is $(a^4 - 4b^2c^2 - b^4)/4b^2c$.

24. A hollow sphere is full of water of weight W, and has a tiny hole at the top. Find the magnitudes of the resultant fluid thrusts on the upper and lower hemispherical surfaces cut off by a horizontal plane through the centre O. [Answer: $W/4, 5W/4$.]

Prove also that the resultant thrust on either of the hemispherical surfaces cut off by a vertical plane through O makes an angle $\tan^{-1}(3/2)$ with the vertical.

25.† A rectangle $ABCD$ is immersed so that AB is in the upper surface of a homogeneous liquid L which rests on a second homogeneous liquid L' with which L does not mix. The depth h of CD below AB is twice that of the surface of separation of L, L'; and the density of L' is twice that of L. Find the depth of the centre of pressure on the rectangle. [Answer: $0.7h$.]

*26. A wall of a tank slopes inwards from the bottom at an angle θ to the vertical, and contains a triangular trap-door ABC of mass M, hinged so that it can open outwards smoothly about the side BC which is horizontal and above the level of A. The trap-door may be treated as a uniform lamina, and the heights of A, B, C above the level of the floor of the tank are a, b, b. Water is poured slowly into the tank. Prove that the trap-door will remain closed, when wholly below the water surface level, provided the depth of the water in the tank is less than $\frac{1}{2}(a + b) + (M/\rho\varDelta) \sin \theta$, where $\varDelta$ is the area of ABC, and ρ is the density of the water.

*27.† A circular area of radius a is just immersed vertically in a heterogeneous liquid whose density at depth z below the surface is kz. Find the magnitude of the thrust on the area and the depth of the centre of pressure. [Answers: $5\pi a^4 kg/8$; $7a/5$.]

28. The length, weight and specific gravity of a uniform thin rod are l, W and s. To an end of the rod is attached a light vertical string by means of which the rod is slowly raised to the vertical from an initial position in which the rod was floating horizontally in the water. Prove that the tension of the string and the length of the rod immersed remain constant during the motion.

Also prove that the work done during the raising of the rod is

$$Wl(1 - s)/\{1 + \sqrt{(1 - s)}\}.$$

29. A uniform rod of length l and specific gravity s, where $s < 1$, can turn smoothly about its upper end which is fixed at a height h (where $h < l$) above the surface of water. If $h/l < \sqrt{(1 - s)}$, prove that the vertical position is unstable, and find the inclination of the rod to the vertical in the stable equilibrium position.

[Answer: $\cos^{-1}\{h/l\sqrt{(1 - s)}\}$.]

30. A right circular uniform solid cone of semi-vertical angle $30°$ floats stably in water with its vertex A uppermost. Find the greatest value k of its specific gravity.

[Answer: 0.578.]

Prove that if the specific gravity has the value k, the cone will float stably with A lowermost.

31. Show that a uniform solid hemisphere whose specific gravity is 0.5 would be in equilibrium, but not stable equilibrium, floating in water with its plane face vertical.

*32. A uniform rectangular block floats in a homogeneous liquid with its upper face $ABCD$ horizontal; $AB = a$, $AD = b$, where $a \leqslant b$, and the vertical dimension is c. Prove that the equilibrium is stable, whatever the densities of the block and the liquid, provided $c < a\sqrt{(2/3)}$.

*33. A cylindrical tank of circular section (radius a) is smoothly pivoted about a fixed horizontal axis passing through its centre of mass which is at height b above the base. Prove that if the tank contains a depth h of water, the upright equilibrium configuration will be stable only if $b > (a^2 + 2h^2)/4h$.

Deduce that if $b < a/\sqrt{2}$, there is instability for all depths.

34. The volume of an inflated tyre is 144 in^3 and that of the barrel of a pump inflating it is 24 in^3. If the valve (assumed light and frictionless) in the tyre opens when a particular downward stroke of the pump is two-thirds completed, find in atmospheres the pressure in the tyre just before and just after the stroke, assuming Boyle's law to hold.
[Answers: 3, 3·17 atm.]

35. A cylinder contains gas at 20 atmospheres pressure. The nozzle is opened, and and it is found that after 40 min the pressure is 11·45 atm. Assuming that the rate of escape varies as the difference between the inside and the outside pressures, and that Boyle's law is relevant, prove that the pressure is about 10 atm after a further 10 min.

36. Masses m, m' of two ideal gases in which the ratios of the pressure to the density are k, k' are mixed at the same temperature. Prove that the ratio of the pressure to the density in the mixture is $(mk + m'k')/(m + m')$.

37. A lift pump has a vertical pipe AB of cross section 3 sq. in. through which water can pass from a water supply whose level is 14 ft below the top end B. At B, there is a valve separating the pipe from the barrel, of cross section 5 sq. in., of the pump. The piston moves inside the barrel from B through a height of 1 ft during each upward stroke. Assuming ideal conditions, find the distance the water rises inside AB during the first upward stroke.
[Answer: About 14 in.]

38. A common hydrometer floats in water with 2·2 cm unimmersed, and in alcohol of specific gravity 0·85 with 0·8 cm unimmersed. Find the specific gravity of a liquid in which it floats with 3·6 cm unimmersed.
[Answer: 1·21.]

39. The specific gravity of a body is found using a hydrostatic balance, the effect of the surrounding air being ignored. Prove that the calculated result s is too great by $\sigma(s - 1)$, where σ is the specific gravity of air (referred to water).

40. A cylindrical diving bell of length 10 ft is lowered into the sea until its top is at a depth of 15 ft. The temperature of the air just above the sea is 97°F, and the height of the water barometer is 32·5 ft. The temperature of the sea-water surrounding the bell is 60°F. Find the depth of the water in the bell when the level has become steady.
[Answer: 4 ft 4 in.]

41. A thin uniform glass tube is bent so as to have four vertical portions with the outer pair open upwards. The tube contains mercury so that the surfaces in the four portions are at the same level, the air enclosed inside being at atmospheric pressure and occupying a length l of the tube. When mercury sufficient to occupy a length x of the tube is now slowly added to one of the outer arms, the level of mercury in the other outer arm is raised through a distance y. If h is the height of the mercury barometer, prove that $x = 4y(h + l + 2y)/(h + 2y)$.

42. A thin uniform U-tube has its arms vertical, one end being open and the other closed. The tube contains mercury to the same level in both arms, and in the remaining length l of the closed arm there is air at atmospheric pressure and temperature $T°C$. The air in the closed arm is now cooled to the freezing-point of water. Prove that the mercury in the open limb falls a distance x, given by
$$2x^2 - (2l + h)x + Thl/(T+ 273) = 0,$$
where h is the height of the mercury barometer.

43. The height of a mercury barometer is 30 in at ground level, and 29 in in an aeroplane above. Treating the atmosphere as an isothermal ideal gas, estimate the height of the aeroplane correct to the nearest 50 ft. [Answer: 900 ft.]

*44.†† Assuming the atmosphere to be an isothermal ideal gas, and that g varies (outside the Earth) inversely as the square of the distance from the Earth's centre, prove that the pressure at height z above the ground is $p_0 e^{-Rz/H(R+z)}$, where p_0 is the pressure at ground-level, H is the height of the homogeneous atmosphere, and R is the Earth's radius.

*45. If the atmosphere were an isothermal ideal gas, prove that it would extend to infinity.

If the atmosphere were an ideal gas in an adiabatic state, with the constant γ (of § 16·65) equal to 1·4, prove that the atmosphere would on the other hand have a finite height equal to $7H/2$, where H is the height of the 'homogeneous atmosphere'.

*46. Treating the atmosphere as an ideal gas in which the absolute temperature diminishes uniformly with respect to height, from 280° at sea-level to 210° at a height of 12 km, prove that if the height of the mercury barometer is 76 cm at sea-level, it is 14·2 cm at the height of 12 km.

*47. Prove that for a given mass of given ideal gas in adiabatic conditions: (a) $T/\rho^{\gamma-1}$ is constant; (b) the work done by boundary forces during a quasi-statical volume change is $MK(T_1 - T_0)/(\gamma - 1)$, where the notation is that of §§ 16·62, 16·65, T_0 and T_1 being the initial and final temperatures.

*48. Prove that the incompressibility of an ideal gas is equal to p, γp, in isothermal, adiabatic conditions, respectively, where p is the pressure and γ is as in § 16·65.

49. The period T of torsional vibrations of a body suspended by a wire is assumed to depend only on the radius a, length l and rigidity μ of the wire and on the moment of inertia I of the body about the axis of the wire. Given that the dimensions of μ are as for stress, prove by dimensional considerations that $T = \sqrt{(I/\mu)} a^{-3/2} f(a/l)$, where $f(a/l)$ is some function of a/l.

50. A thin uniform fixed U-tube, with vertical arms open at the tops, contains mercury, the total length of the mercury in the tube being $2h$. If the mercury is now slightly disturbed longitudinally from the equilibrium configuration, prove that, dissipative influences being neglected, the mercury will oscillate along the tube with period $2\pi\sqrt{(h/g)}$.

†† Variation in g is to be taken into account only in Ex. 44 on this page.

CHAPTER XVII

APPENDIX

"Better is the end of a thing than the beginning thereof."
—*Ecclesiastes.*

The following notes supplement the earlier theory and provide information on some topics of special current interest.

17·1. Vector product. Let **u**, **v** be two vectors whose magnitudes are u, v, and whose directions are inclined at an angle α.

The vector product of **u** and **v**, denoted by **u**×**v** (sometimes by **u**∧**u**, or [**uv**]), is defined as the (free) vector whose magnitude is $uv \sin \alpha$, whose direction is perpendicular to both **u** and **v**, and whose sense is given by the right-hand screw law.

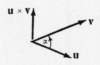

Fig. 222

It can be shown that vector products as thus defined conform to the addition rule of vectors.

In referring to the 'right-hand screw law', we mean that a right-handed screw turning through the angle α in the sense indicated in Fig. 222 would move forward in the sense of the vector **u**×**v**.

We remark here that wherever a reference frame $OXYZ$ is mentioned in the following pages, we understand the frame to be orthogonal and *right-handed*. I.e., a right-handed screw rotating about OZ in the sense from OX towards OY would move forward in the direction OZ. (To an observer looking in the direction ZO, the positive sense of rotation in the plane OXY is thus anticlockwise.)

17·11. Elementary properties of vector products are as follows. Let **u** be any vector, of components u_1, u_2, u_3, parallel to the axes OX, OY, OZ; and similarly with **v**, **w**; let **i**, **j**, **k** denote unit vectors parallel to OX, OY, OZ. Then it can be shown that:

(i) If **u**×**v** = 0, then $u = 0$ or $v = 0$ or **u**, **v** are parallel.
(ii) **u**×**v** = −**v**×**u**.
(iii) **u**×**u** = 0.
(iv) **i**×**i** = 0; ... ; ... ; **j**×**k** = **i** = −**k**×**j**; ... ; ....

(v) $\mathbf{u} \times (\rho \mathbf{v}) = (\rho \mathbf{u}) \times \mathbf{v}$, where ρ is any scalar.
(vi) $\mathbf{u} \times (\mathbf{v} + \mathbf{w}) = \mathbf{u} \times \mathbf{v} + \mathbf{u} \times \mathbf{w}$.
(vii) The components of $\mathbf{u} \times \mathbf{v}$, referred to OX, OY, OZ, are
$$u_2 v_3 - u_3 v_2, \quad u_3 v_1 - u_1 v_3, \quad u_1 v_2 - u_2 v_1.$$
(viii) $\dfrac{d}{dt}(\mathbf{u} \times \mathbf{v}) = \dfrac{d\mathbf{u}}{dt} \times \mathbf{v} + \mathbf{u} \times \dfrac{d\mathbf{v}}{dt}$.

The proofs of (i) to (v) can easily be carried out by the reader; the proof of (vi) involves a geometrical construction which we shall not give; (vii) is proved by writing $\mathbf{u} = u_1 \mathbf{i} + u_2 \mathbf{j} + u_3 \mathbf{k}$, and using (iv), (v), (vi); (viii) is proved in the same way as (xiv) of § 6·171.

17·12. The notion of vector product is specially useful in mechanics wherever moments are involved. (See §§ 12·11, 12·46.)

17·13.
It will be noticed that vector products obey certain rules, e.g. (v), (vi), analogous to rules in elementary algebra; but that the analogy breaks down in the case of certain other formulae, e.g. (ii), (iii).

17·14.
The student may be interested to prove the following important results involving vector products:

(a) $(\mathbf{u} \times \mathbf{v}) \cdot \mathbf{w} = \mathbf{u} \cdot (\mathbf{v} \times \mathbf{w}) = \begin{vmatrix} u_1 & u_2 & u_3 \\ v_1 & v_2 & v_3 \\ w_1 & w_2 & w_3 \end{vmatrix}$.

(b) $\mathbf{u} \times (\mathbf{v} \times \mathbf{w}) = (\mathbf{u} \cdot \mathbf{w}) \mathbf{v} - (\mathbf{u} \cdot \mathbf{v}) \mathbf{w}$.

17·2. Angular velocity of a rigid body. In § 13·22, the angular velocity of a body was considered two-dimensionally. The three-dimensional case is more complicated and requires a vector treatment. This will now be given in outline.

17·21. Consider first the case of a rigid body moving about a fixed point O, and let $OXYZ$ be a reference frame. (All reference frames are taken to be rigid.) To begin with, $OXYZ$ will be treated as at rest.

At times t, $t + \Delta t$, let C, C' denote the configurations of the body. It can be shown that the net change from C to C' is equivalent to a pure rotation, through an angle $\Delta \theta$ say, about some axis through O, say $K'OK$. (In this result Δt and $\Delta \theta$ are not necessarily small.)

Now let $\Delta t \to 0$. The limiting position, $I'OI$ say, of $K'OK$ is called the *instantaneous axis* of motion at time t. We shall take I to be on that side of O which would make the rotation appear clockwise to an observer looking in the direction OI.

The angular velocity, Ω say, of the body at time t is then defined as the (free) vector whose magnitude is

$$\lim_{\Delta t \to 0} \left(\frac{\Delta \theta}{\Delta t}\right),$$

and whose direction is that of OI.

The convention determining the sense of Ω is seen to correspond to the right-hand screw law (see § 17·1), the screw pointing in the direction OI.

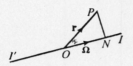

Fig. 223

17·22. Let P be any point of the body of § 17·21, and let $\mathbf{r}$, $\mathbf{v}$ be the position vector and velocity of P at time t. Then

$$\mathbf{v} = \Omega \times \mathbf{r}. \tag{75}$$

To prove (75), let v, Ω be the magnitudes of $\mathbf{v}$, Ω; let PN be drawn perpendicular to OI, and let $\angle IOP = \alpha$. Then, by § 17·1, the magnitude of $\Omega \times \mathbf{r} = \Omega.OP.\sin \alpha = \Omega.PN = v$. Further, the direction of $\Omega \times \mathbf{r}$ is at right angles to and up from the plane IOP in Fig. 223; and so also is the direction of $\mathbf{v}$, taking into account the sign convention stated in § 17·21.

17·23. Consider next a rigid body in general motion, and let O be any point of the body; (in general, O will be in motion). Let $OXYZ$ now be any reference frame with O as origin.

Then all the definitions and results in §§ 17·21, 17·22 continue to hold, provided we understand the words 'relative to the frame $OXYZ$' to hold throughout.

17·24. Let us now stipulate that (relative to some background frame) the directions of the axes OX, OY, OZ of the frame $OXYZ$ in § 17·23 remain fixed (although O may still be in motion). It can then be shown that the angular velocity Ω of the body relative to $OXYZ$ is independent of the particular point O taken in the body, and independent of the particular directions selected for OX, OY, OZ.

Hence we can speak unambiguously of the angular velocity Ω at time t of a rigid body in general three-dimensional motion, relative to a frame, without needing to specify the axes of the frame.

It is necessary to remark finally that angular velocities as described above can be shown to obey the addition rule of vectors.

17·3. Rotating frames and time-differentiation.

Up to this point, frames of reference have been generally assumed to be either at rest, or else to have their origins in relative motion but not the directions of their axes. (See § 6·24, p. 102.)

Now let F (or $OX_0Y_0Z_0$) and F^* (or $OX^*Y^*Z^*$) be two frames in relative rotation about a common origin O. Let $\mathbf{\Omega}$ be the angular velocity of F^* relative to F. In Fig. 224 (which is not necessarily coplanar), OX_1^* and OX_2^* show the positions of OX^* relative to OX_0 at times t and $t + \delta t$.

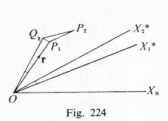

Fig. 224

Let P_1 and P_2 be the positions of a moving point P at these times, and let $\mathbf{r}$ and $\mathbf{r} + \delta\mathbf{r}$ be the corresponding position vectors relative to F. Then the displacement of P relative to F during the time-interval δt is $\mathbf{P_1P_2}$, which we shall denote as $\delta\mathbf{r}$.

Let Q be that point rigidly attached to the frame F^* which coincides with P at time t, and let Q_2 be its position relative to F at time $t + \delta t$. Then the displacement of P relative to F^* is $\mathbf{Q_2P_2}$, which we shall denote as $\delta^*\mathbf{r}$.

By Fig. 224, $$\mathbf{P_1P_2} = \mathbf{Q_2P_2} + \mathbf{P_1Q_2}. \qquad (i)$$

But $\mathbf{P_1Q_2} = \mathbf{v}\delta t$ (to sufficient accuracy), where $\mathbf{v}$ is the velocity of Q relative to F and is, by (75), equal to $\mathbf{\Omega} \times \mathbf{r}$. Hence (i) gives
$$\delta\mathbf{r} = \delta^*\mathbf{r} + \mathbf{\Omega} \times \mathbf{r}\, \delta t.$$

Dividing by δt and letting $\delta t \to 0$, we have
$$\frac{d\mathbf{r}}{dt} = \frac{d^*\mathbf{r}}{dt} + \mathbf{\Omega} \times \mathbf{r}. \qquad (76)$$

Since any free vector, $\mathbf{u}$ say, has the same geometrical properties as a position vector, we have the following important connection between the rate of change ($d\mathbf{u}/dt$) of $\mathbf{u}$ as measured relatively to F and the rate of change ($d^*\mathbf{u}/dt$) measured relatively to F^*:
$$\frac{d\mathbf{u}}{dt} = \frac{d^*\mathbf{u}}{dt} + \mathbf{\Omega} \times \mathbf{u}. \qquad (77)$$

---

* The asterisk on pages 346-349 is a mathematical symbol.

17·31.

Suppose that, in addition to having an angular velocity Ω, the frame F^* has a motion of 'translation', its origin having velocity v_0 (relative to F) at time t. Then, if u is any free vector, it can be shown that (76) still holds, the star still indicating the time-rate of change relative to F^*.

17·32.

The case of a *localised* vector is more complicated. Let its magnitude and direction be given by u, and let its moment about the origin be M. Then in the circumstances of § 17·31, it can be shown that (76) still holds, but that

$$\frac{dM}{dt} = \frac{d^*M}{dt} + \Omega \times M + v_0 \times u.$$

17·4. Centrifugal force, etc.

The results of § 17·3 will now be applied in a dynamical context. The frame F will now be taken to be absolutely at rest, and $OX^*Y^*Z^*$ will be denoted simply as $OXYZ$. We shall take the special case in which $OXYZ$ rotates about OZ, which is taken fixed, with angular velocity of magnitude Ω.

Let P be a particle moving in the plane OXY. Let v^* and f^* be its velocity and acceleration relative to $OXYZ$, and v and f its absolute velocity and acceleration. We proceed to express the equations of motion in a form involving v^* and f^*.

(To fix ideas, consider a small object P moving on the rotating floor, F^* say, of a merry-go-round, and suppose that the only observations of the motion of P are made relative to F^*. The problem is to form equations of motion for P in terms of these observations, the forces on P, and the angular velocity of F^*.)

17·41.

Let r be the position vector of P, and Ω the (vector) angular velocity of $OXYZ$ at time t. We first express v and f in terms of r, v^*, f^* and Ω.

By (76), the result for v is immediately given as

$$v = v^* + \Omega \times r, \qquad (78)$$

where v is the absolute velocity of P, and Ω is the (vector) angular velocity of $OXYZ$.

Differentiating (78) with respect to t, and using (viii) of § 17·11, we have

$$\frac{dv}{dt} = \frac{dv^*}{dt} + \Omega \times \frac{dr}{dt} + \frac{d\Omega}{dt} \times r;$$

i.e., using (76) and (77),

$$f = \frac{d^*v^*}{dt} + \Omega \times v^* + \Omega \times \left(\frac{d^*r}{dt} + \Omega \times r\right) + \frac{d\Omega}{dt} \times r$$

$$= f^* + 2\Omega \times v^* + \Omega \times (\Omega \times r) + \frac{d\Omega}{dt} \times r, \qquad (79)$$

since $v^* = d^*r/dt$ and $f^* = d^*v^*/dt$.

As a particular case of § 17·14 (b), we have
$$\Omega \times (\Omega \times r) = (\Omega \cdot r)\Omega - (\Omega \cdot \Omega)r.$$
So far, all the equations in this section apply three-dimensionally. Now introducing the two-dimensional restriction we are taking, we have $\Omega \cdot r = 0$ since Ω and r are perpendicular. Hence (79) becomes
$$f = f^* + 2\Omega \times v^* - \Omega^2 r + \frac{d\Omega}{dt} \times r, \qquad (80)$$
which is the desired expression for f.

(If, in addition to having an angular velocity, $OXYZ$ has a motion of 'translation', the origin O having velocity v_0 and acceleration f_0, say, at time t, then it may be shown that the terms v_0 and f_0 need to be added to the right sides of (78), (80), respectively, to give the absolute velocity and acceleration of P.)

17·42. Let R be the resultant force on P. Then, by the fundamental equation (26a), and (80),
$$R = m\left(f^* + 2\Omega \times v^* - \Omega^2 r + \frac{d\Omega}{dt} \times r\right);$$
i.e.,
$$mf^* = R + m\Omega^2 r - 2m\Omega \times v^* - m\frac{d\Omega}{dt} \times r. \qquad (81)$$

The equation (81) is interpreted thus: The mass-acceleration of P *as estimated by reference to the rotating frame* is equal to the resultant of:

(a) the resultant R of the actual forces;
(b) the *centrifugal force* $m\Omega^2 r$;
(c) the *Coriolis force* $-2m\Omega \times v^*$;
(d) $mr \times (d\Omega/dt)$.

The centrifugal force is in the direction OP, i.e. radially outwards from O. The Coriolis force is in the plane OXY and is perpendicular to the velocity v^*.

It should be noted that the centrifugal force, the Coriolis force and (d) are not actual forces as hitherto understood, but are effects contributing to the mass-acceleration *relative* to the rotating frame.

17·43. Particular cases.
(i) If the angular velocity Ω of the rotating frame OXY is constant, the term (d) is zero. If, further, P is at rest relative to OXY, then $v^* = 0$, and the Coriolis force is also zero. These circumstances apply in *uniform* circular motion of a particle P with angular velocity Ω about a centre O. If OX is taken as a rotating axis always containing P, and if $OP = a$, where a is constant, then P is in 'relative equilibrium', relative to OX, under the 'centrifugal force' and the actual forces present.

(ii) Consider a particle P of mass m inside a thin smooth straight tube OA rotating about O in a horizontal plane with uniform angular speed Ω, the only (actual) forces on the particle being its weight and the force exerted by the tube. Take OX as a rotating axis always coinciding with the axis of the tube, and let $OP = r$. The term (d) is zero. Also the components in the direction OX of the actual forces and of the Coriolis force are zero. Hence, by (81), the motion of P relative to the tube is determined by the centrifugal force $m\Omega^2 r$, radially outwards.

(iii) An interesting experience of the Coriolis force is that of an airman who moves his head forward when 'looping the loop'. The moving aircraft being his frame of reference, he appears to experience a force on his head in a direction towards or away from the floor according as he loops the loop with his head pointing inwards or outwards.

(iv) It is of interest to re-work Ex. (5) of § 7·6 (p. 132), taking a frame of reference rotating with the Earth. (This example is concerned with the variation of g over the surface of an assumed spherically symmetrical Earth.)

Let P be a particle of mass m lying on the Earth's surface at a point of latitude ϕ. Take a frame of reference fixed in the Earth, rotating about the polar axis NS (Fig. 71) with the angular velocity Ω of the Earth. Relative to this frame, P is in equilibrium under (a) the gravitational force mg_0 to the centre of the Earth, (b) the supporting force of the ground, (c) the centrifugal force $m(R\Omega^2 \cos \phi)$ acting in the direction MP (Fig. 71). The force (b) is equal and opposite to the weight of the particle, and is of magnitude mg and direction opposite to the arrow marked Pg in Fig. 71. The principles of statics applied to these three forces yield the equations (i) and (ii) of p. 132.

17·44. Cartesian form of equations.

The cartesian forms of (78), (80) and (81) may be deduced as follows.

Let $(x, y, 0)$ be the coordinates of P, referred always to the rotating frame $OXYZ$; then $\mathbf{r} = (x, y, 0)$, $\mathbf{v}^* = (\dot{x}, \dot{y}, 0)$ and $\mathbf{f}^* = (\ddot{x}, \ddot{y}, 0)$. Let $\mathbf{R} = (\Sigma X, \Sigma Y, 0)$. Also, $\mathbf{\Omega} = (0, 0, \Omega)$.

Using (vii) of § 17·11, we may then deduce from (78) and (80) that the components parallel to OX, OY of the (absolute) velocity and acceleration are

$$\dot{x} - \Omega y, \quad \dot{y} + \Omega x; \tag{78a}$$
$$\ddot{x} - 2\Omega \dot{y} - \Omega^2 x - \dot{\Omega} y, \quad \ddot{y} + 2\Omega \dot{x} - \Omega^2 y + \dot{\Omega} x; \tag{80a}$$

respectively.

Similarly, we deduce

$$\left. \begin{array}{l} m\ddot{x} = \Sigma X + m\Omega^2 x + 2m\Omega \dot{y} + m\dot{\Omega} y, \\ m\ddot{y} = \Sigma Y + m\Omega^2 y - 2m\Omega \dot{x} - m\dot{\Omega} x \end{array} \right\} \tag{81a}$$

as the cartesian equations corresponding to (81).

Note:—In the foregoing discussion, no discrimination was made between the position vectors of P at time t relative to F and F^* (or $OXYZ$); the same $\mathbf{r}$ was used for both. The necessity for discrimination arises in considering *changes* of position relative to the two frames, as the time changes.

17·45. Radial and transverse components of velocity and acceleration.
Let P be a point moving in a plane, and at time t, let its polar coordinates be (r, θ) referred to an origin O and reference line OA (Fig. 225) fixed in the plane. The radial and transverse components of the velocity and acceleration of P in the plane, i.e. the components along and at right angles to OP (in the direction of θ increasing) can be derived using rotating axes.

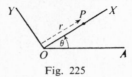

Fig. 225

Let OXY be a frame in the plane of motion, rotating so that P is always on OX. The angular velocity Ω of the frame is then $\dot{\theta}$.

Relative to OXY, the cartesian coordinates (x, y) of P are $(r, 0)$. Hence, by (78a), the components of the velocity of P in the directions OX, OY, are $\dot{r}$, $r\dot{\theta}$; and by (80a), the components of acceleration are $\ddot{r} - r\dot{\theta}^2$, $2\dot{r}\dot{\theta} + r\ddot{\theta}$. These are the required radial and transverse components.

For many purposes, it is convenient to write the transverse component of acceleration as $r^{-1}d(r^2\dot{\theta})/dt$, which is easily seen to be equal to $2\dot{r}\dot{\theta} + r\ddot{\theta}$.

The reader may care to verify, by putting $r = a$, where a is constant, that the formulae include as particular cases the formulae (37) and (39) (page 127) on circular motion.

17·5. Newtonian principle of relativity. Let F_0 be the frame of reference assumed on p. 28 to be absolutely at rest. Let F be a frame which has a motion of translation, but not rotation, relative to F_0, the acceleration of the origin of F being $\mathbf{f}'$. Then the (absolute) acceleration of any particle P is equal to $\mathbf{f}' + \mathbf{f}$, where $\mathbf{f}$ here denotes the acceleration of P relative to F. Hence, if $\mathbf{R}$ is the resultant force on the particle,

$$\mathbf{R} = m(\mathbf{f}' + \mathbf{f}).$$

If, in particular, the velocity of F relative to F_0 is uniform, then $\mathbf{f}' = 0$, and $\mathbf{R} = m\mathbf{f}$. This equation has the same form as the fundamental equation (26a) on which the main structure of Newtonian mechanics is built.

Hence there is no need in Newtonian mechanics to assume that there exists a frame in a state of absolute rest. It is sufficient to refer velocities, accelerations, etc., to any one of a special infinite class of frames, each of which is in uniform motion relative to any other member of the class. This is the Newtonian principle of relativity. The members of the special class of frames are called *inertial frames*.

It also follows that, in Newtonian mechanics, a uniform velocity of translation may be superposed on a system under consideration without any significant effect (provided it is superposed on the *entire* system).

The student who is interested may care to follow out the implications of these remarks in relation to the use that is made of the concepts of *work*, *kinetic energy*, etc.

17·6. Newton's law of gravitation. Let P and P' be any two particles of masses m and m', distant r apart. From Kepler's observations (§ 17·72) of the motions of planets, and other observational data, Newton made the inductive inference that P and P' attract each other with 'gravitational' forces of magnitude Gmm'/r^2, where G is the *constant of gravitation*. The value of G is about $6 \cdot 67 \times 10^{-8}$ g^{-1}-cm^{-3}-sec^{-2}, or $6 \cdot 67 \times 10^{-11}$ kg^{-1}-m^{-3}-s^{-2}.

17·61. The Earth's external gravitational field. Let M be the mass of the Earth, treated as spherically symmetrical. It can be deduced from Newton's law of gravitation that the force of gravitational attraction between the Earth and an external particle P of mass m is GmM/r^2, where r is the distance of P from the centre of the Earth. Let g_0 denote the (absolute) acceleration of a freely falling particle. Then the gravitational force on P is mg_0, and hence by (26a)

$$g_0 = GM/r^2 \qquad (82)$$

for an external particle.

Since the values of g_0 and r at the Earth's surface are observationally known to good accuracy (the mean radius of the Earth $\approx 6 \cdot 37 \times 10^6$ m), (82) enables the mass M of the Earth to be determined when G is known. The result is $5 \cdot 98 \times 10^{24}$ kg. (The first good determination of M was made in this way in 1798 by Cavendish, who used a modified form of apparatus constructed by a brilliant geophysicist John Michell.)

It needs to be realised that in treating the Earth as spherically symmetrical, we are once again using a model representation. The results given in this section are only an approximation to the actual state of affairs.

17·611. External gravitational potential. It can be shown that the inverse-square law of gravitation gives a conservative field. Using the result $\int r^{-2} dr = -r^{-1}$, it can be shown that (in the model representation taken) the gravitational *potential* (i.e. the potential energy of a particle of unit mass) at an external point distant r from the Earth's centre is $-GM/r$.

17·62. The Earth's internal gravitational field.

At a point below the surface of the Earth at distance r from the centre O, it can be shown that the local value of g_0 (the gravitational attraction per unit mass) is equal to GM'/r^2, where M' is the mass of the material inside a sphere of the Earth of radius r and centre O.

With increasing depth below the surface, both M' and r decrease. Thus the value of g_0 at any internal point depends on the variation of density with depth inside the Earth. It happens that for some distance below the surface, M' and r^2 decrease at nearly the same proportionate rate, with the result that g_0 is constant within less than 2% down to a depth of 2400 km. The maximum value of g_0 (a little greater than 10 m/s²) is reached at a depth of 2900 km, where there is a boundary separating the Earth's 'mantle' (on the upper side) from an appreciably denser 'core' below. The value of g_0 decreases steadily with depth inside the core, and is zero at O.

17·7. Motion under inverse-square law of attraction.

Let (r, θ) be the polar coordinates of the position of a particle P of mass m moving in a plane and subject only to a force of magnitude $\mu m/r^2$ (where μ is constant) acting always towards the origin O (taken fixed).

By (26a) and § 17·45, taking resolved parts radially and transversely, we have for the equations of motion

$$\left.\begin{array}{r}\ddot{r} - r\dot{\theta}^2 = -\mu r^{-2} \\ r^2 \dot{\theta} = h\end{array}\right\}, \qquad (83)$$

where h is constant.

(By (57), p. 225, h is the angular momentum per unit mass of P about O. The principle of angular momentum independently shows that $mr^2\dot{\theta}$ is constant, since the only force present acts always through O.)

It is deducible by simple calculus from (83) that

$$d^2q/d\theta^2 + q = 0, \quad \text{where} \quad q = r^{-1} - \mu h^{-2}.$$

This has the same form as the simple harmonic differential equation (43) (§ 8·4) and the general solution is therefore

$$r^{-1} - \mu h^{-2} = q = A \cos(\theta - \alpha),$$

where A and α are integration constants. The solution can be re-written as

$$lr^{-1} = 1 + e \cos(\theta - \alpha), \qquad (84)$$

where $l = h^2/\mu$, and $e \; (= Al)$ is taken as an integration constant in place of A.

The equation (84) is the polar equation of a conic with one focus at O. Thus the path, or *orbit*, of a particle subject to an inverse-square law of attraction to a fixed centre is an ellipse (when $e < 1$), parabola ($e = 1$) or hyperbola ($e > 1$). When $e = 0$, the ellipse is a circle.

As stated in § 17·611, the inverse-square law gives a conservative field, the potential being $-\mu/r$. Hence the mechanical energy of a particle moving in the field is $\tfrac{1}{2}mv^2 - m\mu/r$, where v denotes the speed, and is conserved. Thus

$$v^2 - 2\mu r^{-1} \tag{85}$$

is constant for any particular orbit. It can be shown that the constant is equal to $-\mu a^{-1}$, zero, or μa^{-1}, for an elliptic, parabolic, or hyperbolic orbit, respectively, where a is the semi-major (or semi-transverse) axis of the ellipse or hyperbola.

The value, V say, of $\sqrt{(2\mu/r)}$ is called the 'critical velocity' for a given r. If v exceeds V, the orbit is parabolic or hyperbolic and therefore not a closed curve; and the particle will, in effect, ultimately escape the influence of the attracting centre O. On the other hand, if $v < V$ at any point, the orbit is a closed curve, namely an ellipse which has one focus at O. (In the particular case of a circular orbit, the centre of the circle is at O.)

At points on the Earth's surface or outside, the value of μ is by (82) equal to GM or gR^2, where $g \approx 32$ ft/sec² and R is the radius of the Earth.

For a particle projected from the Earth's surface, the critical velocity at the surface (called the 'velocity of escape') is $\sqrt{(2\mu/R)}$ and hence equal to $\sqrt{(2gR)}$, or about 7 miles per second. (Cf. Exs. III, 9.) (Air-resistance is here neglected.)

17·71. Particular case of circular orbit. When the orbit is a circle, $r = a$ say, the second equation of (83) shows that the angular velocity, Ω say, of the particle is constant. Thus the motion in a circular orbit is uniform, the acceleration being $a\Omega^2$ or v^2/a towards the centre O. Since the mass-acceleration of the particle is equal to the force $\mu m/a^2$ towards O, we have

$$v^2 = \mu/a, \quad \text{and} \quad \Omega^2 = \mu a^{-3}. \tag{86}$$

(The result $v^2 = \mu/a$ is also immediately deducible from the more general expression for v^2 for an elliptic orbit.)

It will be noticed that the value of v given by (86) for a circular orbit is $1/\sqrt{2}$ times the critical velocity for the distance a.

By § 7·4, the period of the motion is $2\pi/\Omega$. Hence, by (86), the period is proportional to $a^{3/2}$.

17·72. Motion of planets, satellites, etc. The theory of § 17·7 applies, on certain simplifying assumptions, to the motions of the Moon and comets around the Earth; to the motions of the planets Mercury, Venus, Earth, Mars, Jupiter, Saturn, Uranus, Neptune and Pluto around the Sun; and to the motions of planetary satellites around parent planets such as Mars and Jupiter.

Included among the assumptions are the following: (i) treating each attracting body as spherically symmetrical; (ii) treating each attracted body as a particle and ignoring deformations of shape, tidal effects, etc.; (iii) ignoring the reaction force on the attracting body (which causes the latter itself also to move); (iv) ignoring the attractions of all other bodies except the pair under consideration; (v) ignoring atmospheric resistance. (The effect of (iii), which can be important, is easily allowed for in the theory by taking the origin at the centre of mass of the pair of bodies.)

In 1619, Kepler's three laws, based on observations of the motions of the first six of the planets mentioned above, were published:

I. Each planet revolves around the Sun in an elliptic orbit with the Sun in one focus.

II. The line joining the Sun to the planet sweeps out equal areas in equal intervals of time.

III. The time of one complete revolution is proportional to the three-halves power of the mean distance from the Sun.

It was principally through these observational results that Newton was led to his inverse-square law of gravitation (§ 17·6). The law I corresponds to the result (84). The law II corresponds to a simple deduction from the second of equations (83); (see Exs. VII (3) where the term 'areal velocity' is mentioned). The law III corresponds to the result in the last paragraph of § 17·71 for the particular case of circular motion.

The orbits of planets around the Sun, of the Moon around the Earth, and of planetary satellites in general, are usually of small eccentricity e. *Comets* are bodies which have elliptic orbits around the Sun with fairly high eccentricities (but short of unity); comets may be out of visual range for times running into years, but they return at regular intervals (unless some other influence intervenes). If a body passes round the Sun with orbital eccentricity exceeding unity, it does not return to the solar system.

17·8. Motion of artificial satellites. The remarks in § 17·72 apply equally to artificial satellites in orbit around the Earth (or any celestial body). In the case of a nearly circular orbit around the Earth, the constant μ is equal to gR^2 (§ 17·7), so that by (86) the requisite speed for an orbit of height h above the Earth's surface is $R\sqrt{\{g/(R+h)\}}$, where $R \approx 6\cdot37\times10^6$ m and $g \approx 9\cdot8$ m/s².

For heights of 300, 1000, 4000 miles above the Earth's surface, the orbital speeds are about 4·7, 4·4, 3·5 miles per second, respectively. For the Moon, which is at a mean distance of about 384,000 km from the Earth, the orbital speed is about 1 km/sec.

In practice, departures from the assumptions stated in § 17·72 cause the orbits of artificial satellites to differ significantly from ellipses, the factors including (*a*) the small (but not entirely negligible) resistance of the atmosphere through which the satellite moves, and (*b*) deviations from spherical symmetry in the attracting Earth. The deviations (*b*) include the Earth's ellipticity of shape and irregularities in the distribution of matter in the outer part of the Earth. (Each element of matter inside the Earth contributes to the resultant gravitational force due to the Earth at an external point, and it is only in the case of full spherical symmetry that (82) holds precisely.)

It has proved possible to separate out the effects of (*a*) and (*b*) to a considerable degree, and to use observations of satellite orbits to provide information both on atmospheric resistance at the high altitudes involved, and on various aspects of the shape and internal structure of the Earth.

One noted result has been to show that the Earth is 'pear-shaped'. (In more technical language, a third-order term in a spherical harmonic analysis of satellite orbits has proved to be statistically significant, and to have indicated that the average distance from the Earth's surface to its centre is about 15 metres greater in the southern hemisphere than in the northern.)

17·9. Weightlessness. We conclude this chapter with an account of 'weightlessness', a topic on which there appears to be much confusion in current writings.

In § 3·3, the weight W of a body of mass m at a point P of the Earth's surface has been defined as mg, where g denotes the acceleration relative to the Earth's surface of a body falling freely in the vicinity of P. The weight W is equal and opposite to the force needed to support the body at relative rest on the Earth's surface at P. If a spring balance provides the force of support, the dial shows the reading W.

Next consider a body inside a conveyance, e.g. the lift in Ex. (2) of § 3·4, moving with acceleration near the Earth's surface. The term 'apparent weight' is commonly used to denote the force equal and opposite to the force F which supports the body at relative rest inside the conveyance. The dial of a spring balance attached to the conveyance and supporting the body at relative rest inside would show the reading F.

In the particular case where the conveyance is moving vertically downward with the acceleration g (the case (*e*) of § 3·4, Ex. (2)), the apparent weight is zero.

The Earth's gravitational force on a body of mass m will be denoted as W_0. Thus $W_0 = mg_0$, where g_0 is as defined in § 17·61 (and as used in § 17·43 (iv) and in Ex. (5) of § 7·6). In contexts where bodies are away from the Earth's surface, W_0 is often called the weight. But, as already pointed out in § 3·31 (vi), W and W_0 are not quite identical, since g is a relative, and g_0 an absolute, acceleration.

There are thus three different usages of the term 'weight'.

Since 1957 October 4, when the first artificial satellite invaded the heavens, the term 'weightless' has come to be used in common speech for objects (and men) at relative rest inside satellites in orbit. What is meant by 'weightless' in this sense is, not that the weight W_0 is zero, but that the apparent weight F is zero. Thus when a body is said to be 'weightless', the state of affairs is simply that there is no force of support on the body, which is therefore moving freely with acceleration g_0 under the sole action of gravity.

(For a body to be really weightless in the sense that the force of gravity on it is zero, the body would have to be either at an infinite distance from any attracting matter, or else at a point where the vector sum of the forces due to all attracting bodies (Earth, Moon, Sun, etc.) is zero.)

Take for example the simplified case of a satellite moving with speed v around the Earth in a circular orbit of centre O and radius a. The satellite has acceleration v^2/a towards O under the sole action of the gravitational force W_0. The satellite is thus 'weightless' in the above sense since no actual external force other than W_0 acts on it. And exactly the same applies to a man at relative rest inside the satellite. But neither man nor satellite is weightless in the sense of being free from the pull of gravity.

(If the appropriate rotating frame of reference is used, it can be said that the satellite (and likewise the man) is in relative equilibrium under the gravity force W_0 and the centrifugal force mv^2/a. This statement is not adequate, however, for the case of an elliptic orbit, because the Coriolis force and the force (d) of § 17·42 would be involved as well.)

17·91. Further remarks.

(i) When a man jumps from the roof of a house to the ground he is (neglecting air-resistance) 'weightless' in precisely the same sense as he is inside a satellite in orbit.

(ii) The physiological effects of 'weightlessness' arise wholly from the absence of an external supporting force. When a man is standing on terra firma, the external forces on him are the force of gravity and the support, F say, of the ground. Let S be a plane horizontal surface imagined as dividing the man into two parts, say at his neck. Then the part of his body below S exerts across S a force equal and opposite to the weight, w say, of his head. Now consider the man in 'weightless' conditions, i.e. when there is no force F present. Then his head (along with the rest of his body) is moving with the acceleration due to the force of gravity, so that, by Newton's second law of motion, the net force across S supporting his head must now be zero, and not of magnitude w as previously. Similar considerations apply to forces across any other section imagined drawn through his body. These forces are all internal to the body as a whole. Thus the immediate physiological effect of 'weightlessness' is a significant rearrangement in the physical stresses normally present inside the man.

"Our state of ignorance is most intriguing."

—E. W. Barnes, late Bishop of Birmingham.

INDEX

Absolute acceleration, 28, 30, 101, 102, 133, 347, 350
— temperature, 330
— velocity, 28, 30, 102, 350
Acceleration, 11, 101
—, angular, 125, 216
— as space-derivative, 12
— due to gravity, 17, 18, 36, 132-3, 351-2
— in circular motion, 127
—, relative, 20, 119, 132-3, 207
—, uniform, 12
Action, reaction forces, 59, 198
Addition of forces, 30, 103
— — free vectors, 94
— — localised vectors, 192, 194, 195
Additivity of mass, 58, 60
Adiabatic conditions, 331
Aerodynamics, 34
Air-friction, 18, 148, 353-5
Altimeter, 334
Altitude, determination of, 332, 334
Amplitude, 137
Angle of friction, 107
Angular acceleration, 125, 216
— coordinate, 125
— momentum, 196, 199, 225
— — about parallel axes, 255, 256
— —, conservation of, 200
— —, principle of, 200-1, 225, 226, 236
— speed, 125
— velocity, 125, 215, 344
Anticlockwise convention, 125, 191, 192, 194, 343
Apparent weight, 35, 37, 38, 319, 355-6
Applied mathematics, 1, 3, 4, 5
Archimedes, principle of, 318, 338
Areal velocity, 134, 354
Artificial satellites, 136, 354-6
Atmosphere, 309
—, homogeneous, 310, 332
—, pressure of, 309, 314
Atwood's machine, 70, 223

Balance, 304
—, hydrostatic, 334
Balloons, 334

Bar, 309
Barometer, 310, 332, 334, 337
Block and tackle, 284
Body force, 306
Boundary force, 306, 323
Bow's notation, 278
Boyle's law, 330
Bulk modulus, 307
Buoyancy, centre and force of, 325

Calorie, 42, 332
Cavendish, Henry, 351
Centigrade, 330
Centimetre, 9
Centre of buoyancy, 325
— — gravity, 197
— — mass, 58, 165, 197
— — —, motion of, 61, 198, 224
— — percussion, 243
— — pressure, 313, 315-6
Centrifugal force, 125, 128, 347-9, 356
Centroid, 168, 197
C.g.s. system, 9, 31
Change of origin, 20, 101, 102, 116, 165
— — units, 34
—, rate of, 11
Changing mass, 79
Charles's law, 330
Chemical energy, 66
Circular motion, 126-33, 348, 350
— orbit, 352, 353-5
Classical mechanics, 4, 5, 28, 350-1
Coefficient of cubical expansion, 331
— — friction, 106
— — restitution, 76, 203-4
Comet, 354
Component, 95
Compound pendulum, 231
Compressibility, 307
Configuration, 58, 196
Conical pendulum, 129, 133
Conservation of angular momentum, 200
— — energy, 47, 48, 66, 113
— — momentum, 62, 199
Conservative (field of) force, 46, 47, 52, 112, 221, 351

359

INDEX

Constant of elastic string, 51
—— gravitation, 89, 134, 351
Constraints, 277, 290
Contact-forces, 106, 221
Continuous friction, 294-6
Continuously distributed mass-systems, 167, 196, 305
Convention, 3, 10, 94
Coordinate, angular, 125
—, position, 10
Coplanar, 87, 190
Coriolis force, 348, 349, 356
Corollary, 3
Correspondence, 5, 35
Couple, 193, 197
—, friction, 218, 226, 235, 272
—, impulsive, 241
Critical velocity, 353
Cubical expansion, 331
Curved surface, thrust on, 321

Dalton's law, 331
Deduction, 1
Definition, 3
Deformable body, 305
Degrees of freedom, 214
Density, 167
— determination, 333
— of air, mercury, water, 310, 330
—, relative, 309
Differential pulley, 285
— screw, 286
— wheel and axle, 285
Differentiation of vectors, 97, 98
Dimensions, 33, 43, 142
Direction-cosines, 120
Discrete systems, 167, 196
Displacement, 12
Dissipation of energy, 48, 66, 69
Distance, 10
Diving bell, 336
Dynamics, 28
Dyne, 31

Earth, central core of, 252, 352
—, density of, 134, 252
—, ellipticity of, 17, 355
—, internal structure of, 17, 252, 338, 352, 355

Earth, mantle of, 252, 352
—, mass of, 252, 351
—, moment of inertia of, 252
—, orbit of, 134
—, ' pear-shaped ', 355
—, pressure inside, 338
—, radius of, 132
—, rotation of, 17, 132, 247
—, variation of gravity in, 352
—, velocity of escape from, 55, 353
Eccentricity, 352, 354
Effective force, 29
— surface, 314
Efficiency, 66, 282
Elastic impacts, 75, 203
— modulus (constant), 51
— potential energy, 52, 69
Elasticity, imperfect, 51, 68
—, perfect, 51, 68, 76, 141
Electrical energy, 66, 89
Electromagnetism, 31
Elliptic orbit, 352-4
Energy, 44, 65, 112, 202
—, chemical, 66
—, conservation of, 47, 48, 66, 113
—, dissipation of, 48, 66, 69
—, elastic, 52, 69
—, electrical, 66, 89
—, heat, 66, 332
— in problems with strings, 52, 54, 69, 114
—, kinetic, 44, 65, 112, 202, 218, 219, 237, 351
—, mechanical, 47, 65, 66
—, potential, 46, 47, 52, 65, 69, 112, 113, 203, 219, 290, 324, 351-3
—, principle of, 45, 47, 65, 112, 113, 203, 219, 221, 324, 353
Enveloping parabola, 151, 155
Epoch, 137
Equilibrium, 30, 103, 197, 258
—, breaking of, 292, 295
—, limiting, 106
— of floating solid, 325
—, relative, 348
—, stability of, 279, 291, 324, 325, 326
—, stable, unstable and neutral, 279
Equimomental systems, 186
Equivalence of forces, 195, 196

INDEX

Equivalent particle, 62, 198
— sets of vectors, 195
— simple pendulum, 144
Erg, 42
Experience, 1, 4, 5, 6
Experiment, 1, 2
Extended angular momentum principle, 200-1, 236
Extension, 51
External forces, 60, 198, 240, 270, 277
Extraneous force, 47

Fahrenheit, 330
Field of force, 46, 112, 351
Flexible, 68
Floating solid, 325
Flotation, plane of, 325
Fluid, 306
— at rest, 305
—, perfect, 323
— thrust, 312-22
Foot, 9, 10
Force, 28, 102, 196, 348
—s, action and reaction, 59
—s, addition of, 30, 103
— at a contact, 106, 221
— at an axis of rotation, 232-3
—, body, 306
—, boundary, 306, 323
—, centrifugal, 125, 128, 347-9, 356
—, conservative, 46, 47, 52, 112, 221
—, Coriolis, 348, 349, 356
— -displacement graph, 52
—, effective, 29
—, external 60, 198, 240, 270, 277
—, extraneous, 47
—, field of, 46, 112, 351
—, impulsive, 40, 41, 62, 66, 199, 241
—, internal, 60, 198, 220, 270, 277, 289
— of buoyancy, 325
— polygon, 260, 263, 278, 279
—, resultant, 29, 30, 102, 196
F.p.s. system, 9, 31
Frame of reference, 28, 101, 344-51
Frameworks, 271-9
Freedom, degrees of, 214
Free vector, 93
Frequency, 138
Friction, 36, 46, 106, 204, 266
—, air-, 18, 148, 353-5

Friction, angle of, 107
—, coefficient of, 106
—, cone of, 107
—, continuous, 294-6
— couple, 218, 226, 235, 272
— of string round post, 294-5
Fulcrum, 283
Fundamental equation of particle dynamics, 30, 102, 350

Gas, 307, 329-33
— constant, 330
— equation, 330
—, ideal, 330
—, work of expanding, 324
Gases, kinetic theory of, 332
—, mixture of, 331
Geometry, 3, 9, 33
Governors, 133, 135
Gramme, 31
— -molecule, 330
— weight, 35
Graph, force-displacement, 52
—, velocity-time, 22
Graphical statics, 278
Gravitation constant, 89, 134, 351
Gravity, acceleration due to, 17, 18, 36, 132-3, 351-2
—, parabolic motion under, 148-55, 237
—, vertical motion under, 17
Gyration, radius of, 178

Heat, 66, 332
Heavy, 31
Helix, 286
Homogeneous, 308
— atmosphere, 310, 332
Hooke's law, 51, 68
Horizontal, 17
Horse-power, 43
Hydraulic press, 309
Hydrodynamics, 323
Hydrometers, 319, 328, 333
Hydrostatic balance, 334
— paradox, 312
Hydrostatics, 305
Hyperbolic orbit, 352-4
Hypothesis, 3

Ideal gas, 330
Impact, elastic, 75, 203
—, inelastic, 67, 76
—, law of, 76, 203-4
Imperfect elasticity, 51, 68
Impulse, 39, 61, 103
—, instantaneous, 40, 103
Impulsive couple, 241
— force, 40, 41, 62, 66, 199, 241
— tension, 69, 75
Inclination, 104, 121
Inclined plane, 104, 107, 155
Incompressibility, 307
Induction (logical), 1, 2
Inelastic impact, 67, 76
— string, 68, 220
Inertial frame, 350
Input of energy, 282
Instantaneous axis (centre), 237, 256, 344
— impulse, 40, 103
Integration of vectors, 98, 99
Intensity of pressure, 307
Internal forces, 60, 65, 198, 220, 270, 277, 289
— kinetic energy, 202, 219, 254
Inverse square law, 17, 89, 134, 351, 352-4
Isolation, method of, 277
Isothermal conditions, 331

Joints, 218, 271-3
Joule, 31, 42

Kelvin scale of temperature, 330
Kepler, 351
Kepler's laws, 354
Kilogramme, 31
Kilowatt, kilowatt-hour, 89
Kinematics, 9, 214
Kinetic energy, 44, 65, 112, 202, 218, 219, 237, 351
— —, internal, 202, 219, 254
— —, rotational, 219
— —, translational, 202
Kinetic theory of gases, 332
Knot (in string), 296
— (unit of speed), 24

Lamina, 167
Lami's theorem, 260

Latitude, 17, 132
Law of gravitation, 17, 89, 134, 351, 352-4
— — impact, 76, 203-4
—s — motion, 31, 60, 88
—s, Kepler's, 354
Lemma, 3
Levers, 283
Light, 31
Like parallel vectors, 193
Limiting equilibrium, 106
Linear momentum, 198
Line-density, 168
— -integral, 99
— of action, 94, 190, 196, 198
Liquid, 307, 329
Localised vector, 94, 190, 347
Logic, 1, 2, 295

Machines, 282
Magnetism, 262
Mass, 29, 35, 58, 60, 102
— -acceleration, 29, 102, 348
—, additivity of, 58, 60
—, centre of, 58, 165, 197
—, varying, 79
Mathematical logic, 2, 295
— model, 3, 4, 6, 18, 28, 51, 68, 214, 305
Mathematics, 1, 2, 3, 4, 5
Matter, 28
Measure, 9, 34
Mechanical advantage, 282
— energy, 47, 65, 66
Mechanics, classical, 4, 5, 28, 350-1
—, quantum, 4
Metacentre, metacentric height, 327
Metre, 9, 10
Michell, John, 351
M.k.s. system, 9, 31
Model, mathematical, 3, 4, 6, 18, 28, 51, 68, 214, 305
Modulus of elasticity, 51
Mol, molecular weight, 330
Molecules, 305
Moment of couple, 193
— — inertia, 178, 219
— — localised vector, 190, 191
Moments, theorem of, 194

INDEX

Momentum, 29, 59, 102, 196
—, angular, 196, 199, 225
—, conservation of, 62, 199
—, principle of, 39, 40, 60, 103, 198-9, 224
Motion of centre of mass, 61, 198, 224
— on inclined plane, 104, 107, 155
— relative to Earth, 17, 36, 132-3, 352-6
— under constant gravity, 148-55, 237

Nautical mile, 24
Necessary and sufficient conditions, 109, 195, 258, 263, 296
Neutral equilibrium, 279
Newton, Isaac, 4, 28, 31, 60, 76, 194, 350, 351, 354
— (unit of force), 31, 34
Newton's law of gravitation, 89, 134, 351, 352-4
— law of impact, 76, 203-4
— laws of motion, 31, 60, 88, 350
Normal component of acceleration, 127
— — — force, 270
— temperature and pressure, 330
Numerical procedures, 6, 7

Observation, 1
Orbit, 148, 352-5
Orthogonal axes, 95, 101
Oscillations, 138, 141, 144, 232, 326
Output of energy, 282

Pappus, theorems of, 174
Parabola, enveloping, 151, 155
Parabolic motion, 148-55, 237
— orbit, 352-4
Paradox, hydrostatic, 312
Parallel axes theorem, 178
— localised vectors, 193
Parallelogram law, *see* Addition of vectors, 94
Particle, 6, 29
Pascal's principle, 309
Pendulum, conical, 129, 133
—, rigid-body (compound), 231
—, simple, 144
Percussion, centre of, 243
Perfect elasticity, 51, 68, 76, 141
— fluid, 323
Period, 128, 138, 142, 144, 232, 326, 353, 354

Perpendicular axes theorem, 179
Phase, 137
Physiology, 356
Pin-joint, 218, 271-3
Pitch of screw, 286
Plane of flotation, 325
Planets, 353-4
Plumb-bob, 145
Polygon of forces, 260, 263, 278, 279
Position coordinate, 10
— vector, 101
Postulate, 3
Potential energy, 46, 47, 52, 65, 69, 112, 113, 203, 219, 290, 324, 351-3
— — associated with weight, 47, 113, 203, 219
— — of elastic string, 52, 69
— —, stationary, 290, 324
—, gravitational, 351, 353
Pound, 31, 36
— weight, 35
Poundal, 31, 34
Power, 43, 110
Practical units, 10
Precision, 4, 5
Premises, 1, 2
Pressure, 307
—, atmospheric, 309, 314
—, centre of, 313, 315-6
— gauge, 334
— in fluid under gravity, 308-9
Principle of angular momentum, 200-1, 225, 226, 236
— — Archimedes, 318, 338
— — energy, 45, 47, 65, 112, 113, 203, 219, 221, 324, 353
— — momentum (linear), 39, 40, 60, 103, 198-9, 224
Probability, 2
Product of inertia, 184
— — scalar and vector, 95
—, scalar, 96
—, vector, 99, 191, 201, 343
Projectiles, 148, 237
Proof, 3
Proposition (logical), 2, 295
Pulleys, 70, 283, 285
Pumps, 335
Pure mathematics, 3

Quantum mechanics, 4
Quasi-statical, 324

Radial and transverse components, 350, 352
Radius of gyration, 178
Range of projectile, 150, 151
Rate of change, 11
Reaction, 59
Reciprocal figures, 279
Relative acceleration, 20, 119, 132-3, 207, 347-8, 350
— density, 309
— equilibrium, 348
— motion, 20, 102, 116, 125, 132, 345-51
— to origin, 11, 101, 102, 345
— velocity, 20, 116, 347, 350
Relativity, 4, 28, 350
Resolved parts, 103
Restitution, coefficient of, 76, 203-4
Resultant force, 29, 30, 102, 196
— of free vectors, 94
— — localised —, 194, 195
Retardation, 12
Right-hand conventions, 343, 345
Rigid body, 68, 214
— — pendulum, 231
Rockets, 80
Rotating axes, 346-9
Rotation, 102, 125, 225, 344-50
—, work during, 226
Rotational energy, 219
Rough, 106
Routh's rule, 183
Russell, Bertrand, 3

Safety-valve, 311
Satellites, 136, 353-5
Scalar, 93
— product, 96
Science, 1
Screw-jack, 286
Screw law, right-hand, 343, 345
Sea-mile, 24
Second, 9, 10
Second moment, 178
Sections, method of, 277
Simple harmonic motion, 19, 137, 326
— pendulum, 144
Siphon, 336

Sliding friction, 106
Slug, 36
Small variations, 145
Smooth, 36, 48, 69, 106
— surface, motion on, under gravity, 113
Solid, 306
Sommerville, D. M. Y., 89
Sound, velocity of, 25
Specific gravity, 309, 333
— weight, 309
Speed, 11
—, angular, 125
Spring, 51
— balance, 35, 355
Stability, 279, 291, 324, 325, 326
Standard units, 10
Statical equivalence, 196, 197
Statics, 30, 258
Stationary potential energy, 290, 324
Steelyards, 283, 303
Stress, 277, 305, 306, 307
Strings, 51, 68, 114, 141, 220, 294
Stroud system, 10
Strut, 68, 272, 279
Submarines, 335
Sun, mass of, 134
Superelevation of rails, 129, 162
Superincumbent liquid, 314
Surface-density, 168
— tension, 306
Symbols, 3
Symmetry, 133, 273, 277
System of particles, 58, 196

Tangential component of acceleration, 127
Temperature, 329, 330
Tension, 51, 68, 277
—, impulsive, 69, 75
Theorem, 3
— of moments, 194
Theoretical units, 9, 10
Thermodynamics, 42, 307, 330
Thick, thin, 167
Three-force theorem, 263
Thrust, 68, 277
— of fluid, 312-22
Tie, 68, 272, 279
Time-differentiation and rotating axes, 346

INDEX

Toothed wheels, 286
Torque, 226
Traction, 305
Translational kinetic energy, 202
Translatory motion, 102, 347, 348, 351
Transverse component, 350, 352
Trigonometrical theorem, 267
Two-dimensional, 190

Uniform, 12, 168
— acceleration, 12
— circular motion, 128, 139, 348
Units, 9, 10, 31, 33
Unit system, 9
— vector, 95
Unlike parallel vectors, 193
Unstable equilibrium, 279

Validity, 2, 4
Valve, 311, 335
Varying mass, 79
Vector, 93, 190
— addition, 94, 192, 194, 195
— couple, 193
— differentiation, 97, 98
—, free, 94
— integration, 98, 99
—, localised, 94, 190, 347
—, position, 101
— product, 99, 191, 201, 343
— subtraction, 94
—, unit, 95

Velocity, 11, 101
—, absolute, 28, 30, 102, 350
—, angular, 125, 215, 344
—, areal, 134, 354
—, critical, 353
— of escape, 55, 353
— of point of rigid body, 216, 217
— of sound, 25
— ratio, 282
—, relative, 20, 116, 347, 350
— -time graph, 22
Vertical, 17, 145
— circle, motion in, 128
Vibration, 137, 141, 144, 232, 326
Virtual displacement, velocity and work, 287
Volume-density, 167

Watt, 43
Watt's governor, 135
Weight, 34, 36, 103, 132, 197, 355-6
—, apparent, 35, 37, 38, 319, 355-6
—, molecular, 330
—, specific, 309
Weightlessness, 355-6
Weston differential pulley, 285
Wheel and axle, 230, 285
Work, 41, 109, 220, 221, 226, 240, 323, 324, 351
—, virtual, 287, 288
Working hypothesis, 3